Human–Animal Interactions in Anthropocene Asia

This book examines the theme of human–animal interactions contextualized against the idea of the Anthropocene.

Focused on China and its immediate Asian borderlands, this interdisciplinary collection provides a powerful and insightful analysis of the ecological challenges that mankind's traditional activities have created. Through in-depth case studies, each focusing on a particular human–animal dynamic, the book contextualizes and advances the understanding of existing environmental and ecological problems faced by local communities in Asia. In particular, the book hopes to transcend the duality of the nature versus culture debate by locating animal-ecological problems in the behavior of human institutions, beliefs, and practices, which are often affected by prevailing cultural proclivities, political ideologies, economic interests, and scientific agendas. Through interrogation of theoretical concepts of Anthropocene and human–animal binary, the volume highlights the controversial debates that follow their usage as well as their empirical utility understanding human–animal interactions historically, thereby engaging a broader interdisciplinary conversation increasingly links these two fields together.

Providing a platform for discussion and dialogue for a wide audience, this book will appeal to students and scholars of environmental history and politics, anthropology, political science and policy studies, China studies, and Asian studies more generally.

Victor Teo is a political scientist specializing in the International Relations of the Indo-Pacific. He was most recently The Cold War Visiting Research Fellow at CRASSH, University of Cambridge, UK; and Wang Gungwu Visiting Senior Research Fellow at ISEAS-Yusof Ishak Institute, Singapore.

Routledge Contemporary China Series

For more information about this series, please visit: https://www.routledge.com/Routledge-Contemporary-China-Series/book-series/SE0768

Human–Animal Interactions in Anthropocene Asia

Edited by Victor Teo

Routledge
Taylor & Francis Group

LONDON AND NEW YORK

First published 2023
by Routledge
4 Park Square, Milton Park, Abingdon, Oxon OX14 4RN

and by Routledge
605 Third Avenue, New York, NY 10158

*Routledge is an imprint of the Taylor & Francis Group, an informa
business*

British Library Cataloguing-in-Publication Data
A catalogue record for this book is available from the British Library

Library of Congress Cataloging-in-Publication Data
Names: Teo, Victor, editor.
Title: Human-animal interactions in Anthropocene Asia / edited by Victor
  Teo.
Description: First edition. | Abingdon, Oxon ; New York, NY : Routledge,
  2023. | Series: Routledge contemporary China series | Includes
  bibliographical references and index.
Identifiers: LCCN 2022044035 (print) | LCCN 2022044036 (ebook) | ISBN
  9781032079219 (hbk) | ISBN 9781032079264 (pbk) | ISBN 9781003212089
  (ebk)
Subjects: LCSH: Human-animal relationships--Asia. | Human-animal
  relationships--China. | Animals and civilization.
Classification: LCC QL85 .H8428 2023 (print) | LCC QL85 (ebook) | DDC
  591.5095--dc23/eng/20221018
LC record available at https://lccn.loc.gov/2022044035
LC ebook record available at https://lccn.loc.gov/2022044036

ISBN: 978-1-032-07921-9 (hbk)
ISBN: 978-1-032-07926-4 (pbk)
ISBN: 978-1-003-21208-9 (ebk)

DOI: 10.4324/9781003212089

Typeset in Times New Roman
by SPi Technologies India Pvt Ltd (Straive)

For my son Jason W. Teo
who so loves animals

Contents

Figures

Tables

Contributors

Dr. Aurore Dumont is an anthropologist. Her research focuses on the nomadic pastoralism and religious practices of the Mongol and Tungus people of the People's Republic of China from the late Qing up to the present. After completing in 2014 a PhD from École pratique des hautes études, Paris, France, she was a postdoctoral fellow of the Centre for China Studies at the Chinese University of Hong Kong (2015–2017) and then at the Institute of Ethnology at Acadamia Sinica, Taiwan (2018–2019). She is currently a Marie Skłodowska-Curie fellow at Groupe Sociétés, Religions, Laïcités (GSRL) in France.

Dr. Keokam Kraisoraphong is Associate Professor at the Faculty of Political Science, Chulalongkorn University, Thailand. She is a member of the executive board and a Senior Fellow at the Institute of Security and International Studies (ISIS), Thailand. Her latest publications include, *inter alia*, "Profits Downstream, Unsustainability Upstream: Illegal Logging and Siamese Rosewood Trade in the Greater Mekong Basin (GMB)" in *Illicit Industries and China's Shadow Economy: Challenges and Prospects for Global Governance and Human Security* (New York: Routledge, 2019); "China, Japan, and the Greater Mekong Basin: A Southeast Asian Perspective" in *China-Japan Relations in the 21st Century* (2017); "Water Regime Resilience and Community Rights to Resource Access in the Face of Climate Change" in *Human Security and Climate Change in Southeast Asia: Managing Risk and Resilience* (London: Routledge Taylor and Francis Group, 2012); "A Water Regime for Human Security: The Lower Mekong Basin" in *Mainstreaming Human Security in ASEAN Integration: Regional Public Goods and Human Security, Volume 1* (2010); "Crossing the Threshold: Thailand's Path to Rethink Security Sector Governance" in *Peacebuilding and Security Sector Governance in Asia* (2014); *Conflict in Southern Thailand: Seed for Security Sector Reform* (Asia Security Initiative Policy Series no. 21, Singapore: RSIS Center for Non-Traditional Security (NTS) Studies, 2013, "Thailand and the Responsibility to Protect," *The Pacific Review* 20(1), 2012.

Dr. Hang Lin is Professor of history at Hangzhou Normal University. He has completed his MA and PhD in Chinese history at University of Würzburg, and a postdoctoral project at University of Hamburg. His major research interests focus on the history, material culture, and ethnic traditions of non-Han peoples in North and Northeast Asia. His recent publications include: "Empress Dowagers on Horseback: Yingtian and Chengtian of the Khitan Liao (907-1125)," *Acta Orientalia Hung* 73.4 (2020); "A Sinicised Religion under Foreign Rule: Buddhism in the Jurchen Jin Dynasty (1115–1234)," *Medieval History Journal* 22.1 (2019); "Re-envisioning the Manchu and Qing History: A Question of Sinicization," *Archiv Orientalni* 85 (2017).

Dr. Bhim Subba is Assistant Professor at the Department of Political Science, University of Hyderabad, Telangana, India and a visiting Research Associate at Institute of Chinese Studies, Delhi. From March 2019 to February 2020, he was affiliated with the Department of Political Science, Sikkim University. Dr. Subba holds a PhD from Department of East Asian Studies, University of Delhi, and an MA in Political Science from Jawaharlal Nehru University. He is a recipient of the Institute of Chinese Studies-Harvard-Yenching Institute (ICS-HYI) China-India Doctoral Studies Fellowship at Graduate School of Arts and Sciences, Harvard University (2015–2017), a Ford Student Fellow conferred by the India China Institute, The New School, New York, and a Confucius Institute Scholarship awardee in 2013–2014. His research interests are comparative politics, Chinese studies, and international affairs. His articles have appeared in peer-reviewed journals, edited book volumes, review blogs, and other online commentaries and analysis.

Dr. Sangay Tamang as Assistant Professor at the Department of Humanities and Social Sciences, IIT (ISM) Dhanbad, Jharkhand, India. He holds a PhD from the Department of Humanities and Social Sciences, IIT Guwahati, Assam, India. His research interest lies at the intersection of ecology and ethnicity in the Eastern Himalayas. He has contributed articles to journals like Economic and Political Weekly, sociological Bulletin, Indian Anthropologist, Ethnicities, Himalaya, European Bulletin of Himalayan Research etc.

Dr. Victor Teo is a political scientist and a lawyer by training. He received his PhD from the International Relations Department at the London School of Economics & Political Science. His main research focus is on the international relations of Indo-Pacific, with particular emphasis on China and Japan. He has secondary interests in the illicit political economy, global governance, and environmental politics in Asia. He was most recently The Cold War Visiting Research Fellow at CRASSH, University of Cambridge, and *Wang Gungwu* Visiting Senior Research Fellow at ISEAS-Yusof Ishak Institute Singapore. From 2007 to 2020, Victor was a faculty member at the University of Hong Kong, where he taught modules on China's modernization, illicit political economy, and Sino-Japanese and US-Japan relations. His most recent publications are "Rethinking the San Francisco System in Indo-Pacific Security: Enduring Legacies, Structural Contradictions and Geopolitical Rivalry" (2022) and "Japan's Arduous Rejuvenation as a Global Power" (2019).

Dr. Carmina Yu Untalan holds a PhD in International Politics from Osaka University. She is a Postdoctoral Fellow at the International Institute for Asian Studies, Leiden University. Her research interests include postcolonial and decolonial studies, critical international relations, international political sociology, and East Asia. She was a recipient of the Japanese Government Scholarship, and is a former graduate student fellow of the Asia Research Institute, National University of Singapore, and Institute of Critical Social Inquiry, The New School, New York.

Dr. Lisa Yoshikawa is Professor of History and Asian Studies at Hobart and William Smith Colleges in New York, USA. Her first book, *Making History Matter: Kuroita Katsumi and the Construction of Imperial Japan* (2017), examined the intersection of Imperial Japan's nation building and the historical field's development. Yoshikawa is currently working on her second book, *The Empire's Menagerie: Mapping Animals in Imperial Japan*, which examines how the empire was reimagined as it expanded and incorporated territories and species, and how the negotiations of political and faunal boundaries impacted human and nonhuman animal relationship to the environment.

Dr. Sungwon Yoon is Assistant Professor at the Programme in Health Services & Systems Research, Duke-NUS. She is a public health researcher and behavioral scientist. Her main research interest lies in understanding individual and population health behavior which may have public health significance. Her current research projects include assessment of post-screening behavior for chronic conditions (diabetes, hypertension, and hyperlipidemia); risk perceptions and health-seeking behaviors in individuals with pre-diabetes; personalized behavioral intervention using mobile technology for chronic disease self-care; health services research pertaining to cancer and end-of-life care; primary care–based integrated community care team intervention; medication adherence; and global health governance. She was a Korea Government Fellow at London School of Economics and Political Science (LSE). She was trained as a medical sociologist at Seoul National University and Ewha Womans University in Korea and subsequently completed a second doctorate in public health at London School of Hygiene and Tropical Medicine (LSHTM) in the United Kingdom.

Acknowledgments

The editor gratefully acknowledges the assistance of many wonderful colleagues that made this book possible. A deep debt is owed to Harvard Yenching Institute for the workshop grant that enabled the project to materialize. The author would also like to thank Professor Robert Weller (Boston University) and Dr. Ma Jianxiong (HKUST) for their insightful comments on the early drafts of the conference proposal for the workshop as well as the comments received from the referees at Harvard. The editor would also like to express his thanks to the University of Hong Kong for additional workshop grant and for making the logistical and personnel support available for the December 2019 meeting in Hong Kong. In particular, special thanks must go Mr. Eddie Lau, Dr. Diao Tiantian and the SMLC Office for the assistance in the organization of the workshop in what was then a particularly challenging and tumultuous environment.

The editor would also like to convey his appreciation to the Academy of Korean Studies for the support (AKS-2016-LAB-2250005) that enabled him to work on this project during his stay at the University of Cambridge as the Cold War Research Fellow. The editor is particularly indebted to Professor Heonik Kwon at Cambridge for all his friendship, advice, and support all these years and for his support in the author's academic endeavors. Additionally, a note of gratitude must be given to the numerous friends and kind informants that the editor met during his numerous fieldwork trips over the years across different field sites: in China (colleagues and friends in Tibet, Qinghai, Xinjiang, Liaoning, Heilongjiang, Jilin, Fujian, Sichuan, Yunnan, and Guangdong provinces), Myanmar, Malaysia, Indonesia, and Thailand. Additionally the editor also wishes to express his sincere gratitude to Miss Stephanie Rogers, publisher at Routledge, for her patience and support. The editor also would like to convey his thanks to the staff members at Routledge, especially Mr. Andrew Leach, Miss Thivya Vasudevan and her team for the preparation of the final manuscript. Finally the editor also wishes to acknowledge his group of lively and supportive friends in Singapore for all the camaraderie and friendship throughout these years, especially through the difficult COVID pandemic period. Special thanks must go to Leong Choy Wai, Ham Yew Kok, Tan Yew Hwa Mark, and Cedric Chan amongst others for all the stimulating conversations, happy hours, long hikes and excursions.

1 Human–Animal Interactions in Anthropocene Asia

Culture, Development, and the Politics[1]

Victor Teo

Introduction

We live in a world where animals, wild or domestic, and sometimes imaginary are an integral part of our lives. Such encounters are often more common than assumed. For most people living in urban settings today, their closest and real interactions with animals are confined to that of pets, zoo exhibits, animal cafes, or the odd wildlife in semi-rural settings nearby. Many of these experiences of animals are therefore of an imaginary nature, mediated through folklore, art, historical narratives, documentaries, or clips on social media. Compared to the earliest ancestors, the difference cannot be more stark. Human–animal interactions are central to the subsistence and everyday experience of the earliest humans. As human civilizations emerged and flourished, the interactions of human–animals too evolved and changed. These interactions, however, have always been circumscribed by two independent, but interrelated, processes. The first is evolution and modernization of human societies, and second, the ecological constraints imposed by the environment and planet. These two processes have affected and circumscribed how human interacts with nonhuman animals from pre-historic societies right through to present day. In the scholarly world, there is increasing interest in the study of human encounters, real or imagined, with animals by scholars from various academic fields (Bulliet, 2005; Demello, 2012; Kalof, 2007; Kemmerer, 2006; Marvin & McHugh, 2014; Turner et al., 2018; Tyler & Rossini, 2009; Twine, 2010). At the same time, there is also a sharp increase in interest in the fields of environmental and ecological studies of the changing conditions on our climate and what this means for the planet and humankind. While these are two separate trends, there is an increasing convergence between these disciplines.

Recently, as scientists announced unofficially that we have entered the Anthropocene era, scholars of human–animal interactions are too increasing interested in how the arrival of the Anthropocene would impact on the animals they study, as well as how it impacts upon human–animal interactions. This volume seeks to contribute to this effort. Each chapter is written around a case study of an animal at the heart of the author's concern, and the problematic explored mostly in an interdisciplinary manner. Make no mistake:

DOI: 10.4324/9781003212089-1

the human–animal binary and the idea of Anthropocene are still hotly debated ideas in the academia, and the way the respective authors are discussing these concepts in the volume attest to this. Nonetheless collectively, this volume provides an important starting point for the scholarly community to weave these concepts together.

Today, many governments worldwide have announced policies to strive for a greener future, often reinforcing what many NGOs working in climate change, wildlife conservation, and other green causes are striving for. Yet, despite the increased recognition and political interest in the green movement and in the protection of animals, there is a plurality of forces that impede this effort. The importance of culture and tradition, the necessity of maintaining global capitalism and harmonious trading relations and the functioning of bureaucracy and governments – all these forces intervene and ameliorate in how humans interact with animals in the Anthropogenic world.

This introductory chapter seeks to set the tone for the book by contextualizing the broad relationship between animals in Chinese and Asian culture. It would then raise some interesting questions and delineate patterns in the evolution of human–animal interactions, before moving on to examine human–animal interactions contextualized against the arrival of the Anthropocene epoch in Asia. The last section of the chapter provides a summary of the various chapters and arguments contextualized against the general theme of the book.

The Centrality of Animals in Asia Cultures: Folklore, Religion, and Culture

In 2017 on the Island of Sulawesi, Indonesia, archeologists discovered a cave with a 44,000-year-old mural depicting human-hybrids hunting pigs and water buffalo. Investigations eventually showed that this turned out to be the earliest form of pictorial record of storytelling and figurative artwork in the world (Aubert et al., 2019). In the central Indian state of Madhya Pradesh, the Bhimbetka rock shelters too showcase paintings of human hunting wild boars that date back to 40,000 years ago. This form of creative early Paleolithic cave art (which in itself is a narrative of some kind found across different continents) has been deemed to continue until period marked by emergence of the Holocene (about 10,000 years ago).

Even though archaeological figments and artifacts suggest that the earliest ancestors of humans moved from Africa (or Europe) (Slimak et al., 2022) to Asia over 2.1 million years ago, we do not have absolute clarity of how they survived or lived on a daily basis. The anecdotal evidence today suggests that for most part, our earliest ancestors lived in caves and survived as hunter-gatherers. While many of these cave art could be inspired by artistic imagination and born out of their daily experiences, they could be drawn for purposes of recounting experiences, record keeping, entertainment, religion, or pure aestheticism. What is clear though is the centrality of animals as a theme, suggesting that human ancestors experienced a kind of shared existence with animals in their daily experiences. Despite their varying ages (from carbon dating), and geographical locality (discovered in every continent of the world), many of these cave paintings are quite similar. Some of these cave-art depicts smaller etchings of human-like figures hunting, battling, domesticating, or working with substantially larger animals or beasts with astonishing clarity. The paintings seem to convey the enormity of tasks facing early humans in confronting these animals that they hunt suggests the angst and anxiety of their daily existence. It laid bare the asymmetricity of their capabilities and the mortal dangers these animals posed to them, particularly of the spartan knowledge they had when they had to face these animals.

Scientists tells us that since the planet entered the age of Holocene just over 10,000 years or so, the earth's planetary conditions became relatively stable, particularly with regard to the planet's overall temperature. This allowed for humans across the world to thrive, grow, and establish sophisticated societies. By 2000 BC, the Egyptians, Huns (Mongolians), Chinese, Iranians, and Hindus had already established civilization of different size and sophistication. Human existence was no longer confined to caves, and the human condition had made great strides forward. These societies made social advancements in communications systems (such as writing and messaging), legal codification, and medium of payment and allowed vibrant economies with crop cultivation and husbandry to grow. Humans became resilient enough to undertake long voyages or grueling expeditions to explore, trade or to fight. Domesticated nonhuman animals by then have become instrumental to these endeavors. In China's Northwestern Gobi desert, the Mogao Grottoes (otherwise known as Thousand Buddha Grottoes) is one of the most important centers of Buddhist arts in the world. First constructed in 4th century AD, the caves today has some of the most well-preserved cave paintings in world depicting vivid imagery of politics, culture, religion, and ethnic and foreign relations of medieval China. Some of the painting depicts camel caravans on traversing the ancient silk route as part of tribute expeditions undertaken then. Compared to the earliest cave paintings, the animals in these paintings have been reduced in size, and often are no longer the sole focus of the art pieces. Instead, they showcased humans' achievements in culture, technology, and society (no matter how primitive), and their power over nature (including nonhuman animals) their cave dwelling ancestors never enjoyed.

As civilizations expanded, human engagement in animal husbandry increased, as people switched from hunting to pastoralism and ranching (Ingold, 1980). The flourishing of cultures in Asia also meant that at different levels, humans' symbolic interactions with animals multiplied exponentially too (Kalof, 2007). Like the ancient Mayans who worshipped jaguars and the Egyptians who assimilated animals as deities (cats and jackals morphed into the Anubis, birds to Horus, and crocodiles to Sobek), Asian cultures too began to integrate various animals into their everyday culture, religion, and mythology. For instance, the Hindu culture considers several animals considered to be sacred, including the monkey (Hanuman), the elephant (Ganesh), the tiger (Durga), the cow, and even the rat (Ganesh). Javanese culture prizes the rooster, and for the Thais elephants for their special cultural significance. Thais revere these magnificent animals role as sacred animals in Buddhism, reinforced by a folk belief that the elephants bring luck and prosperity. This form of belief that associates human destiny with animal reverence permeates in most Asian cultures.

The Chinese zodiac has since time immemorial provided a cosmological narrative to the lunar calendar, providing meanings and understandings of time in accordance to Chinese and East Asia folklore. Using the following (real or mythical) animals – rat, ox, tiger, rabbit, dragon, snake, horse, goat, monkey, rooster, dog, and pig – each year is linked to an animal and its attributes. Astrologers and fortune tellers would often ascribe a person's personality and life events to the supposed divine influence in accordance to this classification. Chinese parents delight in having a baby born in the year of dragon, while many will refrain from having a "tiger-year" child often due to the anticipated personality issues and/or bad luck associated with these horoscope signs. Regardless of religion, Chinese couples often wonder about their mutual compatibility of their horoscopes as if these animal classifications of time affected their personalities. Just as modern Fengshui masters in Hong Kong and Greater China rely on the Zodiac calendar for their work, advising businesses how to "position" their furniture and interior layout to maximize their good luck ahead, ordinary folks too rely derive the forecast of fortunes and pitfalls each year in spring. This Sinic obsession with animals is seen in Asia across the board. Japan adopted this during the Edo period, and the zodiac animals are known as *Junishi*. Each Zodiac animal is actually associated with the patron saints in Japanese Buddhism. Likewise, the Vietnamese, Thai, the Koreas, and Burmese have similar zodiacs with "local" adjustments. The Vietnamese Zodiac has the "cat" replacing the "rabbit," while Thai horoscope replaces the "dragon" in the Chinese horoscope with "naga" (Sanskrit for "cobra"), a serpent-like creature that purports to guard against evil spirits in Thai temples.

In Indonesian culture, the aforementioned "naga" is a creation of the Indic influence and is manifested as a deity associated with water and fertility. In Japan, in Inari Shrines that worship the Kami (God) of agriculture and fertility, one would find fox deities standing guard as Inari's messengers within. An outstanding example would be the Toyokawa Inari Shrine, where thousands of fox statues stand guard on its grounds. Visitors to China will no doubt be greeted with stone statutes of lions and/or *Kirins*, a mythical creature (麒麟). Whether it's the Temple of Heaven in Beijing, or homes across the Taiwanese island of Kinmen/Jinmen (Figure 1.1) or outside the doors of HSBC Bank in Hong Kong, these stone lions and Kirins provide luck, magnificence, and prosperity signifying a cosmic balance of yin-yang forces in life. Ponds in Chinese and Japanese gardens are usually filled will *Kois*. The colors and shape of *Kois* are reminiscent of the Yin-Yang symbol, signifying prosperity and longevity for the owners.

Figure 1.1 A small statue of the Wind Lion God (风师爷) in Jinmen Island, Taiwan (photograph by the author). Local Taiwanese people believe the totems of these stone lions have the power to ward off spirit, bring good fortune, enhance career, or bring peace.

If the various folk beliefs that imaginary association with certain animals can bring luck and prosperity is valid, then surely association for other animals can equally bring misfortune, trouble, or uncanny experience. Whether one calls it occult or superstition, many believe that such encounters are just as bad as chancing upon ferocious animals (tigers, bears, snakes) in the wild. In South Korea, black crows are seen as bad luck, often ominously as the harbingers of death. Black cats have always been considered to be Halloween superstition and represent bad luck in India, British Commonwealth states, and the United States. This "black cat" myth traces all the way back to the Catholic church in the 13th century. In Vietnam, the cats don't even have to be black to bring bad luck. In Tibetan culture, the "nagas" in Tibetan culture (again half-human, half cobra derived from Hindu culture) are seen to be nasty-dwelling spirits that bring disease and misfortune.

Strangely enough, however, Tibetans consider vultures sacred despite their off-putting appearance and habits. The appearance of vultures on Tibetan hillsides would invariably signal the passing off someone. As the deceased's

Figure 1.2 Vultures feeding at a Tibetan Skyburial Ceremony in China
(photograph by the author).

loved ones congregate to pray and honor the deceased's life before the final
send-off in Tibet's traditional sky burial ceremony, the vultures congregate at
the sidelines ready for a meal. After the rites and prayers, the Tibetan monks
would cut up the body for the vultures instead of cremation or burial (see
Figure 1.2). By the end of it, usually nothing is left, not even bones. It's
shocking to outsiders, but a form of mutualism that exist between the Tibetan
nation and the vultures that exist till today. This practice registers the symbolic and symbiotic relationship their people have had with these animals of
part of Tibetan culture. As with religions, Tibetan Buddhism have treated
animals as central to its conceptual repertories to help its followers negotiate
their lives in various aspects (Ohnuma & Ambros, 2019).

In addition to rituals and religions, the centrality of animals in Asian cultures is reinforced by long history of literati and artistic work that emphases
and celebrate their existence. Often in Asian cultures, there is much interplay
between humanistic creativity. Writers and story tellers often conjured of
human–animal interactions that humanize the animals with attributes and
traits that speak to the philosophical and normative aspirations of the communities. Every Chinese child has that magical moment of being enthralled
by the 16th-century folklore *Wu Chengen*'s "Journey to the West" (西遊記).
The story offers a blend of fantasy and reality, detailing the story of a monk,
Xuanzhang, who travelled westward (toward modern-day Central Asia and
India) in search of scriptures. The story is told and retold across generations,
replete with flesh-eating demons and monsters and in turn protected by three
deity disciples – personified by a monkey, a pig, and water-buffalo, and a

horse which is actually a dragon prince transformed. For centuries, this legend has been one of the central stories at the heart of Chinese culture and identity. This genre of stories is not unique to just China. Most Asian cultures would have some form of these stories. Young Korean children are often treated with stories and tales about their national animal – the tiger (*Horangi*). The tiger is such an important animal for the Koreans that it was used as their official mascot for the 1988 Seoul Olympics. Russian and Japanese children are often treated to tales of bears, while the panda is often reserved for this role in young children's literature in China.

Many of these stories are regaling tales of morality, ambivalence, and uncanny encounters. These narratives often take on themes that appear to be impossible in real life, and posit choices that readers in reality would have to struggle with. In Indonesia, every young child learns about the story of the Dragon Princess of Komodo. The mythical earliest leader had twins – a healthy boy while his sister took the form of a Komodo dragon, who ran off when she was a child. Years later while on a hunt, the young prince almost accidentally killed his own sister. He finally reconciled with the sister after learning that that their mother was herself a Komodo dragon that had taken on human form. This legend today is used to backstop and explain the narrative of the harmonious coexistence between men and lizards on the island of Komodo in Indonesia today.

Young Chinese children today learn from (蒲松齡) *Pu Songling*'s 聊齋誌異 (*liaozhai zhiyi*) – that foxes, ghosts, and demons can coexist together in our realm. The Legend of the White Snake (白蛇传) that has been adapted for plays, TV series, and movies tells of a moving story between an immortal white snake falling in love with a protagonist. These stories allude to the taboo relationship that men cannot have with beings not in their realm, as evidenced by the reincarnated white snakes. This kind of stories also highlights the moral conflicts and stereotypes in the human world. In recent years, a South Korean K-Drama entitled "My Girlfriend Is a Nine-Tailed Fox" (*Nae Yeojachinguneun Gumiho*) was a considerable fictitious tale that gave a modern twist to the East Asian myth of a fox spirit who takes on human form.

One of the most popular *Wuxia* Novelist in modern China, Louis Cha aka *Jinyong* (金庸), often employs animals in his novels. In the Legend of the Condor Heros (射鵰英雄傳), *Guo Jing*, the main protagonist, grew up in the Steppes of Mongolia and became skilled in the hunting and shooting of condors. The condor imagery is present throughout the novel, suggesting that it is a kind of metaphor for natural chivalry, grace, and positive yang energy that *Guojing* is imbued with throughout the novel. In another subsequent, novel, *The Return of the Condor Heros* (神鵰俠侶), *Jinyong* promoted the Condor to become an actual person, who was able to nurse the main protagonist, *Yang Guo*, to health and practiced kungfu with him. There are other novels with animal themes in them. *The Heaven Sword and Dragon Saber* (倚天屠龍記) has characters with special abilities akin to these animals: lions (金毛獅王), bats (青翼蝠王), dragons (紫衫龍王), and eagles (白眉鷹王).

Each of these characters is imbued with animalistic traits in their personality and kungfu they profess to practice. The realities of these alternate cosmological universes are often borrowed and translated into everyday realities in popular culture. TV adaptations and large-screen movies such as *A Chinese Ghost Story* (倩女幽魂 1987; 2011); *Painted Skins I and II* (画皮 2008; 2013), and *Madame White Snake* (白蛇傳說之法海 2011) convince audiences of the interconnection of reality and the spiritual realm. *Jinyong*'s novels have been made and remade across several decades on Hong Kong, Chinese, and Taiwanese cinemas.

The utilization of animals' imagery in Arts is astonishing commonplace too and permeates the daily life of Asian people. A causal visit or stroll through any museums in Hanoi, Jakarta, Taipei, Kyoto, or elsewhere will attest to this. The ceramics, stone carvings, jade and metal artifacts with an animal based design are just too numerous to count. These are, however, not the only genres. From Ming dynasty painter *Bian Jingzhao*'s (边景昭) "Bamboos and Cranes" to *Xu Beihong*'s (徐悲鸿) "galloping horses," Chinese artists have utilized various animatic themes such as tigers, horses, flamingoes, and cranes in their work. This is done across various medium – from bronze pieces to porcelain, from Beijing opera to contemporary Hong Kong movies. In Japan, there are numerous animal-themed artifacts (such as fox, dogs, cats) in the collections of various museums. For instance, the Tokyo National Museum, in collaboration with Japan Foundation and the National Gallery of Art in DC and LA County Museum of Art, was often able to put on roving exhibitions, such as the one based on "The Life of Animals in Japanese Art," showcasing the Japanese people's unique relationship with animals. The symbolic importance of animals in Asian cultures grew as human settlements expanded and civilization became more sophisticated.

In Service of the Sovereign and the State

Beyond culture and religion dimensions, animals also began to play a central role, both metaphorically and literally in the political life of Asian nations. Politically, the ease of appeal to the masses endowed them with an unprecedented place in political symbolism and rituals. These animal-related symbols are monopolized by the ruling elites, often to register an unquestioned celestial endorsement, either to protect or to legitimize their households for perpetual rule. These interpretations have often been accepted wholesale and unopposed by the people, often because they are integrated and fused into national folklore and popular culture, and transmitted across generations. Take Thailand for instance. Elephants are often used to pay homage to the crowning of new monarchs. On the 7 May 2019, ten elephants painted in white marched to Bangkok's grand palace to honor Thailand's newly crowned King Maha Vajiralongkorn, a day after his coronation ended (Kittisilpa, 2019). Five months later, North Korean leader Kim Jung-un was filmed riding a white horse up the slopes of Mount Paektu, Korea's holiest mountain, in a not so subtle alluding to the similarities he shares with DPRK's founding leader, Kim Il-sung, and his personal prowess and defiance to the West (Smith, 2019).

State institutions often adopt animal-related imagery. The outstretched bald eagle in the Great Seal of the United States is well known throughout the world. In Indonesia for instance, many of the national emblem and state crests too are designed around the eagle, while in Singapore, the most famous symbol synonymous with the nation is the merlion – a lion with a fish tail. National emblems help consolidate political recognition, and instantly confer importance and legitimacy on any institution or person bearing the emblem. This legitimating function is important for societies to have that ideological glue to rally the masses to support both the country and the crown. Generations of Chinese emperors have regarded themselves to be true divine beings, often drawing inspiration and legitimacy from associations with mythical dragons (龍) and phoenixes (鳳凰) that are perceived to be at the top of the animal hierarchical structures. This understanding, arguably from the 16th century, has underpinned epistemological of how Chinese people view the political universe. The robes of the emperors and empresses, whether Manchu or Han or Mongol, are replete with the embroidery of dragons and phoenixes, fictitious animals with heavenly properties in the imagination of the Chinese nation. The political legitimacy of heavenly mandate is mediated by a cosmological balance derived from Buddhism from India. Such a mandate underpins the belief system and worldview of the rulers and the ruled: that forces of nature demands cause and effect, karma and retribution. If the emperor has the mandate of Heaven, then the emperor's realm would be prosperous and peaceful. If the Heaven decries the legitimacy of a ruler, then the realm will be plagued by natural disasters, famines (as crops are affected by locusts 蝗蟲), and diseases (such as plague from rats 鼠疫). The influence from India, particularly Buddhism on Chinese culture, is strong: animals are what they are because of the sins they conducted in their previous life; and likewise if humans do not behave appropriately in this life, they run the risk of becoming a cow or a horse to toll their next life (做牛做馬). Chinese culture also appropriately assures sinners that the purgatory is overseen by guards with cows head and horse faces (牛头馬面). Beyond just imaginaries, history has also shown that Chinese politics have also utilized real animals in their statecraft. Elephants have always been revered in China and Southeast Asia because they are a symbol of power and prosperity as the Chinese name 象 (xiang) often rhymes with good fortune 祥. Chinese emperors have therefore always been keen on the notion of a peaceful order under heaven, indicative of their benevolence and legitimacy (太平景象).

Animals have requisitioned for the sovereign as an important instrument in politics and diplomacy. The first giraffes to China were shipped back by Admiral Zheng He from present-day Kenya and Somalia and were mistakenly received in the Chinese court as the mythical Kirins. The Chinese court has always been happy to receive elephants or other exotic animals as tributary or diplomatic gifts as long as they provide the appropriate celestial representation. In premodern Southeast Asia, many of the political entitles made tributes that included rare animals to the regional hegemon, China. These political tributes included gifts such as live shells, peacocks, horses, and monkeys, and often had economic aims as well. Tributary countries expect reciprocal gifts from the Chinese court in value that far exceeds what they have given, and they do this earnest also as a way to earn revenue, and access to goods and resources they otherwise would not have. Gifts of animals went out the other way too. The Chinese Empress Wu reportedly sent Pandas to Japan as diplomatic gifts.

This practice of political animal tributes (or exchanges) was not just limited to China. Elephants were also imported from Vietnam via China to Nagasaki, where they walked to Edo to be presented to the Japanese court. Naturally, the Japanese were all enthralled and amused by this. The rarity of the elephant as an exotic animal made them an extremely important and valuable present to the Japanese rulers as well. Like in the Chinese case, the circulation of "rare" animals then as economic resource is not just confined to tribute exchanges. The prevalent historical narrative on the origins of ornamental kois today suggests that the Japanese had imported kois from China, with the intent to breed them for food initially. The farmers in *Niigita* then isolated the most beautiful orange kois to be kept as pets, only for them to be farmed and reintroduced to China on a larger scale. This tale is interestingly similar to the story of the silkworms: although the Chinese were first to employ silkworms in the production of silk, it was the Japanese who imported silk, improved Chinese technology, and produced better silk for East Asia.

In our contemporary Asia, the same patterns of animal gifting for diplomatic purposes continue. During the Cold War, pandas were gifted by Beijing to 64 countries, including former Soviet Union (1957 and 1959), North Korea, the United States (1972), France (1973), the United Kingdom (1974), Mexico (1975), Taiwan (2008), Malaysia (2014), Japan (1972 and 1982), Spain (1978), and West Germany (1980) (Lam, 2016; Mok, 2018). From 1984 onward, China decided it would only loan out the pandas due to its declining numbers. Today, despite the hostility China faces internationally from the developed West and Japan, many countries are still very keen to for China to gift (or, more accurately, lease) a couple of pandas to them. In 1953, Vietnamese leader Ho Chi Minh gave China's Mao Zedong two Asian elephants, as did Sri Lanka in 1973. Indonesia President Suharto gave Komodo dragons to Singapore in the 1980s, and to the United States in the 1990s. Mongolia has a habit of presenting dignitaries with horses and Australia Koalas (Lam, 2016). North Korean leader Kim Jung-en had two pure breed Korean hunting dogs gifted to South Korea as a token of friendship after he met President Moon Jae-in of South Korea. Kim's father, Kim Jong-Il, had done the same in 2000 after a significant meeting with South Korean President Kim Dae-jung. Former Prime Minister Abe too had tried to gift Russia

Figure 1.3 The famous hotspring monkeys in Nagano, Japan
(photograph by the author).

President Vladimir Putin an Akita dog, but the gift was declined. Japan too has always been very good at promoting herself using animals as a tourist destination. Anyone who has been to Kansai would have definitely had some wonderful memories of feeding the deers in Nara, experienced the interaction with the monkeys in the hot-springs of Nagano (Figure 1.3) or whale watched in Hokkaido.

The Evolvement of Human–Animal Interactions in Contemporary Everyday Lives

The culture of human–animal interactions in Asia today is deeply linked to the way animals have become instrumental to lives of the people in the region. While it is not possible to exhaustively discuss all relevant types of human–animal encounters and interactions, some important patterns of these interactions have evolved and emerged in recent times.

First, with the advent of technology, knowledge, and progress, the human–animal relationship has grown asymmetrical, with humans coming to wield extreme power over the natural world, animals included. Whether it's blue whales, crocodiles, tigers, or the humble chicken, humans are capable of enslaving, subjugating, and/or hunting them down. The evolution of humans as the true apex predator and dominator in the animal kingdom is unprecedented in history.

Second, animals have become indispensable in our everyday activities pertaining to subsistence and consumption, social identity and political economy, companionship and pleasure. We have evolved and accumulated much scientific knowledge and experience with animals to the point that human domination cannot be rolled back. This has in turn created problems for Mother Nature's ability to regulate and heal with disastrous consequences.

Third, over the course of time, each animal has come to possess an intrinsic and derived value or worth that determines its importance to humans. This "worth" to humanity is often contingent on their appeal as food or processed product; value as instruments for work or transport; or their rarity, unique characteristics or endearing value to humans. The "worth" of each animal therefore can fluctuate and is therefore relative. This means that animals can be exchanged, bartered, traded, or gifted in the human world. This, of course, does not negate the fact that humans often can appreciate animals in their own right for their magnificence and beauty, with many even form their own intimate relations with animals in their lives, often outside of the owner–pet relationship.

Fourth, this perceptual difference allows humans to differentiate animals into groups of desirability. This idea has been developed in a sociozoologic scale that depicts the "worth" of animals through a "ranking" according to their place or use in human society rather than by species (Arluke & Sanders, 1996). The four prime categories envisaged by this model are: friends, tools, vermin, and demons. Some animals would be deemed good by us (because of their utility as pets or food) while others hated because of the harm (cockroaches and mosquitos). Yet, others will be regarded different according to their circumstances. For instance, humans might think that bears are magnificent when watching them on TV documentaries, but balk at the idea of ever running one into real life.

Fifth, certain categories of animals would therefore receive privileged treatment because humans feel that these categories of animals are desirable, useful and highly protected, while others are on the receiving end of disdain, neglect, or outright disrespect. Pandas, for instance, are highly protected species, and elicit much public affection and sympathy for them. Consequently, the government in China, for instance, has mandated very strict legislation to protect them, and according much concern for the species whether in captivity and wild. Few other animals are accorded the same kind of protection and concern from the State and people in China, even when there are many species that are at risk from going extinct. Not all animals that are labeled "desirable" would receive good treatment. Some species are bred with the best treatment, with them becoming our food in mind. Think of those beer-guzzling music-listening wagyu cows in Kobe, chestnut-eating black-pigs on Jeju Island, or the free range chickens in Chiang Mai's artisanal farms. All these animals are all destined for the dining table, with the only caveat that the diners must be able to afford them. One might argue they have a good life before they are dispatched.

Sixth, it would therefore appear that human–animal interactions today involve a great deal of hypocrisy, particularly when our treatment of or attitude toward the animals depends entirely on where we locate them vis-à-vis our utility for them. Those animals that are of no use or refuse to be used by us, or considered undesirable, are often neglected or subject to be abuse or ill-will. Snakes, rats, mosquitoes, and cockroaches, for instance, are often considered as "undesirable," and hunted down. This sort of attitude is not only prevalent in Asia, but often is deeply embedded in cultures around the world. This has often led to a contestation over what is "best" for conservation in and outside of Asia. The privileging of dogs and cats has generally, for instance, led to a number of Western NGO campaigns in Asian countries where the consumption of cats and dogs is seen to be a grotesque and outdated practice. The defenders of these practices often point to the double standards and the hypocrisy of such criticisms, creating what appears to be an unbridgeable gap.

Seven, the aforementioned points impact how we think about the categorization of animals, the loss of biodiversity, and conservation practices. Today, layman (i.e., non-scientists) categorize animals in accordance to where they live (oceans, jungles, or dessert-based) or how they move (fly, crawl, swim) or perhaps whether are nocturnal or not. Others employ a more widely used binary: the wild–domesticate divide. Yet others class animals based on whether they are "exotic," and needless to say, this might connotate different things for different people. The list of categorization is long, and these are just some examples. For most part, these categories are too often general, artificial, and often not useful for conservation or the enhancement of human–animal relations. Take the wild–domesticate divide for instance. This binary category fails to capture the existence of a class of animals that exist somewhat in between these two categories. As often is the case, these categories are blurred in reality. Are tigers (and other animals) bred in captivity not "wild" and therefore not "pristine" enough to be protected? Are the deers that live in urban areas (say those deers outside the temples in Nara Japan) considered domesticated or not if they do not live in the forests? For species that populate areas outside their ordinary habitat or origin, are they no longer "pure" and deserving of protection? What about cockroaches, rats, and stray dogs that reside in urban rubbish dumps? These questions are not random or "academic" in nature, as humanity have often been confronted with real-life situations which require our judgment. Today scuba divers in Southeast Asia and elsewhere are racing to rid their coral reefs of lion fishes because these fishes are regarded as "predatory" as they feed on corals, often destroying stretches of reefs at a time. We also often hear of efforts to destroy invasive species of animals introduced into new environments as a result of stowaway or accidental release. Examples abound: the Asian carp in the

United States (name changed to Copi); invasive green crabs or the Mitten crab from China that is now invading River Thames.

More questions are raised: If human–animal interactions result in new circumstances for certain species, what is this "balance" (how do we define it) that we seek to restore? Are conservationists correct in the sense that we should "rewild" as much as possible? When does the line between alien and native blurs? Do we go all out to wipe out a certain species because it is out of place? If endangered species that are bred in captivity are not released back in the wild, do they not count as the population of the said species too? Humans have always been a part of the evolution process, but now, they too are becoming an arbitrator in Darwinian process in the animal world (deciding what must be wiped out and what is to be preserved) – is this correct? The new patterns we see today in these interactions show the ingenuity as well as the domination of human kind across all the realms of interaction, but with these success raises more questions instead of answers. In the following sections, we highlight some interesting issues and dilemmas that have cropped up across the different roles animals have played in human societies, and discuss some of the dilemmas and questions in human–animal interactions over five realms in humankind's daily lives.

Subsistence and Consumption – The earliest humans viewed animals both as a competitor for food, and in being a food source in themselves. Approximately 10,000 years ago, humans began to first domesticate certain species of animals (initially sheep) for food in Mesopotamia, which is current-day Syria, Iraq, Turkey, and Iran (Gibbons, n.d.; National Geographic, 2022). This shift from hunting gathering to animal husbandry allowed for human settlement to expand and grow quickly. By the mid-20th century, humankind had developed large-scale industrial farming, reaping unprecedented harvests, and providing adequate and affordable food for the first time in history for the masses. The advances in food production technology, while beneficial as a whole for humankind, also desensitize humans to the suffering of the animals that are being farmed. The modern man often does not need to think of the cow or chicken one consumes as "live" animals when looking at a piece of burger or chicken nuggets on his plate. Such desensitization has enhanced the exploitation of breed species to unprecedented heights (Nibert, 2017). Many issues have cropped up over the past two decades, notably to do with the ethics concerned with animal cruelty (Harari, 2015; Halteman, 2011) in industrial farming. This in turn has raised questions in how burgeoning consumption of poultry, meat, and fishery stocks have impacted upon our environment and health (Human League, 2021; Rao, 2013).

The range of animals that has been consumed in Asia has extended beyond the "cultivated" or "domesticated" species that are bred. They often include wildlife that people of other countries do not eat. Often these demands are fuelled by traditional cultural or folk beliefs, but also often justified on the medicinal or rehabilitative or aphrodisiac of these animals. Indeed, the news (photos or reports) of these markets (in China, Indonesia, and Indochina) are often deeply upsetting or worrying for Western audiences.

Figure 1.4 Crocodile as a delicacy in seafood restaurant in China
(photograph by the author).

The motivation for this form of consumption is often traced to local cultural traditions or egoistic consumption. This is most evident in China. As the local saying goes, "food is the most important thing for citizens" (民以食为天) – this means that Chinese citizens revere food as the most important thing in the world. There is a demand for wildlife often not commercially available, and that the demand for these exotic meats remains high. Across Asia, everything from crocodiles (Figure 1.4) to snapping turtles to civet cats and even tigers are eaten. In some border areas, these practices become even more pronounced. The perceptual differences in what could be eaten in the West versus those opinion in some Asian cultures often create a huge but remarkable gulf in the narrative on how Asians (or particular ethnicity group in Asia, e.g., Chinese or Indians) treat certain species of animals.

The various nationalities of China have a fascinating culinary relationship with the natural world, consuming much exotic plants and animals for medicine and food. For instance, the Chinese have had a fascination for caterpillar fungus (冬虫夏草), extracted only in the wild from the Qinghai-Tibetan plateau. Most Chinese today believe it is some sort of worm which changes to plant in winter (derived from the Chinese name). In reality, these are actually mummified moth larvae killed by the fermentation of fungus growing within. The caterpillar fungus is harvested from the wild, and is an incredibly lucrative industry for the otherwise poor nomadic people of the Tibetan plateau.

Its role and centrality in Chinese culture sustain the industry: the Chinese believes that the caterpillar fungus is anti-aging and cures a variety of ills from erectile dysfunction to cancer. Even though this has yet to be scientifically verified, the belief of its potency is so strong in Chinese culture that almost all Chinese households who could afford this would buy it for consumption. The caterpillar fungus is one small subset in a complex political economy of trade that involves other animals or animal products (see Figure 1.5) in Chinese culture: from rhinoceroses' horns to pangolin scales, from bear bile to all sorts of penises, animals are hunted for traditional Chinese medicine (or folk medicine), for their pelts as garments (e.g., wolf pelts); for ornaments and accessories (wolf tooth worn by Mongolians) or for food (e.g., civet cats that so famously was responsible for the SARS epidemic in China). In certain parts of China and Indochina, cats are considered delicacies and could be eaten. Other countries such as Korea consume dog meat. In Cambodia and the border provinces of Myanmar as well as in Sumatra, animals like bats, snakes, and monkeys are eaten. The consumption of these wildlife (or non-domesticated animals) directly impinges on the continued viability of the species in question, as well as the question of animal-related diseases finding their ways into the human world. The 2003 SARs and 2019 COVID pandemic are cases in point.

Some consumption cultures in Asia pose particularly difficult challenges for conservationists. The culture of wearing fur (in Russian Far East,

Figure 1.5 Taxidermized wildlife for sale in Kashgar, Xinjiang Province, China (photograph by the author).

Northern China, Koreas) sustains an industry that is considered extremely barbaric. Many conservationists are outraged at the remarkable cruelty inflicted on the animals raised for the fur industry. In Japan and Korea, some raw fish restaurants serving sashimi wrap a wet towel with ice around fish's head and gills as the chef work to slice up the fish, before serving the entire fish (its mouth still gasping and eyes blinking on the plate along with the raw flesh slices) to customers who have a perchance for "fresh" food. This type of consumption practice again would elicit outrage from Western cultures as well as from the conservationists. Amid the criticisms, there will often be a cultural backlash: why is it alright for cows or sheep to be eaten but not cats or dogs? In the same light, there have been questions asked with regard to tunas – why is it that "dolphin friendly" labels are found on tuna-cans, but we have never seen "tuna-friendly" dolphins for sale?

As various Asian societies are becoming more influenced by Western (often global) norms, more questions today are being about the treatment of animals in how Asians process them for food and consumption: what animals should and could be eaten; how are they acquired and processed (whether they are treated right; if they are stored or killed mercifully; the question of sustainability)? These broad conversations have provoked deep soul searching in Asia. The good news is by and large, there is ample evidence that practices in developing Asia are slowly changing to converge with the practices we see in developed world, even though there will be pockets of resistance and exception.

Political Economy and Social Identity – Due to their intrinsic value, animals have always had a role in the in the political economy of the human world. In premodern societies, where currency distribution is limited or non-existent, domesticated animals serve the function of value retention. The powerful and rich often showed off their wealth and influence by the size of their herds of livestock or the kind of animals they kept as pets. Mongol chieftains that ruled over large tracts of the steppes historically rewarded their gallant soldiers with plunder and horses in exchange for support, fight, or sacrifice. For most of the pastoral societies across Asia, nomadic or sedentary, the domesticated animals are prized as a source of both wealth and status. In contemporary Tibet for instance, a mature yak is worth about USD 2,000 (RMB13,000). Some well-to-do pastoral local families would have at least 80 yaks, while the richer families could have as many as 500 yaks, making their net-worth approximately USD 160,000 to USD 1 million alone. This excludes their land, tents, and other husbandry such as sheep or goats. Be that as it may, these Tibetan pastoralists do not see themselves as being "rich." They are comfortable in the sense that they know that will never go hungry. Most of them rely on their herd to put food (meat, milk, and butter) on the table, and, when there is abundance of food, distribute to their urban dwelling friends and families to upkeep social relations. Their excess wealth allows them to afford creature comforts, such as SUVs for men or gold jewelry for the womenfolk. However, for most of them, the true value of these domesticated animals lie in their working relationship with the animals that speaks to their productivity as pastoralist, herders or famers (Kowner et al.,

2019). These animals also often solidify their sense of identity as part of a larger nation. Very often, this affinity is tied closely to either the ethnicity or regional affiliation, as much as it is to the geographical and environmental conditions that define the people. From camel herders in Inner Mongolia to snake farmers in Guangxi, from mule caravans in highlands of Yunnan to donkey ranches in Xinjiang, these selected animals provide a key anchoring point in the individual's, tribe's and nation's self-identification. For the inhabitants of the Himalayan states, yaks played a vital role in enhancing the livelihoods of ethnic groups. Working tribes have long survived the climates and geographical terrains they face by relying on their animals. One of the most salient problems facing many groups today is an ideational crisis when the livelihood they make off tending or herding the animals is no longer economically viable or environmentally sustainable. Often, state policies compound the difficulties of these tribes. The difficulty of these modern-day challenges for the social identity of these cultures cannot be understated.

Animals, however, still play important roles in the modern work place (Figure 1.6) particularly in Asia. In India and Thailand today, both camels and elephants continue to be important tools for transportation of heavy goods in rural and sometimes urban areas. Likewise in rural areas such as Vietnam, the Philippines, and Indonesia, farmers use buffalos to plough rice fields. Modern laboratories across Asia use animals such as monkeys, rabbits,

Figure 1.6 North Koreans tending to their fields with the help of their cows in North Hamgyong Province, DPRK

(photograph by author).

or rats to experiment on as medical or test subjects, all in the name of science and technology. Rats, however, are not always regarded as vermin or pests – in Cambodia, these rats play somewhat of a heroic role as they are trained in helping to clear land mines (Figure 1.7). Over the recent years, the list of ethical transgressions involving animals has increased with the emergence of scientific techniques, such as genetic engineering and breeding as well as the cloning of animals (Grandin & Whiting, 2018).

While this is not at all unusual, norms and expectations toward animals between developed and developing Asia appear to be widening. In most developed countries, there are animal welfare laws, animal ethics, and derived practices well established and in place. In developing countries, the standard of these institutions varies. For most of Asia, the culture of respect toward animals certainly still falls short. Particularly in the developing economies, there are continuing reports of elephants, camels, donkeys and horses being beaten, starved and abused. In some cases, the animals are often neglected and overworked to the point of collapse, where they are killed for consumption after. Indeed if animals could only speak, what would they say about this sort of abuse that humans hold over nonhuman animals? The lack of

Figure 1.7 Land mine clearing rat taking a break in Siam Reap, Cambodia (photograph by the author).

agency on the part of animals as well as changing expectations means that today, Asian societies are awakening to have a close hard look at human–animal interactions in the realm of their work.

Companionship

It would be a mistake to assume that most of the experiences that animals have in Asia are all negative ones. Historically, the evolution of human societies also meant that some animals have become close to living among humans as pets. Research indicates that the earliest pet dogs appeared about 33,000 years ago in East Asia, and eventually these dogs migrated out toward the Middle East and Europe (Wang et al., 2016). Unlike farmed animals, the lives of household pets often come close to being a loved family member. Some also often act as indispensable aides in our daily lives. Guide dogs help the blind to navigate. In Central Asia and the Mongolian grassland, pet hunting falcons help to capture prey for their masters. These animals are allowed to cohabit in homes of humans, where they often are able to build incredible personal bonds with their human owners. Many of these pets are able to play important roles in humans' lives, particularly in a protective capacity. Pet dogs, for instance, are expected to alert us to intruding snakes or bears or other unwelcomed creatures just cats living in our midst are expected to catch mice, roaches, or scorpions that intrude into our home.

Our experiences with household pets today should be contextualized against the large numbers of animals that live as strays or urban "wild." Here we are, of course, referring to those animals that live in close proximity with the humans, but are neither "pet-worthy" nor considered "wildlife" by many. For instance in Southeast Asia, iguanas and monkeys roam much of urban or semi-rural habitats freely as do wild-boar and pythons, often living an existence where they share a mutual coexistence or proximity with humans. Then there are those of the "domesticate" species that are allowed to roam wild and free. Dogs and cats, for instance, can live and often elk out an existence as strays on the streets. This begs the question: how did such a binary of wild–domestic categorization come about, and what are the implications of us thinking about animals this way?

Such binary categorization often dictates which animals are safe for us to coexist with, and which ones are not. Often we draw such knowledge blindly from existing social conventions and norms, at least until they are challenged or excepted of course. The ideas of pets should only be dogs or cats or parrots sit warm and fuzzy until we read of the ridiculous reports. Druglords and dictators, of course, delight in keeping a tiger or two. Saddam Hussein and his son Uday kept lions, cheetahs, bears, and monkeys in the palace (Constable, 2003); Libya's Gaddafi's son, Saadi, kept nine lions at Tripoli's main zoo (Harding, 2011). The infamous Columbian druglord Pablo Escobar smuggled four hippos (Daniels, 2021) which have now grown into an eighty-strong herd in Columbia. Some Nigerians (Figure 1.8) keep hyena as pets (Siddons, 2018).

Figure 1.8 Keeping hyenas as pets in Nigeria.

(Copyright Peter Hugo; Creative Commons License 2.0 https://commons.wikimedia.org/wiki/
File:Gadawan_Kura.jpg).

These "exotic" pets are not just a proclivity of the despotic or the crim-
inally minded. In Asian cities, "exotic" pets are commonplace: cobras,
pythons, iguanas, chameleons, elephants, and, of course, even tigers are
common place (Yan, 2015). This phenomenon has led to the rise of a
lucrative illicit trade in exotic animals for pets. In the rural areas of
Indochina, there are often wildlife markets where various species of animals captured from the jungles by the natives are brought out for sale –
some are endangered species. Indonesia has foiled a number of smuggling
of animals destined for exotic pets market. In 2016, four men were charged
for stuffing 41 endangered white cockatoos and 84 eclectus parrots were
discovered squashed into plastic piping from the North Maluku islands
destined for the Philippines (AFP, 2017). In 2019, Indonesian authorities
foiled a gang in East Java from smuggling Komodo dragons after selling
them on Facebook for USD 1,400 each (BBC, 2019). Indonesian police in
Riau arrested a resident for smuggling 4 African lion cubs, a leopard cub,
and 58 Indian star tortoises from Malaysia (Jakarta Post, 2019). Ego,

profits, and a perchance for the exotic drive this trade forward. The desire to protect the animals, as well as the need to keep certain ferocious animals away from population center has led to various states introducing wildlife laws. As per market forces, illegality often boosts demand as supply is constricted for "exotic" pets. The higher the demand, the greater the profit and the stronger the market for these pets grow. For every one case of smuggling that goes detected, we cannot be sure how many cases were successful. The point here is this: markets and states have real consequences on the bearing of how we calibrate our treatment of and relationship with animals in various capacities today. Economic forces and political power are the two most important drivers of the trade in animals today – and often these forces do not recognize this wildlife–animal divide. The wildlife–domestic binary just does not provide adequate analytical utility to understand the complexity of current human–animal relations – and often this divide creates problems for human society, particularly when they create obstacles in conservation (e.g., rehousing of captive animals or animals born in captivity is often frowned upon).

Figure 1.9 Animal products sourced from Siberia, Russia, for sale at *Heihe*, a border town in Heilongjiang Province across from Blagoveshchensk in Amur Oblast. *Many of these products are smuggled into China by Chinese, Mongolian, or Russian traders and are then resold to retailers, bars, restaurants, and households across China.*

(photograph by the author).

Amusement – Beyond the pleasure of keeping pets and developing an intimate relationship with them for company and intimacy, even those people who do not own pets would enjoy interactions with animals. As early as 252 BC, ancient Romans used to hold battles (known as Venationes) in the Colosseum that saw venatores (like gladiators) fighting animals such as lions, bears (from Hungary, Austria), tigers (from Persia), crocodiles, elephants and rhinos (from India) to amuse the masses. Roman emperors ranging from Titus to Trajan to Commodus all found these mass animal killings politically beneficial. Royalty in Asia often organized large-scale hunting both as a political specter and as a veiled attempt to conduct local inspections (Allsen, 2011). It was also from ancient Rome that the first idea of a performing circus involving animals was conceived and popularized in the 17th-century France. Today, these practices continue on a smaller scale. There are big game hunters who pay vast sums of money to shoot lions or tigers. Across Asia and Latin America, there are still many activities which involve pitting animals against animals. Cockfights (Figure 1.10) are regular social activities in Indonesia, Indochina, and Mexico, as are fights involving cobra and mongoose in Thailand.

Beyond these sadistic activities, there have long been records of humans establishing intimate relations with animals. These are often outlawed and often taboo. Nonetheless, there is increasing number of people investigating this taboo subject of bestiality, of sentiment or sexual relations between human and nonhuman animals (Bakke, 2009). Today, most people have more broadly found pleasure in watching and interacting with animals in controlled settings. May it be safari parks, zoos, circuses, bird parks, aquariums, animal (e.g., owl) cafes or farmstays, humankind has been able to create opportunities for

Figure 1.10 Cock fight in Bali, Indonesia
(photograph provided by Jason Teo).

limited human–animal interactions for the masses, particularly for those that live in modern metropolises where interactions with animals are rare.

From a human perspective, all these interactions are pleasure-oriented. The perspective from the animals' point of view however might not be the same: getting shot at, imprisoned, and gawked at for life is surely not at all pleasant. At the heart of these interactions again is the important question of agency (McFarland & Hediger, 2009) and consent.

In developing Asia, most of the interactions that allow humans to come up close to animals under controlled environments fall into this category. The motivation of those providing such experiences are sometimes interest-based, but most are commercially oriented (for profit in whole or part). Examples of the former are rescue sanctuaries, and for the latter, pet cafes or for-profit zoological gardens. The problem, however, is the perspective from the animals. No one would agree, for instance, that animals need to be consulted when they are being paraded around (even in zoos) or put in enclosures for viewing for the duration of their natural life. A zebra, for instance, might have a better life in the zoo than out in the plains of Africa when he/she risks her life daily due to hungry lions; there is no evidence to suggest that the zebra concerned might like it better than a zoo. These "experiences," however, are collective efforts that subjugate the animals often into alien environments that are suited more for humans and provide a romanticized version of the animals that mask the cruelty that belie such activities (Board, 2021; Santos, 2022).

Over the long course of history, human–animal interactions have become increasingly rarer and more circumscribed as humans move and congregate into cities. Even as the symbolic role they play in Asian cultures increased drastically, the number of people having contact with animals beyond those that are considered "domesticated" among us has decreased. As urbanization rapidly transforms Asia's landscape, the impact of dwindling forests and habits mean that for most part, encounters with "wild" animals are even more unlikely. Up until most recently, the nature of human–animal interactions had predominantly been asymmetrical, exploitative, and often underreported and unappreciated. Often we hear of human–animal conflict (Dabas, 2021; Wallen, 2022), but the reality is that most of these incidents are often underreported. There are, of course, instances where humans are able to coexist with animals (AFP, 2021; Mukherjee, 2022; Xinhua, 2020).

It would be prudent to highlight that the human–animal binary should not convey the idea that humans would always be able to dominate and enslave the rest of the nonhuman animals. It might serve us well to recall the numerous incidents where humans have been killed or injured by non-human animals – there are still reports of shark attacks off the coasts of the United States, Australia, and Egypt. In recent years, Tiger attacks have increased in Nepal as their numbers in the wild had increased because of the protection put in place for them. In India, elephant and leopard attacks on humans are still a yearly occurrence, just as bear attacks are reported in Alaska, Montana, and parts of Russian Siberia. Human beings are as much part of the animal kingdom and natural world whether we like it or not – as we coexist, live, cultivate relations (master–slave, companion, sexual partner), eat (or sometimes eaten by) these animals. The dichotomy the human–animal binary presents might be a mirage in reality. If anything, the recent COVID-19 pandemic has seen the demise of a large segment of the human population across the world. It serves to remind us that regardless of the political and cultural boundaries humans have erected, humans are still not immune to the viruses that animals might carry. The interdependency between human world and animal is underestimated and understated.

As a member of the animal kingdom, humans are not only capable at harming other animals, but also very adept at fighting themselves. This is more so now with the technology at our disposal. From the days of dueling with clubs and knives, human societies have grown so powerful that they can now eradicate each other from the face of the earth with their weapons of mass destruction.

Everyday we hear of innovation in military technology in the press (read: innovation in killing other humans). From drones to lasers to genetic weapons, it is now possible to kill others with a press of a button and watch them die on the screen. Technology has not removed the primal instincts of desire, rage, envy, and greed from humans. At this point, one may well wonder if humankind's mastery of technology and knowledge has enabled humans to become the true apex predator on our planet. Some of us might regard this is the case, but certainly, we are also more capable of killing each other too much like many of the other nonhuman animals. As alluded to earlier in the chapter, the problem with the different binary categorization is that they do not capture the complexity of nature world we live in. In this instance, one therefore should be aware that this construction of human–animal binary itself as a concept is problematic and debated, and would require much more theoretically exploration.

The Arrival of Anthropocene Era in Asia: Impact on Human–Animal Interactions

The development of human–animal interactions in Asia and the world is intricately tied to how successful humankind has become in its mastery of nature and technology. The entrepreneurial and ingenious spirit of humankind incessantly drove the improvement of the human condition. By the end of the 20th century, the idea of modern nationhood, alongside the development of modern technology, has spread from Europe to other parts of the world through colonization as well as the principal processes of globalization. The result is that more than ever before the world converges on the understanding of "modernity." This has not come without costs and injuries to our planet and environment.

According to the International Union of Geological Sciences (IUGS), the planet technically is still in the Holocene epoch that started about 11,700 years after the last ice age. Fundamentally, the Holocene is an era that is defined by a "balance" in planetary perimeters that has produced the most stable climate system on earth. At the core of this stability is a relatively mean stable temperature that sees a fluctuation of plus or minus one degree Celsius every decade. This is drastically different from the era before, as ice cores and rock sediments have revealed that in the period of 100,000 years since the first appearance of humans, the mean temperature was fluctuating over plus/minus ten degrees a decade. Scientists have labeled this period of stability in an unprecedented interglacial age. The Holocene's relative stability in temperature has therefore created the conditions for a warmer era, with four seasons and predictable weather, allowing for human civilizations the world over to emerge, grow, and thrive. From hunter-gatherers living in caves, humans began sedentary lifestyles in rural settlements and townships, cultivating crops and undertaking animal husbandry. By the mid-19th century, with the accumulation of knowledge, experiences, and skills, humans began to embark on a journey that saw them adapting, harnessing, and conquering the nature world. As humans became more knowledgeable, they also became more effective and efficient at overcoming obstacles, including those posed by Mother Nature, to achieve their aims.

Recently, scientists have been debating whether we have entered the age of Anthropocene, otherwise known as the age of humans. Collectively, proponents of the Anthropocene single out the fact that humans have become a force in their own right to drive geological change. Most pointedly, these changes stem from human activities, mostly from their modernization and have impacted the planet in an unprecedented world. The idea of Anthropocene is one of the most important concepts and has attracted the attention of scholars working in diverse fields, ranging from humanities (Bladow & Ladino, 2018; Deloughrey, 2019; Jagodzinski, 2018; Lo & Yeung, 2019; Oppermann, 2017; Vince, 2014), social sciences (Chandler, 2018; Clement, 2021; Dalby, 2020; Figueroa, 2017; Fremaux, 2019; Gillson, 2015; Lorimer, 2015; Purdy, 2015; Watanabe & Watanabe, 2019) and the sciences (Bhaduri et al., 2014; Birkeland, 2015; Levin & Poe, 2017; Voigt & Kingston, 2016). The concept of Anthropocene is still contested and debated (Ammarell, 2014; Hilton, 2014; Hudson, 2014; Philip, 2014), much like the human–animal binary. Scholars focused on Asia are concerned that the Anthropocene concept is too Eurocentric particularly as the starting period of the Anthropocene era is focused on important dates in the European history calendar. Did the Anthropocene begin in Europe from the 16th century or perhaps from the beginning of the Industrial Revolution? If this was indeed the case, the focus would therefore be on the activities of the Global North – the rich developed countries today leading the charge against environmental conservation. Alternatively, it is entirely plausible that the Anthropocene actually began from 1945 when the first atomic bombs were dropped in Hiroshima and Nagasaki in Japan, or in 1950 when the Great Acceleration which saw a dramatic increase in activities affecting the planet took off (National Geographic, 2019). A clear-cut watershed timeline is controversial, and no one can be reasonably sure what happened, even though population, carbon, and methane levels all spiked after the 1950s (Myer, 2019).

The importance of focusing on not just the Global North, but also the Global South, with all the costs and consequences (intended or otherwise) is an important theme for Asianists, just as they are keen to highlight the uneven cost–benefit differentiation and spatial distribution of the environmental ills and planetary damages that the idea of Anthropocene has managed to mask. Notwithstanding the fact that the Anthropocene as a concept is still contested, amorphous, and ill-defined, there is some basic consensus among advocates. The gravity of the consequences of the Anthropocene era, however, is not lost, with some claiming that it distills down to a question of security (Dalby, 2020) and survival (Beeson, 2019; Seploski, 2000). The sense of urgency is not lost as scholars, policymakers, and much of humanity recognize the impending human-induced environmental crisis that we will soon face (Shao et al. 2014; Tsing et al., 2017), and humankind needs to act in concert in order to save the planet.

By and large, the working hypothesis today among many scientists is that

> human activities, rather than natural processes are now the dominant driver of change on the Earth's surface – that carbon pollution, climate change, deforestation, factory farms, mass die-offs, and enormous roadworks have made a greater impact than any other force in the past 12,000 years.
>
> (Myer, 2019)

From the standpoint of those that inhabit our planet, this is particularly problematic. If the Holocene's climate and environmental stability provided the conditions for human societies to flourish, then surely the eroding of these conditions would cause chaos for human societies and for the animals that inhabit this planet with us. For much of the period since the industrial revolution, most of human societies have been so focused on the pursuit of modernity and development that they have overlooked the actual cost to the environment and the planet that sustain us. The issues emerging from the impact of humankind's activities on the planet first appear to be isolated in nature, but today these issues are recognized as being intricately interlinked (Fieldman, 2021; Noone, 2013).

We have converted much of the world's arable land for agriculture and for rearing livestock. The United Nations Food and Agriculture Organization (FAO) estimates that five billion hectares, or 38% of global land surface, are used as cropland, while two-thirds are used for grazing livestock. This has led to livestock becoming the single largest anthropogenic user of land (FAO, 2006: xxi). As the world's population doubled between 1961 and 2016, there is increasing pressure on land use, with global cropland area per capital decreasing from 0.45 ha per capital to 0.21 ha per capital in 2016 (FAO, 2020). For the period 2007 to 2016, China has the largest agricultural land extent (500 million hectares or Mha) followed by the United States and Australia (400 Mha each), and finally Brazil (278 ha). India has around 179 Mha of cropland. Over the years, China has become one of the world's largest importers of food, as domestic production is unable to cope with growing consumption on the back of economic growth, but also because of declining productivity of the arms due to dispersion of farmland and soil contamination. Likewise, in India, over 100 million hectares have been considered degraded, and most of these degraded land are either rain-fed farmland link to food security or forest land that offers the best defense against climate change (Sengupta, 2021). In much of Southeast Asia, particularly in Indonesia and Malaysia, pristine primary forests have been replaced by palm oil cultivation. Human development is therefore the primary reason for the changes to our biosphere. The advanced state of deforestation has directly led to the loss of habitat for the animals that live in the wild, with a significant impact on biodiversity. The loss of natural habitat also impacts their immediate well-being, and long-term natural behavior. Livestock also contribute to a great deal of greenhouse gas emissions, through their carbon dioxide respiration and the release of methane (Gerber et al., 2013; Goodland & Anhang, 2009). The FAO estimates that livestock accounts for 9% of the anthropogenic carbon dioxide emissions, and 37% of methane, 65% of nitrous oxide, and 64% of ammonia. Additionally they also account for 8% of human water use, and contribute to reduction of replenishment of freshwater through degrading soil (FAO, 2006: xxiii). While this might be seen as a boon for the cattle population, very often the reduction in natural forests and jungles for grazing land reduces the habitats of many other species.

Siberian tigers in Northeast China and Russian Far East have had their "range" drastically reduced and divided by urban areas, making it hard for them to breed. Orangutans, which have 97% of their DNA similar to humans, have habitats mostly in the tropical rainforests of Borneo and Sumatra. Due to the expansion of palm oil industry, their natural habitats are being decimated. This has had grave consequences for them as they are forced to migrate to less ideal territories where they often starve and experience decreased reproductive rates (Gutknecht, 2020). There are also instances where expanding settlements intrude into the animals' natural habitats. In Mumbai, 250,000 residents have homes established within the boundary of the Sanjay Gandhi National Park, often putting the residents in conflict when the leopards intrude into their settlements looking for food. Between 1991 and 2013, 176 reported leopard attacks took place; 84 occurred between 2002 and 2004. Nine people were killed by leopards in the month of June 2004 alone (Soumya, 2014). Such human–animal conflicts are not rare (Dabas, 2021; Wallen, 2022) but are rarely reported simply because in most cases, humans have asymmetrical powers over these animals that intrude. Only in instances where the animals are ferocious (such as tigers, bears) do they make the news, accounting for a miniscule of the incidents that occurred.

More significant for the survival of humans and animals is the question of global warming. The impact of human activities emitted sufficient heat-trapping greenhouse gases as well as carbon dioxide that have made the planet warmer. NASA estimated that the average global temperature has increased by 1.1 degree Celsius since the 1880s, with the majority of warming occurring since 1975, at roughly 0.15–0.20 degrees per decade (NASA Observatory, n.d.). This means that over a span of 60 years or so humanity has raised the earth's temperature single-handedly by a degree.

Global warming impacts multiple eco-subsystems of the planet that have helped to keep our climate stable for human development. Most significant of this is the nitrogen cycle (Contiff, 2017; Erisman et al., 2013; Ng et al., 2016). In our time, humankind has made the single most significant impact on the nitrogen cycle in 2.5 million years. Higher temperatures trapped by the carbon dioxide and pollution are melting our ice caps. Through this process, our polar caps are less able to deflect the sun's energy back into space, further contributing to global warming. This cycle of global warming is perpetuated by the fact that many of our subsystems are not working well. The consequence for the Anthropocene world is one where there are actually increased instances of unstable temperatures and extreme weather conditions. The effects of these are mutually reinforcing on other aspects of the planetary ecological balance – and global warming will impact the negative biosphere changes (such as accelerating deforestation, increasing loss in biodiversity, causing disruptions to the hydrological cycle and nutrient cycle); depleting the ozone layers further and degrading our marine spaces through sea-level rise, pollution, and acidification.

The rise of sea level presents a direct threat to coastal cities and island nations as rising sea level would mean that over time, many areas would be subjected to flooding or submergence (Mimura, 2022). Asia stands to be drastically affected by this (Conca, 2020). To make matters worse, some of the cities (Jakarta, Tokyo, etc.) are sinking even as sea level rises (Eco-business, 2022; Gardner & Greenpeace, 2021). Recent research has shown that the previous generation of technology has actually underestimated the number of areas that are at risk to rising sea levels (Board, 2021).

Water makes up about 70% of the earth's surface, and over 50% of our oceans and marine spaces are being overfished, often at an unsustainable rate. The direct threat to the ocean dwelling animals in our Anthropocene era comes from both the direct and indirect activities of man. The most direct and visible consequence is the loss of biodiversity brought on directly and indirectly by human activities, particularly in their pursuit of economic gains (Seddon et al., 2016). Many animals have been driven to extinction or set on path to decline by mankind through the alteration of their natural state of environment – both on land (Chiaverini, 2019) and in water (Center for Biological Diversity, 2013). Through his grasp of technology, humankind has been able to extract marine resources at an ever increasingly alarming pace with remarkable ease. Overfishing has been cited as one of the most important threats to the biodiversity of our oceans especially with large-scale trawling that indiscriminately hauls in a large amount of by-catch. For instance, research has shown that overfishing and the associated by-catch problem is causing one third of sharks and rays species to head toward extinction (Pacoureau et al., 2021). This problem appears to be worsening despite pledges by nations to stop it (Woody, 2019). Demands for certain species of fish have seen the stocks of these fishes dwindle over the last few decades (WWF, n.d.). We have indirectly through climate change and pollution put coastal communities at risk because of changing sea levels, and threatened fragile ecosystems along the coast, such as coral reefs, mangrove swamps, and river and freshwater systems.

In a short span of a few decades, humankind has pushed themselves out of a "stable" state that both humans and animals have relied on for the last 10,000 years or so. The stability that Holocene period has provided for men, animals, and plants depends upon to survive, grow, and thrive will disappear in this new era. The Anthropocene era today threatens to destroy and upend the world that we know slowly but surely over the next few decades (Fioritti, 2020). It is therefore not surprising that we are seeing various narratives of apocalypse endings for humankind (Huff, 2021) associated with the Anthropocene. The most drastic of these narratives of course ends with the sixth extinction event (otherwise known as the Holocene Extinction). Unlike the previous extinction events on earth that are largely driven by natural phenomena, this one would be unique as it is driven largely by man's activities, with some arguing that three quarters of the animals would vanish or extinct within 300 years (Gibbons, 2011). Even though the idea of the Anthropocene is largely contested, there is no denying that implicit in many of the writings on the Anthropocene is that this is a "boiling a frog" slowly type of situation that humankind is facing.

While there are diverse views where humankind stands in the age of Anthropocene, the recent work on planetary boundary concept (Steffen et al., 2015) is gaining much traction. This model (Figure 1.11) aimed to define the environmental limits within which humanity can safely operate. This easily understood conceptual model paints an easily comprehensive picture on the important indicators necessary for the planet to be self-sustaining, self-healing, and healthy, that is, a Holocene-like condition that can support contemporary human societies. If any of indicators are exceeded due to humankind's activities, it would drastically alter the planetary balance, possibly beyond a point of no return as it would impact the other boundaries as well, as illustrated in Figure 1.11. These stipulated boundaries are critical for anthropogenic perturbation of critical Earth-system processes. As the model shows, the planet is reaching alarming levels with regard to the level of extraneous nitrogen and phosphorus in the ecosystem. The planet is at high risk insofar as biodiversity is concerned as many species of plants and animals have gone extinct. Regardless of whichever variable is affected, it would impact directly and indirectly both humans and the animals that share our planet. Conversely, if each variable stays within the safe boundary, the planet would remain in a zone where the changes in the sub-ecosystems remain manageable.

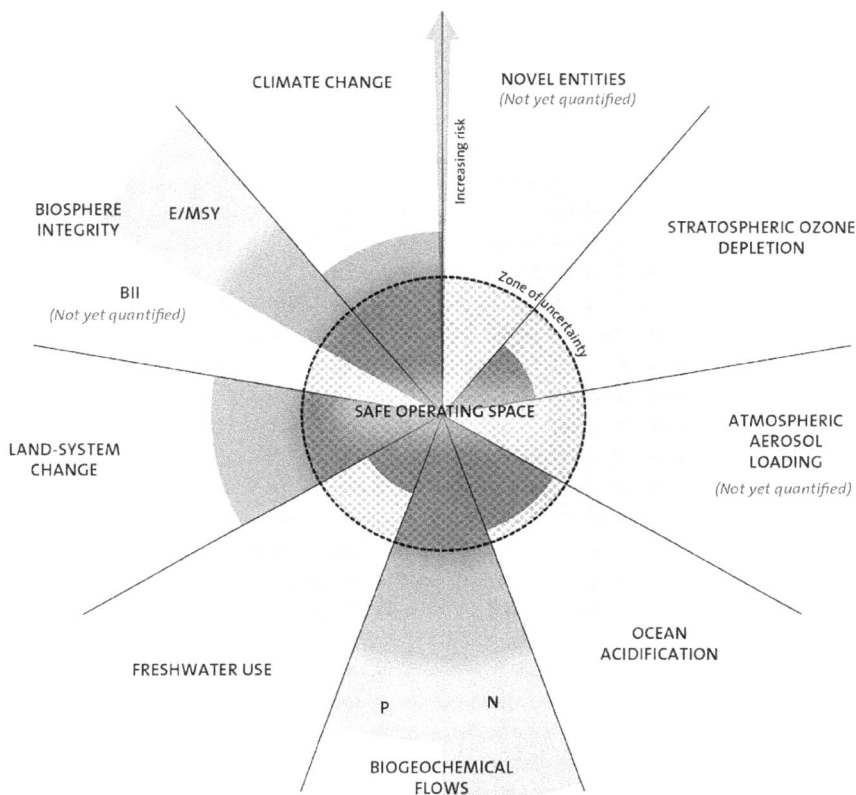

Figure 1.11 Planetary boundaries model illustrated in black and white

(Credit: J. Lokrantz/Azote based on Steffen et al., 2015; CCBY 4.0; https://www.stockholmresilience. org/research/planetary-boundaries.html).

Human Agency over Animals and the Responsibility to Protect: Transcending Symbolism to Action-Oriented Approaches

The concept of Anthropocene was criticized early on by scholars of Asian studies as the concept seems to mask the question of responsibility as to "who" was to be held responsible for the damage to our planet. The bulk of the literature today seems to indicate that most of the environmental problems are the transgressions of the Global South, seeing how the advanced countries such as those in Europe have the best environmental practices today. Yet if the planetary damages are the sum total of all the exploitation that has gone on before, then surely the Global North should shoulder the greater share of the responsibility and burden today, if the South is just playing catch up (even if they are doing more damage in their developmental strategies along the way).

The Anthropogenic consequences for the animal world and human–animal interactions are equally devastating. Our obsession with development and modernization has left a negative impact and legacy on the animal world because humankind has seldom taken the animals' rights and welfare into account for much of its recent history. Humans are often ignorant of how their actions directly or indirectly impinge on the ecosystem.

Just like the problematic nature of the human–animal binary, the concept of Anthropocene is also manifestly unsuited when we try to apportion responsibility or conceptualize remedies. Anthropocene advocates certainly suggest that humankind is culpable. Anthropocene as a concept how just doesn't allow us to attribute which country should do more or in what ways. While it is true that some Chinese businesses are, say, logging trees in Myanmar illegally or that their fishing fleet are often engaging in expeditions far from PRC's shores (hence the charges of illegal fishing or overfishing), developed nations too are not completely sin-free in such enterprises. The global consumer demand for affordable (and hence disposable) goods fuel the "made in China" export phenomenon, just as greater demand for cheaper seafood in affluent Asia would welcome cheap, no-questions-asked fishery imports, never mind if the fish is harvested off the coast of Somalia or Chile. Humankind's capitalistic preference for cheap and quality products often supersedes any ethical considerations. The ability of hegemonic states, groups, or powers to monopolize the discourse, draw lines, call out, and action anyone outside the group or line is equally worrying. Like the human–animal binary, the concept of Anthropocene often masks the power realities that underlie global development. In short, we cannot blame the Eskimos for the melting ice caps just because they live in the polar vicinity just as we cannot blame the rural poor for cutting trees for fuel when they have nothing to keep them warm.

Insofar as human–animal interactions are concerned, the advent of the Anthropocene discourse is not all bad news. It has galvanized communities across the world to take ecological change seriously. Today, the idea of conservation is no longer a fringe idea but is a growing movement across the world where the younger generation are definitely a lot savvier and more aware than the older generations about conservation. Even at the preschool level, the conservation message has gotten across. Insofar as human–animal interactions are concerned, again most people understand that humankind transcends the approaches undertaken by their ancestors. There is growing consensus that human–animal interactions need to move beyond traditional (and often practical) themes related to food, work, or amusement.

The first step, of course, is the recognition of the role of human agency, and its importance in the welfare of animals. Nonhuman animals are unable to act in the same way as humans do, and their lack of agency is a question that most of us do not think of easily in undertaking daily interactions with them (McFarland & Hediger, 2009). The ideas of exploiting animals for man's benefits in all respects – food, work, amusement, and pleasure – are now increasingly being challenged by ideas of what humankind can do to protect animals from harm. Increasingly, the emerging generation today is more informed about the importance of conservation to mitigate the adverse consequences of the Anthropocene era, and more aware of the responsibilities of people, groups, and nations involved. The constitution of the remedies, however, still lags behind – as conservation is hardly advocated as an international public good – and those countries and companies benefiting from the damages done in the third world are certainly not playing their part enough to mitigate the various environmental disasters far away.

Notwithstanding this, what is also true is that today's generation are a lot more conscious about conservation than the generations before. The advent of this consciousness cannot have happened without the rise of a dedicated class of "green" warriors. While environmentalism can be traced to the earliest thinkers in the late 18th century with the industrial revolution, the arrival of the idea of Anthropocene in our discourse meant that more than ever, we are becoming more conscious of the planet that we live in. The rise of global nongovernmental organizations has have played a remarkable role in advocating the essential protection of our environment and planet at large. Prominent examples are organizations such as Greenpeace, Conservation International, The Nature Conservancy, Ocean Conservancy, World Wildlife Fund, and World Resources International. Certainly in Asia, there are very few communities that have not been affected by this green movement.

For the contemporary generation, the conservation and protection of animals during the any human–animal interactions certainly becomes an important consideration, regardless whether it is for pleasure, work, companionship, or even food. Re-examining how humans relate to animals is now an important consideration across the board in most societies in Asia, not an afterthought. This comes in three important respects.

First, the development of fields of animal ethics, law, and welfare is a big step forward. In the West, and now increasingly Asia, various nations are devoting more time, resources, and intellectual energy toward the development of these fields in their jurisdictions. These, however, are not just academic ideas or discussions. Second, much of these narratives have been translated into policies and laws, and these in turn compelled authorities to establish a plurality of agencies to oversee and to enforce these laws, at both national and local levels. Third, beyond the formal codification of law, various communities are seeking to integrate conservation practices into their daily routine and practices. Even the most disadvantaged of communities are increasingly seeing that care and incorporating good conservative practices as part of their daily routine is critical to their own well-being. Rural communities are also starting to understand that the surest way to protect their future generations is to ensure that their environment is secured as a form of heritage that their children can inherit.

The arrival of the idea of Anthropocene has therefore galvanized our communities into action in ways unseen before. In Asia, most societies have therefore begun to take initiatives to tackle the environmental challenges they have, albeit in differing degrees. This include the protection of the environment in all aspects, transitioning to non-fossil fuels, reducing pollution, and protecting each country's natural resources – from animals to plants. The idea that humans have a responsibility to protect and care for the planet's resources including the animals that inhabit earth with us has given rise to a whole new discourse that is often in agreement and synchronous in principle, but often contentious, conflicting, and lagging in practice. The reason is simple. The challenges posed by the anthropogenic issues are often beyond the wisdom, foresight, and capacity of individual countries, let alone human and subnational groups. Beyond that, there are almost always competing priorities between actors of differing power attributes, interests, and attitudes. When the interests or the issues are transnational boundaries, they often prove to be intractable.

The Role of the State: Tackling the Challenges as Anthropocene as Public Interest Project

The state therefore becomes an extremely important actor in the resolution, if not amelioration, of many of the environmental challenges stemming from the advent of the Anthropocene era. Owning to increased concern for animals and wildlife, whether it is in Southeast Asia, China, Indian sub-continent, or Northeast Asia., laws are being legislated and policies enacted to accord some sort of protection to both domestic animals and the wildlife in their jurisdiction. Laws are also drawn up in the more advanced states to combat cross-border environmental issues, especially those involved in animal smuggling and trade. In China for instance, with the passage and enforcement of 1988 Wildlife Protection Law, a list was drawn up in the subsequent year, in 1989, to protect the various species. This has helped in reality to reverse the fortunes of some endangered animals, such as the Siberian tiger and the panda. Various Southeast Asian countries have also put in measures to prevent the theft of their resources such as trees, wildlife, corals, and marine animals. At the same time, China's laws have also been criticized severely as they were alleged to still offer inadequate protection to wildlife (Bale, 2016), just as the Chinese government has been accused of violating biological ethics standard. The importance of legislation has therefore become more carefully debated and calibrated.

Like most policymaking in reality, even though good intent is often present, the actual progress in enacting and enforcing such legislation is in reality a large mixed bag of results. Far too often, the actions of the state remain spotty, uneven, or ineffectual. In other instances, it's the state that actually reinforces the structures (whether ideational or material) that impede the resolution of the problem. This could be due to a combination of factors: political indecisiveness, bureaucratic inertia, lack of scientific expertise, inadequate resources, vested competing interests at different levels, or war and conflict. Often it is also the authorities that prevent conservationists from doing their jobs to enhance animal protection. For instance, Chinese activists have lamented the government's inability to update the wildlife list has in fact impeded the protection to certain species (such as the pangolin) for a long time. Only with recent updates decades later did this species receive official state protection. The legislative and administrative power of the state instrumental to the protection of animals only works when it is used appropriately in a timely manner. When it works against these goals, the State might be singularly the most significant impediment toward the protection of animals.

Regardless of the realities on the ground, national governments particularly in Asia have largely adopted the discourse of ethics, rights, and protection of animals. The State often mediates between their own (or more accurately the politicians') interests with the interests of other societal groups that often do not take into account the welfare or rights of animals (real estate developers; timber industry; factory developers) as well as other stakeholders (indigenous people, jungle/rural villagers; animal protection activists and NGOs). In most instances, the interests and welfare of the animals were never the primary consideration. When land has to be cleared to make way for highways or cities in the name of development or profit, those that are most affected are often never consulted. The animals cannot speak for themselves, and often those humans who do speak for them do not have enough clout to protect either themselves or the animals. In undertaking their broad responsibilities, it is therefore incumbent upon governments to undertake self-reflection and awareness as to the extent of their responsibility not only toward those the electorate or domestic constituents but to pay special attention to the environment and the conservation of animals. The real challenge is that while there is no doubt increasing number of people that would agree animals should be protected, it is questionable if as many would agree that animals should share in our entitlement and command over the earth's resources at humanity's expense. It is a romantic and idealistic thought, though not a practical one.

The evolving culture of conservation and animal protectionism is therefore an important milestone in human–animal interactions in contemporary Asia. This culture is increasingly adopted by individuals and communities across the board. It aides in the constant negotiation between those that work or assist in the protection of the ecosystem and all living beings, against other individuals and groups that privilege other economic priorities such as development or economic growth. The playing ground is hardly even. Many those of who are at the forefront of working to protect the ecosystem, flora, and fauna are often grassroots and communities on the ground. They very often rely on nature to make a living, and often bear the costs from the Anthropogenic forces. These communities are often marginal and powerless, and lived on the peripheral areas in each country. Often, their priorities are subsumed under more powerful ones. The history of global development is replete with histories of mistakes, out of negligence, or recklessness to the extent that one author suggests that Anthropocene are "ghosts" and "monsters" of the past (Tsing et al., 2017).

Challenges in Human–Animal Interactions in Anthropocene Asia

The authors of this book therefore collectively speak to how the communities they study have come together to face and negotiate the challenges of the Anthropocene era. In many ways, the cases speak of the "ghosts" and "monsters" of past mistakes and neglect that each community in the cases faced. They are all unique because each case speaks to an important animal that is closely related to people concerned, and the dangers that the animal face in the Anthropocene era. Each chapter also challenges us to think about how useful the constructs of the human–animal binary and the Anthropocene are in reality, and highlights how much work needs to be done to improve our ideas and policies toward the protection of the nature world. The first section of this book presents four cases which illustrate fully the challenges local communities living with animals face in this Anthropocene era, and the complexity of the ongoing human–animal interactions for the future of the region and the communities at stake. Two of the cases are drawn from the Indian subcontinent, while the other two are from the peripheral areas of the People's Republic of China.

Sangay Tamang's chapter introduces us to the challenge of conservation work in the Himalayas, particularly of the Red Pandas. This chapter firstly illustrates the challenge of the idea of the Anthropocene as an analytical concept. The red panda is the centerpiece of conservation in many Himalayan states and is undergoing serious anthropogenic threats that resulted in habitat loss, landscape transformation, and changing human–animal interaction. However, the idea of "Anthropocene" is not universal. This chapter will focus on "different kinds of people" categorized as different form of Anthropocene in the Himalayas. This chapter, by using the case of red panda conservation in the Singaila National Park (SNP), Darjeeling Hills, will discuss different aspects of wildlife conservation and its effect on human–animal relationship in the region. Based on ethnographic fieldwork, this chapter will highlight on changing trajectories of landscape, economy, and local communities' relationship with the wildlife after the establishment of National Park in the erstwhile grazing ground. Since, the SNP is one of the famous tourist destinations in Darjeeling Hills, this chapter, by using the case of tourist inflows data, shows how the notion of "tourism" leads to changes in ecosystem, landscape, and local economy, which many conservationists tend to overlook it. Thus, the chapter urges for rethinking the idea of "Anthropocene" by situating local contestation over the term.

Sangay's chapter is followed by three chapters that examine the evolvement of human–animal interactions of herding communities. All three chapters are fascinating case studies of how environmental challenges have accentuated the already difficult conditions that these herding communities face in managing their livestock amid deteriorating political-economic and social conditions. Even as they faced extenuating circumstances such as melting glaciers, political turmoil, bureaucratic struggles, interstate conflict, and identity loss over time, each community had to come up with ways to mobilize their traditions and their way of life.

Aurore Dumont's chapter is a fascinating study of camel herders based in China's loosely populated Inner Mongolia Autonomous Prefecture. Dumont skillfully provides an ethnographic account of how these camel herders have redefined their practices to suit the ecological zones of both the desert and the steppe over time, and negotiate their own practices and lifestyle vis-à-vis the State's development strategies. Once an important mechanism of transport, the importance of the camel was eroded in its connectivity role as China's development journey took off. The fortune of the camel declined compared to other livestock. Dumont's fascinating case study illustrates how the camel in the Inner Mongolia Autonomous region switched from being a beast of burden in nomadic pastoralism to a racing animal used in contemporary recreational activities actively promoted by the Chinese authorities. In particular, the pastoralists have adapted to this major change and utilized camel racing, as well as other "invented traditions" had slowly reshaped domestication practices and relationships between humans, the camels, and their natural environment. The chapter invites us to rethink our understanding of human–animal experience by placing the camel and its herders within the larger context of environment, indigenous skills, and politics. By exploring the intersections between these realms, the chapter discusses how local societies faced the changing paradigm of the Anthropocene over the past seven decades or so in their attempt to preserve their own culture, identity, and developmental trajectory.

The next chapter focuses on the reindeer herders. **Hang Lin** examines the Ewenki of Aoluguya, of Inner Mongolia, who has for generations relied on the reindeer as their prime source of livelihood. In 2003, the community were relocated to a purpose-built settlement as "ecological migrants," justified on the grounds of environmental protection and social development. Although many Ewenki herders are increasingly attracted to the lifestyle offered by regional centers of urbanization, others interpreted the relocation as an attack on the traditional lifeworld, with a number of the Ewenki moving back to the forest. Together with the changing way of living and the increasing importance of tourism as a revenue source, indigenous cultural practices have declined, including shamanistic performances, traditional medicinal use, and traditional dress, whereas the incidence of alcoholism has increased. This chapter analyzes the specifics of reindeer herding and domestication, concentrating in particular on how reindeer shaped the economic and religious lifeworld of the Ewenki. Focusing on the impact of the relocation on the Ewenki, it explores how their distance to the reindeer and the increasing importance of tourism changed their indigenous way of economic, social, and religious living. Through this, it probes into the multidimensional interaction between environment, human, and animal, and by doing so probes into the complex relationship between environmental change and adaptability of ethnic culture.

Bhim Subba's chapter is a study set in the Sikkim Himalaya that sits on the border region between South and East Asia. Sikkim is an ecological transboundary hotspot nestled between India and China, and for centuries has thrived with a substantial yak population. The herder communities and their yaks have become an intricate part of the social and cultural imagination of the alpine region since time immemorial. However, once a shy beast of burden, yak have become the new ruler of alpine socioeconomic landscape: a transition from cultural and nomadic lifestyle to changing social and economic rationale with changing state policies and vis-à-vis environment and security priorities. In the process, other animals were forced out, resulting in equity concerns and also leading to consolidation of yak ownership in the mountains. In addition, the emerging geopolitical landscape in the Himalayas between India and China, climate change, and yak herding lifestyle and sustainability have become a concern for the alpine inhabitants. This chapter thus illustrates these issues from local dynamics, existing environmental and ecological laws and security dynamics that have reshaped the yak herding in the changing social and cultural lifestyle in the alpine borderland region.

These latter three cases make for a very interesting comparison. All three chapters somehow took a long durée approach to examine their communities' relationship and interactions with their respective animal herds in relation to the larger social and natural environment. The challenges the three cases face are relatively similar, particularly with regard to how environmental changes have increasing posed difficulties for the communities' way of life. Faced with the grim prospects of continuing their lives in their traditional ways, they adjust their strategy to ensure and protect their communities' heritage where their interactions with the animals under their charge occupy a central and critical role. A key concept here closely related to is notion of how modernization and development in their broader social context have led to changes in both the natural world and the socioeconomic world. In all three cases, the communities had to adjust to challenges and limitations posed by the arrival of the Anthropocene era, with how the State, manifested through the authorities' actions, handled environmental and economic problems to protect and preserve their traditional identity. In all three cases, the State played an intrusive and somewhat negative role in impacting upon the traditional human–animal interactions of these communities. The chapters also illustrated how these communities negotiated with the State (manifested through the authorities' actions) handled environmental and economic problems to protect and preserve their traditional identity.

Dumont's chapter demonstrates how alongside the rise of the modern transportation system in China, modernization had reduced the utility of camels as a work and human transport over time. Dumont's chapter illustrates the difficulties that community faced, and the arduous journey they took to overcome the problems they faced. In particular, Dumont highlights how the community adopted new forms of human–animal interaction through the emphasis of "ecological knowledge" through their organization of innovative tourism events that helped the community reinvent and rejuvenate their social vitality and economic sustainability. Dumont's chapter highlights the importance of looking at the shifting usage of the camel by the pastoralists and the authorities alike across different context and periods of time.

In Hang Lin's chapter, the Ewenki community too defended their tradition and lifestyle by adapting to the plans the State has put in place for them in accordance to the blueprint for socialist development and modernization. As a nation, they worked hard to negotiate the environmental problems they faced as herders and fight the adverse consequences of their actions to ensure the maximum continuity of their traditions, identity and way of life. The reality is grim, as the Ewenki herders have been left with little space to negotiate their preference for their way of life. The arrival of the Anthropocene has clearly impacted the traditional "symbiotic domestication" Ewenki has had enjoyed traditionally with their reindeers. As the community struggled to cope with State policies that had resulted in partial alienation from their reindeers, the Ewenki set on an ever-struggling journey to cope with life between the taiga and the forest, reindeer herding and tourism, ethnic culture, and modernization.

Subba's chapter argues that experience of Yak herding has transformed to become more exclusive and discriminatory for marginal herders/farmers due to state policies. The effects of the Anthropocene era are felt by melting glaciers, as well as an overall degradation of the yak habitat due to man-made activities, such as the harnessing of hydropower. For the Himalayan population, Subba's chapter highlights the importance of the Yak in their daily lives – that the Yak is not only a beast of burden, an important player in the Alpine economy and an important symbol in the cultural imagination and identity of the people. Yaks are increasingly endangered not only because of the erosion of their traditional alpine habitat, but also because of the problems associated with inbreeding due to the fact that political borders and security zones have limited their movements in the frontier era. This creates problems for the preservation and conservation of the yak gene pool and affects the viability and sustenance of the species in itself.

The question of how the State in itself can become a problem rather than a solution in addressing the challenges of the Anthropocene era is illustrated in the Sikkim case jarringly as security tensions between China and India in the border area became a significant issue for yak herding. In the same light, **Camina Untalan**'s chapter is another interesting case study that examines human–animal interactions across national boundaries. Using the case of Philippine-China South China Sea disputes, it demonstrates how geopolitics' anthropocentrism and fixation toward sovereignty have thwarted promising attempts to put environment at the forefront of foreign policy. It argues for the necessity of recognizing animal agency in international politics as a way of nurturing humanity's common bond with the planet and of reorienting our understanding of world politics, from the politics of destruction to a politics of planetary survival, where big and small states, humans and animals could and should mutually coexist. Her chapter suggests that an Anthropocene-inspired world outlook could help explore possible ways of integrating nonhuman animal life in doing and thinking about international relations.

In the same vein, **Lisa Yoshikawa**'s chapter is an insightful and provocative one that offers readers much food for thought about how modern politics affect the way nation-states and governments go about interpreting problems and framing solutions related to conservation in the Anthropocene era. Yoshikawa's chapter examines the conservation of the Asian giant salamanders in the longue durée. In May 2018, a series of headlines hit the newsstand: "The Adorable Chinese Giant Salamander is Slithering Toward Extinction"; "5 Giant Salamander Species Identified—And They're All in Danger"; "How a Pyramid Scheme Doomed the World's Largest Amphibians." Yoshikawa argues eloquently that these stories provided familiar snapshots of the so-called Anthropocene. Humans, here the Chinese, were threatening the living fossil: through their farming practices to meet culinary and medicinal demands, ineffective government environmental protection, and more. Yet, the history of human-giant salamander relation reveals its complexity that spans two centuries and players beyond China. As at today, the IUCN red-lists two Asian giant salamander species: the Chinese and the Japanese. Until recently, scientists debated these demarcations due to limited technology and politics: competitive empire and nation buildings that shaped species taxa categorization, research trajectories, and conservation policies. It argues that the amphibian crisis is a regional manifestation of a historical and global phenomenon; demanding changes in Chinese human–animal interaction alone repeats the imperialist attitude that helped create the problem in the first place.

Across the world and particularly in Asia, the difficulties faced by the people living in the region in combating the effects of Anthropocene, as well as in overcoming the adverse consequences that the environment's impact on human–animal interactions, have so far been a mixed bag of success. The last three chapters of this book illustrate this through their case focus on China. In some instances, China lags behind global practices because segments of her population remain reticent about certain traditional beliefs and attitudes, particularly in their demand and consumption of certain animals. In other cases, the Chinese nation exhibits particular wisdom toward moving to a model that befits the global movement that is emerging to combat planetary ills. To that extent, the Chinese State has in fact acted more decisively than most governments around the world, even if the developments within China do not appear so.

Keokam Kraisoraphong's incisive chapter examines the fate of the honey-bees in China, and what this tells us about human–animal interactions in the Anthropocene. Starting out with the premise that regardless whether the Anthropocene began with the process of agricultural expansion some 7,000 or 8,000 years ago, or with the industrial revolution of the 18th century, China's ecological footprint has been very visible in this global process due to the Chinese people's domination of the ecosystem. Over the last three decades, China accelerated its speed of economic development by rapidly increasing control and technological mastery of nature. Viewed on the one hand as an achievement, China's inventiveness to push back the limits of nature and replace ecosystems function by human technology is particularly worrying. Contextualizing China's human–honeybee interactions by tracing China's path to becoming the world's leading honey producer and exporter, this chapter shows that China's developmental journey has not gone unchallenged from the perspective of *Anthropocene* as climate and ecological crises. Discussing China's honey trade, the chapter discusses the importance of a paradigm shift for China to become a redeeming agent – when China embraces the epochal consciousness and recognize the two complimenting conceptions of human – as a cultural and social being and as a biological species – to redefine their relations with honeybees, and realize that the fate of one determines that of the other.

Karisoraphong's chapter induces us to think more about the relationship between humans and the food we consume. At the earliest stage of human civilization, the relationship between human and animals has always had competitive and complementarity dimensions. Humans, like many animals, have relied on the consumption of other animals for sustenance. The binary of "wild" versus "domesticated" has thus been a concept that has evolved with humankind's ability to domesticate, farm or grow animal crops for food. Chicken husbandry, for instance, has been ascertained by bioarchaeologists to have spread from Thailand some 3,500 years ago to Europe (Gorman, 2022). Before they were "domesticated," chicken were "wild" animals that flew up to the trees in forested areas. The question of what is considered domesticated is therefore often malleable between cultures. Today, it remains a fact that crocodiles are farmed – but can they be truly considered "domesticated"?

In between cultures, what can be eaten or consumed are also often controversial. Even as humans have largely began to farm and cultivate certain species of animals for large-scale mass consumption in the earliest times, the division between "wild" and "domesticated" animals has always been somewhat unclear and often culture-specific. After centuries of evolvement, what can be considered "domesticated" and acceptable for food is largely homogenous in the developed world. Beef, poultry, and pork constitute the main staples for most. In the developing world, other than the aforementioned categories, there are often exotic animals that are consumed not just as food, but for (purported) medicinal, cultural, or egoistic reasons. The governments and laws here for most part are usually ambiguous as to what can be consumed or eaten. Even in areas that are not, the consumption of the animals considered "wild" or "exotic" often poses major public policy and transnational global challenges from smuggling of exotic animals, the establishing of illegal wildlife markets (UNDP, 2022), and, in the worst-case scenario, global health pandemics.

Yoon Sungwon's chapter highlights the importance of public health considerations in human–animal interactions in the Anthropocene era. The emergence and re-emergence of pandemics that occurred in recent decades underscore the role that human–wildlife interface played in precipitating the spread of infectious diseases. It is reported that 75% of all infectious diseases are potentially zoonotic and of wildlife origin. Outbreaks of Ebola virus infection, Middle East respiratory syndrome (MERS), and severe acute respiratory syndrome (SARS) virus infections are some of the examples that clearly illustrate the adverse consequences of the incursion of humans into wilderness. Although the role of human–wildlife interface in predicting emergent zoonosis is well documented, little attention has been paid to the cultural dimension of zoonosis. In other words, cultural variation in attitudes toward wildlife species as well as perceived risks when interacting with animals has not been well understood. Drawing on zoonotic infections originated from China, this chapter aims to examine the cultural and behavioral milieu in which human consumption of wildlife species as appetite and aspiration takes place in China's market capitalism and how such context contributes to the pathogenic crises. Surveys of wildlife markets in Guangzhou, China, invariably describe the diverse range of wildlife species such as masked palm civets, ferret badgers, barking deer, wild boar, bamboo rats, endangered leopard cats, and various species of hedgehogs, foxes, squirrels, gerbils, and snakes. In particular, the civet cats figure prominently in Chinese cuisines as the main ingredient in the exotic wildlife dish, posing significant challenges for China's ecology and environment. Speaking to the theme of human–animal interactions in Chinese culture, the chapter will discuss the complex relations that involve symbolic construction of civet cats as a substitute in Chinese culture, socioeconomic drivers, contemporary everyday life, and anthropogenic implications.

Yoon's chapter highlights the importance of the idea of the concept of "One Health." Zoonotic diseases emerge and spread when the resilience of people and the environment they live in provide the optimum conditions for the pathogens to spread. The recent COVID-19 pandemic has not only caused unprecedented chaos, but has claimed lives and destroyed economies across the world. Despite the investments that many countries have put into their health care systems, the COVID pandemic still managed to overwhelm hospitals and lock down entire cities. The perils of ignoring zoonotic diseases are very real, and the planetary impact of this disease showed that human civilization is not as resilient we make it out to be. The health of animals, humans, and the environment is more intimately linked than we imagined it to be. If we intrude into the spaces of the animals, enslave or capture these animals for food, we can set off a chain of events that would end up destroying our own environment or species (Hitchens & Johnson, 2020). It is vital that humankind stops destroying nature and eroding various ecosystems for their own developmental needs. The development growth model needs to be strenuously rethought and reconceptualized in many countries, particularly in Asia. At the same time, a more strenuous and united global response is required to fight the planetary crisis we are facing. This is, of course, easier said than done, remaining a distant dream given the practical state of affairs regarding the tussle not only between environmentalists and conservations and their enemies, but often because of the differences within each camp.

In the case of tiger conservation in China, this is clearly seen. China has been one of the countries which have been at the forefront of passing legislation to actually protect wildlife (or more accurately selected wildlife) within its territorial borders. Yet today, the question of tiger conservation has attracted much controversy and criticism, particularly from global green movements and wildlife interest groups based in the West. **Victor Teo's** chapter examines the institution of the private zoo in China, particularly for tigers. These private zoological gardens featuring captive tigers have come under heavy criticism for detracting as opposed to furthering professed conservation goals. These for-profit establishments are accused of profiteering, raising genetic monsters, and diverting resources away from the real tiger species that needed help. While this might be true under circumstances, the chapter argues it is important for everyone to look at the particular circumstances and difficulties faced in tiger conservation, and take the interests of the other social groups into consideration. By interrogating the human–animal relationship in China, the chapter argues for a middle path forward where the Chinese government can draw lessons from these criticisms to enhance their partnership with the private zoological parks to enhance tiger conservation. The chapter also argues that conservationists should also take existing practices and practical steps help to them improve. As the only player with sufficient clout and authority, the Chinese government too can play a mediating role between green advocates, commercial entities, and societal interests. Given that the State is possibly the only actor with sufficient clout to strike a balance, its role should be encouraged but monitored in the efforts to restore the tiger population in China. Conservationists could be more successful if they are more accommodating and stay in tune with the development aspirations of the nation. Fueled by traditional culture and folk beliefs, the illicit demand for tiger products can be countered only by education and persuasion supported by the rigorous and intelligent application of the law. The chapter concludes with some recommendations going forward.

As countries strive toward a global solution to combat the threat we are seeing from the Anthropocene, it is vital that communities, governments, and the people work collaboratively to combat the ills ailing our planet. As the case studies will show, it is often the government that is standing in the way of people trying to combat anthropogenic forces. It would be foolhardy for us to think that given the magnitude of the anthropogenic forces, it is only down to the governments that could and should work collaboratively to combat the Anthropocene. Much of the burden falls on ordinary folks to change their traditions, way of life, and day-to-day outlook. There needs to be a recognition of private and grassroots efforts as well in this fight.

Insofar as animals are concerned, Asians today need to correct many assumptions and traditions that they hold of animals in order to ensure that human–animal interactions return to a more even keel and respectful relationship. Recognizing that the human–animal binary is an artificial term and that human beings are just as much of the animal world (and the reverse is just as true) is a good start.

Underlying everything is the critical understanding that animals are our coinhabitants on this planet, and that their welfare must be taken into account not only because humankind has a moral duty to act, but also it is in our self-interest to do so. If we are interested in securing the future for our children, we should also understand that animals are also a form of "heritage" that must be protected so that our children and their children can enjoy them as well.

Asian societies must evolve with times and adopt and adjust with important developments in animal laws and ethics suited to their societies. Tradition and culture cannot be used unconditionally to defend practices that are detrimental to the conservation of animals. Conservationists would be more successful if they could understand the importance of respecting local cultures and working with social groups to seek win-win solutions. Changes to cultural practices are best done through persuasion and education, often taking much time and effort. This could help reframe humanity's relationship with animals and nature in the long run. Conservation and development can and should coexist. Undertaking both enterprises in a sustainable manner is the key to combating the problems of the Anthropocene, thus ensuring that human–animal interactions become more positive in the long run.

Note

1 The author wishes to thank the initial four reviewers at Harvard Yenching Institute as well as the three anonymous referees contacted by the publisher for their insightful comments. The author also wishes to thank Professor Lisa Yoshikawa for all the insightful comments on draft of this chapter.

Bibliography

AFP, "Indonesia Smugglers Stuffed Exotic Birds in Pipes: Police", 16 November 2017, https://phys.org/news/2017-11-indonesia-smugglers-stuffed-exotic-birds.html

AFP, "Chinese Villagers Co-exist Uneasily with Resurgent Hungry Elephants", *SCMP*, 19 August 2021, https://www.scmp.com/news/china/science/article/3145620/chinese-villagers-coexist-uneasily-resurgent-hungry-elephants

Allsen, Thomas, 2011, *The Royal Hunt in Eurasian History (Encounters with Asia)*, Philadelphia: University of Pennsylvania Press.

Ammarell, Gene, "Whither Southeast Asia in the Anthropocene?: Comments on the Papers from the 2014 Roundtable 'JAS at AAS: Asian Studies and Human Engagement with the Environment'", *The Journal of Asian Studies*, Vol. 73, No. 4, November 2014, pp. 1005–1007.

Arluke, Arnold & Sanders, Clinton R., 1996, *Regarding Animals*, Philadelphia, PA, Temple University Press.

Aubert, Maxime, Lebe, Rustan, Oktaviana, Adhi Agus, Tang, Muhammad, Burhan, Basran, Hamrullah, Jusdi, Andi, Abdullah, Hakim, Budianto, Zhao, Jian-xin, Geria, I. Made, Hadi Sulistyarto, Priyatno, Sardi, Ratno & Brumm, Adam, "Earliest hunting scene in prehistoric art", *Nature*, Vol. 576, 2019, pp. 442–445, https://doi.org/10.1038/s41586-019-1806-y

Bakke, Monika, 2009, "The Predicament of Zoopleasures: Human-Nonhuman Libidinal Relations", in Tyler, Tom & Rossini, Manuela (eds.), *Animal Encounters*, Leiden and Boston, MA: Brill.

Bale, Richard, "Five Ways China's Wildlife Protection Law Will Harm Wildlife", *National Geographic*, 2 February 2016, https://www.nationalgeographic.com/animals/article/160201-China-wildlife-protection-law-conservation

BBC, "Indonesia Foils Komodo Dragon Smuggling Gang", 28 March 2019, https://www.bbc.com/news/world-asia-47729022

Beeson, Mark, 2019, *Environmental Populism: The Politics of Survival in the Anthropocene*, Singapore: Palgrave.

Bhaduri, Anik, Bogardi, Janos, Leentvaar, Jan & Marx, Sina (eds.), 2014, *The Global Water System in the Anthropocene: Challenges for Science and Governance*, Cham: Springer.

Biermann, Frank & Lovbrand, E., 2019, *Anthropocene Encounters: New Directions in Green Political Thinking*, Cambridge: Cambridge University Press.

Birkeland, Charles (ed.), 2015, *Coral Reefs in the Anthropocene*, Heidelberg, New York and London: Springer Science.

Bladow, Kyle & Ladino, Jennifer, 2018, *Affective Ecocriticism: Emotion, Embodiment, Environment*, Lincoln: University of Nebraska Press (Literature).

Board, Jack, "Millions More in Southeast Asia Face Sea Level Rise Risks Than Previously Thought: Satellite Imagery Study", *Channel News Asia*, 2 July 2021, https://www.channelnewsasia.com/sustainability/sea-level-rise-southeast-asia-satellite-imagery-climate-change-1989406

Board, Jack & Promchertchoo, Pichayada, "Beasts of Burden: Hooks, Chains and Pain – How Thailand's Elephants Have Become Symbols of Despair", *Channel News Asia*, 28 September 2019, https://www.channelnewsasia.com/asia/thailand-elephant-shows-circuses-abuse-cruelty-training-858356

Bulliet, Richard W., 2005, *Hunters, Herders and Hamburgers: The Past and Future of Human-Animal Relationships*, New York: Columbia University Press.

Center for Biological Diversity, "Deadly Waters: How Rising Seas Threaten 233 Endangered Species", December 2013, https://www.biologicaldiversity.org/campaigns/sea-level_rise/pdfs/Sea_Level_Rise_Report_2013_web.pdf

Chandler, David, 2018, *Ontopolitics in the Anthropocene: An Introduction to Mapping, Sensing and Hacking*, Abingdon and New York: Routledge.

Chiaverini, Luca, "Biodiversity in the Anthropocene: Threats to Biodiversity in Southeast Asia", *The Oxford Institute of Population Ageing Blog*, Entry 23 October 2019, https://www.ageing.ox.ac.uk/blog/biodiversity-in-the-anthropocene

Clement, Sarah, 2021, *Governing the Anthropocene: Novel Ecosystems, Transformation and Environmental Policy*, Cham: Palgrave.

Conca, James, "Why Climate Change Hits Asia the Hardest", *Forbes*, 30 November 2020, https://www.forbes.com/sites/jamesconca/2020/11/30/climate-change-hits-asia-hardest/?sh=2bf35b430aa8

Constable, Pamela, "The Cats of War", *The Washington Post*, 21 July 2003, https://www.washingtonpost.com/archive/politics/2003/07/21/the-cats-of-war/d3a9de55-7da4-4e41-9a2c-e71f542cc2ec/

Contiff, Richard, "The Nitrogen Problem: Why Global Warming Is Making It Worse", 7 August 2017, https://e360.yale.edu/features/the-nitrogen-problem-why-global-warming-is-making-it-worse

Dabas, Haveer, "Man-Animal Conflict Turns Ugly, Villagers Lynch Leopardess in Bijnor", *The Times of India*, 29 January 2021, https://timesofindia.indiatimes.com/

city/meerut/man-animal-conflict-turns-ugly-villagers-lynch-leopardess-in-bijnor/articleshow/80515354.cms

Dalby, Simon, 2020, *Anthropocene Geopolitics: Globalisation, Security and Sustainability*, Ottawa: University of Ottawa Press.

Daniels, Joe Parkin, "I Was Terrified: The Vet Sterilizing Pablo Escobar's Cocaine Hippos", *The Guardian*, 23 October 2021, https://www.theguardian.com/world/2021/oct/22/vet-sterilising-pablo-escobar-hippos

Deloughrey, Elizabeth M., 2019, *Allegories of the Anthropocene*, Durham: Duke University Press (Literature & Ecology).

DeMello, Margo, 2012, *Animals and Society: An Introduction to Human-Animal Studies*, New York: Columbia University Press.

Despret, Vinciane (Translated by Brett Buchanan), 2007, *What Would Animals Say If We Asked the Right Questions*, Minneapolis: University of Minnesota Press.

Eco-Business, "Asia's Coastal Cities 'Sinking Faster Than Sea Level-Rise'", 27 April 2022, https://www.eco-business.com/news/asias-coastal-cities-sinking-faster-than-sea-level-rise/

Ellis, Erle C., 2018, *Anthropocene: A Very Short Introduction*, Oxford: Oxford University Press.

Elverskog, Johan, "Asian Studies + Anthropocene", *Journal of Asian Studies*, Vol 73, No. 4, November 2014, pp. 963–974.

Erisman, Jan Willem, Galloway, James N., Seitzinger, Sybil, Bleeker, Albert, Dise, Nancy B., Petrescu, A. M. Roxana, Leach Allison, M. & de Vries, Wim, 2013, "Consequences of Human Modification of the Global Nitrogen Cycle", *Philosophical Transactions of the Royal Society Bulletin*, Vol. 368, No. 1621, http://doi.org/10.1098/rstb.2013.0116

Fioritti, Nathan, "The Interconnectedness of Human, Animal and Environmental Health", *Pursuit* (University of Melbourne), 2 June 2020, https://pursuit.unimelb.edu.au/articles/the-interconnectedness-of-human-animal-and-environmental-healths

FAO, "Livestock's Long Shadow: Environmental Issues and Options", *Food and Agriculture Organisation*, 2006, https://www.fao.org/3/a0701e/a0701e.pdf

FAO (Food and Agriculture Organisation of the United Nations), "Land Use in Agriculture by the Numbers", News Entry, 7 May 2020, https://www.fao.org/sustainability/news/detail/en/c/1274219/

FAO (Food and Agriculture Organization of United Nations), "News – Land Use in Agriculture by the Numbers", 7 May 2020, https://www.fao.org/sustainability/news/detail/en/c/1274219/

Fieldman, Anthony, "Rebalancing the Earth Is Dead Simple", *Medium*, 20 December 2021, https://gen.medium.com/rebalancing-the-earth-is-dead-simple-616bb47c4b37.

Fremaux, Anne, 2019, *After the Anthropocene: Green Republicanism in a Post Capitalist World*, Cham: Palgrave Macmillan (Politics).

Gardner, Dinah & Greenpeace, "Five Facts about Sea-Level Rise in Asia That Will Surprise You", 1st July 2021, https://www.greenpeace.org/eastasia/blog/6679/five-facts-about-sea-level-rise-in-asia-that-will-surprise-you/

Gerber, P. J., Steinfeld, H., Henderson, B., Mottet, A., Opio, C., Dijkman, J., Falcucci, A. & Tempio, G., 2013, *Tackling Climate Change through Livestock: A Global Assessment of Emissions and Mitigation Opportunities*, Rome: Food and Agriculture Organization of the United Nations (FAO), https://www.fao.org/policy-support/tools-and-publications/resources-details/en/c/1235389/

Gibbons, Ann, "Are We in the Middle of a Sixth Mas Extinction?", *Science*, 2 March 2011, https://www.science.org/content/article/are-we-middle-sixth-mass-extinction

Gibbons, Ann, "The Evolution of Diet", *National Geographic* (Feature), 2013, https://www.nationalgeographic.com/foodfeatures/evolution-of-diet/

Gillson, Lindsey, 2015, *Biodiversity Conservation and Environmental Change: Using Palaeoecology to Manage Dynamic Landscapes in the Anthropocene*, Oxford: Oxford University Press.

Goodland, Robert & Anhang, Jeff, "Livestock and Climate Change: What If the Key Actors in Climate Change Are Cows, Pigs and Chickens?", *World Watch*, November/December 2009, https://awellfedworld.org/wp-content/uploads/Livestock-Climate-Change-Anhang-Goodland.pdf

Gorman, James, "Before Chickens Were Nuggets, They Were Revered", *The New York Times*, 7 June 2022, https://www.nytimes.com/2022/06/07/science/chicken-domestication-origin.html?referringSource=articleShare&fbclid=IwAR0hl ToPbwE0CpBfcBGyFjGlo85NsGrYo0cVysopxYw2BIYzrKF0JhYqjRs

Grandin, Temple & Whiting, Martin, 2018, *Are We Pushing Animals to Their Biological Limits? Welfare and Ethical Implications*, Oxfordshire: CABI Publishing.

Gutknecht, Danielle, "The Plight of the Orangutans: Finding Hope in an Uncertain Future", *Earth.Org*, 6 March 2020, https://earth.org/the-plight-of-the-orangutans-finding-hope-in-an-uncertain-future/

Halteman, Matthew C., "Varieties of Harm to Animals in Industrial Farming", *Journal of Animal Ethics*, Vol. 1, No. 2, 2011, pp. 122–31, https://doi.org/10.5406/janimalethics.1.2.0122

Hamilton, Clive, Bonneuil, Christophe & Gemenna, 2016, *The Anthropocene and the Global Environmental Crisis*, Abingdon and New York: Routledge.

Harari, Yuval Noah, "Industrial Farming Is One of the Worse Crimes in History", *The Guardian*, 25 September 2015, https://www.theguardian.com/books/2015/sep/25/industrial-farming-one-worst-crimes-history-ethical-question

Harding, Luke, "Gaddafi's Son Abandons His Lions in Flight from Tripoli", *The Guardian*, 30 August 2011, https://www.theguardian.com/world/2011/aug/30/gaddafi-son-abandons-lions-tripoli

Hilton, Isabel, "Crossing the Bridge between Specialized Knowledge and Breadth of Vision in Regard to Climate Change and Asia", *The Journal of Asian Studies*, Vol. 73, No. 4, November 2014, pp. 1001–1004.

Hitchens, Peta Lee & Johnson, Christine K., "Don't Blame the Pagolin (Or Any Other Animal) for Covid 19", *Pursuit*, 8 April 2020 (University of Melbourne Publication), https://pursuit.unimelb.edu.au/articles/don-t-blame-the-pangolin-or-any-other-animal-for-covid-19

Hudson, Mark J., "Placing Asia in the Anthropocene: Histories, Vulnerabilities, Responses", *Journal of Asian Studies*, Vol. 73, No. 4, November 2014, pp. 941–962.

Huff, Leah, "Anthropocene Debates: Talks of the Apocalypse and a Way Forward", *Currents: A Student Blog – Exploring the Intersections of Water, People and the Environment*, 8 March 2021, https://smea.uw.edu/currents/anthropocene-debates-talks-of-the-apocalypse-and-a-way-forward/

Human League, "Factory Farming and the Environment: Impacts on the Planet", *The Human League*, 4 March 2021, https://thehumaneleague.org/article/factory-farming-and-the-environment

Ingold, Tim, 1980, *Hunters Pastoralists and Ranchers: Reindeer Economies and Their Transformations*, Cambridge and New York: Cambridge University Press.

Ingold, Tim (eds.), 1994, *What Is an Animal?*, London: Routledge.

48 *Victor Teo*

Irvine, Leslie, 2009, *Filling the Ark: Animal Welfare in Disasters*, Philadelphia, PA: Temple University Press.

Jagodzinski, Jan, 2018, *Interrogating the Anthropocene: Ecology, Asthetics, Pedagogy and the Future in Question*, Cham: Palgrave Macmillan Springer Nature.

Jakarta Post, "African Lions, Leopard Smuggled in Boxes from Malaysia", 16 December 2019, https://www.thejakartapost.com/news/2019/12/16/african-lions-leopard-smuggled-in-boxes-from-malaysia.html

Joe, Wallen, "Elephants Kills Woman before Returning to Trample Her Corpse at Funeral", *The Telegraph*, 13 June 2022, https://www.telegraph.co.uk/world-news/2022/06/13/elephant-kills-woman-returning-trample-corpse-funeral/

Kaiwen, Su, Jie, Ren, Jie, Yang, Yilei, Hou & Yali, Wen, "Human-Elephant Conflicts and Villagers' Attitudes and Knowledge in the Xishuangbanna Nature Reserve, China", *International Journal of Environment Research and Public Health*, Vol. 17, No. 23, 30 November 2020, p. 8910, https://doi.org/10.3390/ijerph17238910

Kalof, Linda, 2007, *Looking at Animals in Human History*, London: Reaktion Books.

Kemmerer, Lisa, 2006, *In Search of Consistency: Ethics and Animals*, Leiden and Boston, MA: Brill Publishing.

Kittisilpa, Juarawee, "Elephants March in Thailand to Pay Respects to Newly Crowned King", *Reuters*, 7 May 2019, https://www.reuters.com/article/us-thailand-king-coronation-elephants-idUSKCN1SD0RL

Kohn, Eduardo, 2013, *How Forests Think: Toward an Anthropology Beyond the Human*, Berkeley, Los Angeles & London: University of California Press.

Kowner, Rotem, Bar-Oz, Guy, Biran, Michal, Shahar, Meir & Shelach-Lavi, Gideon, 2019, *Animals and Human Society in Asia: Historical, Cultural and Ethical Perspectives*, Cham: Palgrave Macmillan.

Lam, Lydia, "Putin Turns Down Japan's Dog Gift: 5 Types of Animal Diplomacy", *The Straits Times*, 12 December 2016, https://www.straitstimes.com/world/putin-turns-down-japans-dog-gift-5-types-of-animal-diplomacy

Levin, Phillip S. & Poe, Melissa (eds.), 2017, *Conservation for the Anthropocene Ocean: Interdisciplinary Science in Support of Nature and People*, London & San Diego: Academic Press Elsevier.

Lo, Kwai-Cheung & Yeung, Jessica, 2019, *Chinese Shock of the Anthropocene: Image, Music and Text in the Age of Climate Change*, Singapore: Palgrave Macmillan.

Lorimer, Jamie, 2015, *Wildlife in the Anthropocene: Conservation after Nature*, Minneapolis and London: University of Minnesota.

Marvier, Michelle, Kareiva, Peter & Lalasz, Robert, "Conservation in the Anthropocene: Beyond Solitude and Fragility", *Breakthrough Journal*, Fall 2011, No. 2, 1 February 2012, https://thebreakthrough.org/journal/issue-2/conservation-in-the-anthropocene

Marvin, Garry & McHugh, Susan, 2014, *Routledge Handbook of Human-Animal Studies*, Abingdon and New York: Routledge.

McFarland, Sarah E. & Hediger, Ryan (eds.), 2009, *Animals and Agency: An Interdisciplinary Exploration*, Leiden and Boston, MA: Brill.

Mimura, Nobuo, "Rise in Sea Level and the Asia-Pacific Region", *Ocean Policy Institute*, 5 January 2022, https://www.spf.org/en/opri/newsletter/34_2.html?full=34_2

Mok, Laramie, "5 of China's Most Memorable Diplomatic Gifts to World Leaders", *South China Morning Post*, 30 April 2018, https://www.scmp.com/magazines/style/people-events/article/2143649/5-chinas-most-memorable-diplomatic-gifts-world-leaders

Morton, Timothy, 2016, *Dark Ecology: For a Logic of Future Co-Existence*, New York: Columbia University Press.

Mukherjee, Sugato, "Where Humans Don't Fear Leopards", 5 May 2022, https://www.bbc.com/travel/article/20220504-jawai-where-humans-dont-fear-leopards

Myer, Robinson, "The Catalysmic Break That (Maybe) Occurred in 1950", *The Atlantic*, 17 April 2019, https://www.theatlantic.com/science/archive/2019/04/great-debate-over-when-anthropocene-started/587194/

NASA Earth Observatory, "Global Temperatures", n.d., https://earthobservatory.nasa.gov/world-of-change/global-temperatures

National Geographic, "Domestication" – Encyclopedic Entry, Updated May 2022, https://education.nationalgeographic.org/resource/domestication

National Geographic Society, "Anthropocene" – Encyclopedic Entry, June 2019, https://www.nationalgeographic.org/encyclopedia/anthropocene/

Ng, Ee Ling, Chen, Deli & Edis, Robert, "Nitrogen Pollution: The Forgotten Element of Climate Change", *The Conversation*, 4 December 2016, https://theconversation.com/nitrogen-pollution-the-forgotten-element-of-climate-change-69348

Nibert, David (ed.), 2017, *Animal Oppression and Capitalism: The Oppression of Nonhuman Animals as Sources of Food*, Santa Barbara, CA and Denver, CO: Praeger.

Noone, Kelvin J., "Problem Solving in the Anthropocene", *Eco-Business*, 8 October 2013, https://www.eco-business.com/opinion/problem-solving-anthropocene/

Ohnuma, Reiko & Ambros, Barbara, 2019, "Buddhist Beasts: Reflections on Animals in Asian Religions and Cultures", *Religions*, Vol. 10, Special Issue, ISSB 2077-1444, https://www.mdpi.com/journal/religions/special_issues/beasts#published

Ong, Sandy, "Covid Casts Asia's Exotic Pet Trade in Harsher Light: From Thailand to Japan, Health Risks Add to Animal Welfare and Biodiversity Concerns", *Nikkei Asia*, 23 November 2021, https://asia.nikkei.com/Spotlight/Asia-Insight/COVID-casts-Asia-s-exotic-pet-trade-in-harsher-light

Oppermann, Serpil, 2017, *Environmental Humanities: Voices from the Anthropocene*, London and New York: Rowman & Littlefield.

Pacoureau, Nathan, Rigby, Cassandra L., Pollom, Riley A., Jabado, Rima W., Ebert, David A., Finucci, Brittany, Pollock, Caroline, Cheok, Jessica, Derrick, Danielle, Herman, Sherman, Samantha, C., VanderWright, Wade J., Lawson, Julia M., Walls, Rachel H. L., Carlson, John K., Charvet, Patricia, Bineesh, Kinattumkara K., Fernando, Daniel, Ralph, Gina M., Matsushiba, Jay H., Hilton-Taylor, Craig, Fordham, Sonja V. & Simpfendorfer, Colin A., "Overfishing Drives over One-Third of All Sharks and Rays toward a Global Extinction Crisis", *Current Biology*, Vol. 31, No. 21, November 2021, pp. 4773–4787, https://doi.org/10.1016/j.cub.2021.08.062

Philip, Kavita, "Doing Interdisciplinary Asian Studies in the Age of the Anthropocene", *The Journal of Asian Studies*, Vol. 73, No. 4, 2014, pp. 975–987.

Purdy, Jedediah, 2015, *After Nature: A Politics for the Anthropocene*, Cambridge: Harvard University Press (Environmental Law / Politics).

Rao, "Animal Cruelty Is the Price We Pay for Cheap Meat", *Rolling Stone Feature*, 10 December 2013, https://www.rollingstone.com/interactive/feature-belly-beast-meat-factory-farms-animal-activists/

Santos, Maureen, "Animals Asia Uncovers Animal Abuse in Vietnam's Circuses and Amusement Parks", *ASEAN Economist*, 16 April 2022, https://www.aseaneconomist.com/animals-asia-uncovers-animal-abuse-in-vietnams-circuses-and-amusement-parks/

Seddon, Nathalie, Mace, Georgina M., Naeem, Shahid, Tobias, Joseph A., Pigot, Alex L., Cavanagh, Rachel, Mouillot, David, Vause, James & Walpole, Matt, "Biodiversity in the Anthropocene: Prospects and Policy", *Proceedings of the Royal Society B. Biological Sciences*, 2016, http://doi.org/10.1098/rspb.2016.2094, https://royalsocietypublishing.org/doi/10.1098/rspb.2016.2094

Sengupta, Rajit, "Land Degradation in India Hurts Farmers and Forest Dwellers the Most", 27 August 2021, https://www.downtoearth.org.in/news/environment/land-degradation-in-india-hurts-farmers-and-forest-dwellers-the-most-78701Gal

Seploski, David, 2000, *Catastrophic Thinking: Extinction and the Value of Diversity from Darwin to the Anthropocene*, Chicago & London: The University of Chicago Press.

Shao, Wanyan, Keim, Barry D., Garand, James C., & Hamilton, Lawrence C., "Weather, Climate, and the Economy: Explaining Risk Perceptions of Global Warming, 2001-10", *Weather, Climate, and Society*, Vol. 6, No. 1, 2014, pp. 119–134, https://journals.ametsoc.org/view/journals/wcas/6/1/wcas-d-13-00029_1.xml

Siddons, Edward, "Pieter Hugo's Best Photograph: The Hyena Men of Nigeria", *The Guardian*, 19 July 2018, https://www.theguardian.com/artanddesign/2018/jul/19/pieter-hugo-best-photograph

Slimak, Ludovic, Zanolli, Clement, Lewis, Jason E. & Metz, Laure, "New Research Suggests Modern Humans Lived in Europe 10,000 Years Earlier Than Previously Thought in Neanderthal Territories", 9 February 2022, https://theconversation.com/new-research-suggests-modern-humans-lived-in-europe-10-000-years-earlier-than-previously-thought-in-neanderthal-territories-176648

Smith, Josh, "Defiant Message' as North Korea's Kim Rides White Horse on Sacred Mountain", *Reuters*, 16 October 2019, https://www.reuters.com/article/us-northkorea-kimjongun-idUSKBN1WV08G

Soumya, Elizabeth, "Leopard of Mumbai: Life And Death Among City's 'Living Ghosts'", NDTV Mumbai News Feature, 27 Nov 2014, (Originally published in The Guardian) https://www.ndtv.com/mumbai-news/leopard-of-mumbai-life-and-death-among-citys-living-ghosts-704734

Steffen, Will, Richardson, Katherine, Rockstrom, Johan, Cornell, Sarah E., Fetzer, Indigo, Bennett, Elena M., Biggs, Reinette, Carpenter, Stephen R., De Vries, Vim, Folke, Carl, Gerten Dieter, Heinke, Jens, Georgina, M. Mace, Persson, Linn M., Ramanathan, Veerabhadran, Reyers, Belinda & Sorlin, Sverker, "Planetary Boundaries: Guiding Human Development on a Changing Planet", *Science*, Vol. 347, No. 6223, 15 January, 2015, https://doi.org/10.1126/science.1259855

Stuart, Diana & Gunderson, Ryan, 2020, "Human-Animal Relations in the Capitalocene: Environmental Impacts and Alternatives", *Environmental Sociology*, Vol. 6, No. 1, pp. 68–81, https://10.1080/23251042.2019.1666784

Sueroa, Adolfo, 2017, *Economics of the Anthropocene*, Cham: Palgrave Macmillan.

Tonnessen, Morten, Oma, Kristin Armstrong & Rattasepp, Silver, 2016, *Thinking about Animals in the Age of Anthropocene*, London: Lexington Books.

Tsing, Anna, Heather, Swanson, Gan, Elaine & Bubandt, Nils, 2017, *Arts of Living on a Damaged Planet: Ghosts and Monsters of the Anthropocene*, Minneapolis: University of Minnesota.

Turner, Lynn, Sellbach, Undine, Broglio, Ron, 2018, *The Edinburgh Companion to Animal Studies*, Edinburgh: Edinburgh University Press.

Twine, Richard, 2010, *Animals as Biotechnology: Ethics, Sustainability and Critical Animal Studies*, London and Washington, DC: Earthscan.

Tyler, Tom & Rossini, Manuela, 2009, *Animal Encounters*, Leiden and Boston, MA: Brill.

UNDP, "Wave of Solidarity: Combatting Wildlife Trade, Protecting Biodiversity and Wildlife and Protected Areas", 19 May 2022, https://undp-biodiversity.exposure. co/wave-of-solidarity

Vince, Gaia, 2014, *Adventures in the Anthropocene*, London: Milkweed Editions (General Engagement; Environmental Sciences).

Voigt, Christian C. & Kingston, Tigga, 2016, Bats in the Anthropocene. In: Voigt, Christian, and Kingston, Tigga. (eds.) *Bats in the Anthropocene: Conservation of Bats in a Changing World*. Cham: Springer, https://doi.org/10.1007/978-3-319-25220-9_1

Wang, Guo-Dong, Zhai, Weiwei, Yang, He-Chuan, Wang, Lu, Zhong, Li, Liu, Yan-Hu, Fan, Ruo-Xi, Yin, Ting-Ting, Zhu, Chun-Ling, Poyarkov, Andrei D., Irwin, David M., Hytönen, Marjo K., Lohi, Hannes, Wu, Chung-I, Savolainen, Peter, & Zhang, Ya-Ping, "Out of Southern East Asia: The Natural History of Domestic Dogs across the World, *Cell Research*, Vol. 26, 2016, pp. 21–33, https://doi.org/10.1038/cr.2015.147

Watanabe, Toru & Watanabe, Chiho, 2019, *Health in Ecological Perspectives in the Anthropocene*, Singapore: Springer Nature.

Woody, Todd, "The Sea Is Running Out of Fish, Despite Nations' Pledges to Stop It", *National Geographic*, 8 October 2019, https://www.nationalgeographic.com/ science/article/sea-running-out-of-fish-despite-nations-pledges-to-stop

WWF, "Overfishing", WorldWildlife Foundation, n.d., https://www.worldwildlife. org/threats/overfishing

Xinhua, "Across China: China Defuses Human-Elephant Conflict", 19 August 2020, http://www.xinhuanet.com/english/2020-08/19/c_139302139.htm

Yan, Alice, "Chinese Officials Fined for Keeping Tigers as Pets after One Leapt to Death from 11-Storey Building", *South China Morning Post*, 18 March 2015, https://www.scmp.com/news/china/article/1740866/three-chinese-lawmakers-fined-keeping-endangered-siberian-tigers-pets

2 "Rearing the Rare"

History, Coexistence, and Contestation of Wildlife Conservation in the Singalila National Park, Darjeeling, India

Sangay Tamang

Introduction

The most admirable, yet rare, mammal the red panda (*Ailurus fulgens*) has in modern time evolved as a flagship of conservation in Himalayan states like China, Nepal, Bhutan, and various parts of India. Some conservationists consider red panda as "umbrella species" "that means that, ideally, conservation efforts put into place to protect them also will protect other animals within their geographical areas" (cited from Red Panda Network eNews Vol. 17, April 7, 2020). Although it is extremely difficult to sight this mammal in the wild, the red panda in the zoo has become a popular sight of both conservation and tourism. Writing about the red panda in this chapter is perhaps accidental. It was never in my mind that I would be researching some nonhuman species as part of my research work; however, the significance of the red pandas in the making of landscape, politics, and environmental laws in the Eastern Himalayas made this "other species" as an inevitable part of my research. In late 2017 and early 2018, when I first visited the forest villages situated in the Indo-Nepal border primarily to understand the historic evolution of colonial forestry and contemporary forest management as a part of my doctoral thesis, I was amazed by the everyday stories that locals illustrate about this particular species (red panda). It was stories of relationship, companionship, and coexistence that human and nonhuman (red panda) shared within the contested site of forest management in Darjeeling Hills. This chapter aims to explore these relationships of human and nonhuman within the political ecology of wildlife conservation in the Singalila National Park. However, this relationship between human and nonhumans (in this case "red panda") is both contested and harmonious determined by structural rules and laws governing the nature, and "evolved from co–evolutionary histories, from rich process of co–becoming" (Dooren et al., 2016: 2). Co-becoming has been an important aspect of regional political aspiration in postcolonial society in which nonhuman became a symbol of regional identity politics.

Conservationists believed that the increasing anthropogenic pressure on land resulted into the habitat loss of the red panda further aggravated by the poaching in the open international border between India and Nepal in the Singalila.[1] Locals, on the other hand, claim to have been residing in the land

DOI: 10.4324/9781003212089-2

since time immemorial and further complain of being a victim of both conservationists and law enforcing authority after the establishment of the
Singalila National Park.[2] Nonetheless, our engagement with both local communities and forest department staffs has made us believe that apart from
anthropogenic factors, there are several other additional factors contributing
to the anthropogenic problems in wildlife protection in the Singalila which
we often tend to overlook. Hence, this chapter engages with the evolution of
the park and human relationship with the landscape that underwent massive
transformation in recent times. Thus, locating within the realm of critical
political ecology that challenges the exclusionary logic of the state-driven
conservation approach, this chapter attempts to engage with the contestation, challenges and conflict in red panda conservation in the eastern
Himalayas and tries to show how nonhuman has become an integral part of
intersection between environment and ethnicity in the Eastern Himalayas.

A large corpus of scholarship on conservation has successfully argued that
the establishment of protected areas since colonial times has led to tensions
over resources ownership and human mobility (Saberwal et al., 2001), thereby
excluding a large mass of common from nature. However, over the last two
decades "trends in conservation have found new ways of framing both the
relationship between parks and people and people and environment, moving
from top-down to participatory forms of management" (Rasmussen et al.,
2019: 01). This advocacy for community participation in nature conservation
has its own limitation (Agrawal and Gibson, 1999); however, it has gained
significant proponents concerning the ethos of democracy in accessing the
use of natural resources. This chapter attempts to show how tensions in conservation get manifested within the discursive elements of community formation and further illustrates how wildlife (red panda) conservation became a
contested site of regional politics.

Another important dimension that makes the state conservation project
controversial in the Singalila National Park (hereafter SNP) is the presence
of an open political border between India and Nepal (see Figure 2.1). This
border often plays a crucial role in the enforcement of the law, presence of
state authorities, and community involvement. Since it is almost difficult to
demarcate the line between these two nations, two contrasting environmental
laws often played a significant role in determining the ecosystem in the SNP.
The protected space with the rigid colonial rule on the Indian side is con-
trasted by the "community forestry" in Nepal, and this leads to cross border
conflict over the settlement of laws and management of the park. The Indian
state views the expansion of agriculture on Nepal side as one of the serious
threats to red panda habitat (as mentioned in the management plan) while
communities claim to have historically coexisted with red panda before the
national park came into existence. Thus, this chapter with ethnographic
exploration would attempt to reflect on such contestation over the politics of
conservation in the SNP.

The chapter is organized as follows: the second part would reflect on the
historical evolution of the national park in the Singalila Range of the Eastern

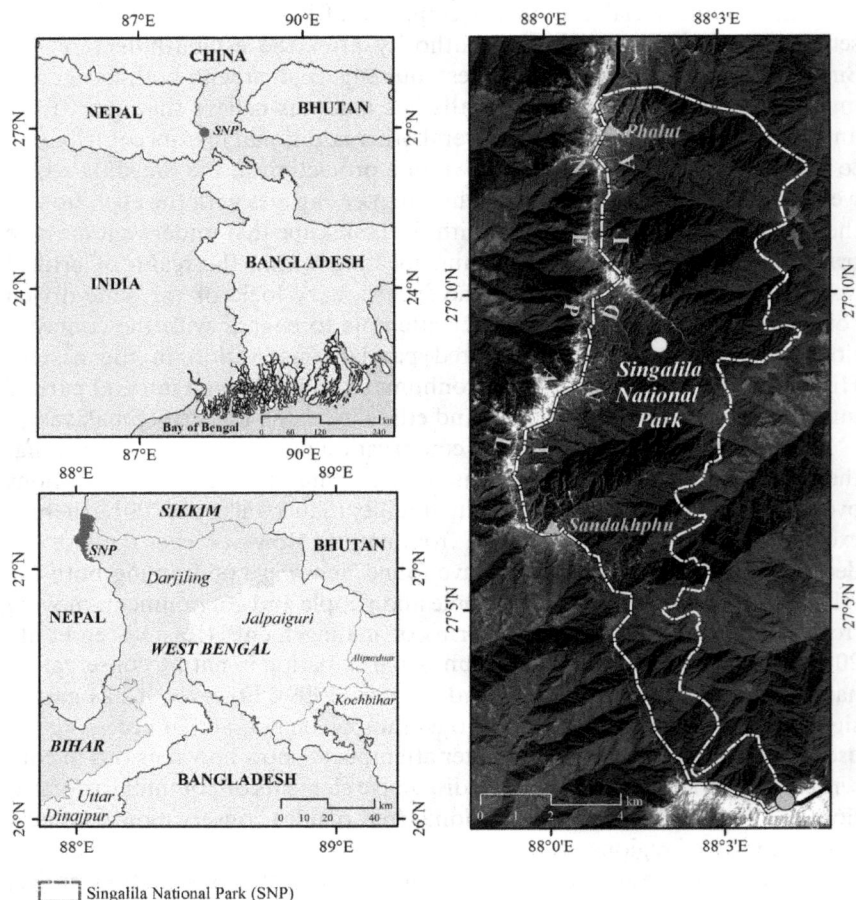

Figure 2.1 Location of the Singalila National Park
(figure created by author).

Himalayas. It would further reflect on the regional conservation agenda that had conditioned the establishment of the park in the region. By engaging briefly with the red panda's conservation the third section intends to explore the contradiction over the idea of "Anthropocene" in which different kinds of people are categorized differently and classified. This coexistence between human and red panda needs to be critically analyzed while considering the "kinds of people" (Hacking, 2007). What are the kinds of people that cause more environmental and cultural damages and how this affects wildlife conservation in the region? The fourth section discusses the role of other agencies like bamboo in red pandas' conservation that invites rethink on "more than human histories" (O'Gorman and Gaynor, 2020) and/or "multispecies Solidarities" (Dooren, Kirksey and Munster, 2016).

In a recent DNA study conducted in China, it is found that Chinese red pandas and Himalayan red pandas are two separate species based on physical features. "Chinese red pandas have redder fur and striped tail rings, while Himalayan pandas have whiter faces,"[3] and the study urges that the conservation action plan needs to be specific to the two different species in its contextual, regional, and local levels. However, the history of red panda and the reading of some conservationist writing in China and India suggest some historical linkage among the family of the species. Although two species might differ scientifically and also might not have coexisted together in the same space, however, to understand human and wildlife coexistence at the local level needs much deeper and critical engagement with the local history of not just species but also the landscape in which it thrives.

The SNP (see Figure 2.1) located in the isolated Himalayan landscape with large part covered by blankets of the alpine forest is home to Himalayan red pandas as well as to various highland communities such as Sherpa, Lepcha, Bhutia, and micro ethnic communities of larger Nepali heritage. Due to the complex geographical landscape spread through different elevation and difficult hilly terrain, this study was conducted through multi-sited ethnography (Marcus, 1995) in villages lying in both the core and buffer zone of the SNP. As the area is difficult to access due to its widely scattered villages, we approached the place through "walking ethnography" (Cheng, 2014; Chettri and McDuie-Ra, 2018). Similarly, by considering the strategic location of the park along the international and national border, walking ethnography helps in capturing the mobility of everyday life, environmental flows, and cultural connectivity in and between different spaces produced by the concept of borders and national park. However, we deviate a little away from walking ethnography due to the fact that the route to the SNP is famous for trekking and there are differences in general walking and doing trekking. Even though walking is an integral component of doing trekking, however, differences in terms of its accessibility of time and space often render different meaning between the two. Walking, in general, is not based on any stipulated time or destination unlike in trekking where a predefined plan of covering a certain distance conditioned the trekkers. One can walk at any point of time (day or night, month or year) as per his/her convenience; however, trekking has a certain specific season and is mostly conducted during the daytime.

Further, the SNP remains closed for visitors from June 15 to September 15 of the year due to the breeding season of wildlife; hence, we conducted fieldwork in three different phases: the first phase was conducted on the eastern side of the park in May 2018 and villages – mostly *Gorkhay*, *Samenden*, and *Ramam* – were visited; the second phase was conducted on the western side of the park in late May and early June 2018, and the third phase was conducted in the entire park during October and November 2018 and May 2019. However, we spent a good amount of time in the lower periphery of the park to understand the dynamics of everyday mobility within and outside the park.

Evolution of the Singalila National Park in Darjeeling Hills

Most of today's national park and wildlife sanctuaries in the country used to be hunting grounds for British officials or princes or landed aristocracy (Beinart and Hughes, 2007; Mackenzie, 1997; Rangarajan, 2015). However, the SNP evolved primarily not as a hunting reserve[4] for white officials, rather as a site for adventure, science, and sports. Thus, the story of the Singalila shares different historical narratives. The area that is today designated as the SNP in Darjeeling Hills was purchased by the British Government from the descendants of Chebu Lama[5] in 1882–1883 (O'Malley, 1907 [1999]: 91) and used mainly as an approach route to Mount Everest and Mount Kanchenjunga expedition by early colonial officials like Jules Jacot-Guillarmod and the famous occultist Aleister Crowley in 1905. Even Tenzing Norgay used routes from Singalila during several expeditions (Norgay and Ullman, 1955).

Owing to its high elevation and difficult terrain that constrain accessibility to forest product, timber logging from the Singalila was not an easy task, and hence, the region was used mostly for white men's desire for expedition, science, and sport. The Himalayan club established on 6 October 1927 described that the British interest in the Himalayas is "to encourage and assist Himalayan travel and exploration, and to extend knowledge of the Himalaya and adjoining mountain ranges through science, art, literature and sport" (Corbett, 1929: 2). Similarly, the Singalila was reserved for the experimentation of white's knowledge on expedition and science.

Notably, the higher ridges forest of Darjeeling was not a desirable place for colonial hunting unlike those in other parts of Assam, Central, and Western India and hence, the British forest department was less concerned about the protection of wildlife.[6] The Singalila also served as an important destination for various natural history museums to collect specimens of floras and faunas of the Eastern Himalayas. In 1883, a famous British botanist, Sir Joseph Dalton Hooker, visited the region of Singalila for his scientific exploration of floras in the Eastern Himalayas. However, in 1908 the Singalila was brought under systematic management of modern forestry by Mr. Trafford and attempts were made to create systematic group felling and plantation. Similar efforts were also made by another colonial forest official, Grieve, in 1912. However, all these efforts proved to be inefficient considering the remote location and inaccessible track.[7]

Apart from the colonial desire for expedition, the Singalila was an important transit point during the Gorkha invasion of Sikkim in the late 18th and early 19th centuries. In 1788–1789, they invaded eastward across the slope of the Singalila and advanced up to Rubdentze West Sikkim, which was once the capital of the Rajah of Sikkim (Samanta, 2018: 08). Most importantly, the landscape was an important grazing ground for trans-human grazers like Bhutias, Sherpas, and some others nomadic groups who traveled upward during summer and move downward during winter[8] and moved around Tibet, Bhutan, Nepal, and independent Sikkim. Even after the British took control over the forest, the upper part of the Singalila was kept open for grazing and the grazers used to establish goths (herder huts/cattle stations) and settle down either temporarily or permanently. In fact, the Singalila with its richness in Silver Fir, Birch, and Rhododendron with large tract covered by bamboos (dominantly *Maling*) often reflects a landscape deeply entrenched by human–nature interaction that cohabitated human and wild animal since ages.

Nevertheless, the postcolonial state control over wildlife in India generates considerable significance of wilderness and, as argued by Rangarajan, that "wildlife in free India remained a subject to be dealt with by foresters" (2001: 39). This power bestowed to foresters to protect nature and wildlife was not as vigilant till 1970 as the old colonial and princely ethos of hunting continued to shape wildlife in India. It was only in the early 1970s when Indira Gandhi[9] addressed a major international conservation conference in New Delhi, there emerged a new dimension of wildlife (ibid.). In 1972 when the wildlife protection act was passed, the significance of the Singalila diverted toward protecting Himalayan fauna that was increasingly gaining popularity after the establishment of Himalayan Zoological Park or commonly known as the Darjeeling Zoo in 1958. The same year in 1972, the Darjeeling Zoo became a registered society, with an agreement that maintenance costs would be shared by the central and state governments. In 1975, the then Prime Minister of India, Indira Gandhi, renamed the zoo as Padmaja Naidu Himalayan Zoological Park in memory of the late Smt. Padmaja Naidu, ex-governor of West Bengal. The establishment of the zoo in the urban area constantly shaped the image of wild space and wildlife, and hence, the Singalila was reconfigured outside of its excessive isolation and represented it as the hub of Himalayan endangered animals like snow leopard, Himalayan wolf, and red panda.

The 1980s witnessed more stringent rules to protect forest and wildlife especially after Indira Gandhi lectured in 1982 on "how the country needed some really 'hard measures' to halt denudation of hill catchments" (Rangarajan, 2015: 174). Similarly, the forest bill of 1982 that proposed to bestow magisterial power to forest officers created great resentment among many Adivasi or forest dwellers in the country. This state control over forest and wildlife was marked by the exclusionary logic of conservation that perceives human as one of the essential threats to wildlife and keeping people outside the areas of conservation interest was considered as a critical need. In 1986, the Singalila was declared as a wildlife sanctuary. The same year, the region witnessed a massive political turmoil unfolding over the issues of the Gorkha identity crisis and the demand for separation from West Bengal. This has greatly shaped the idea of community in the region, and hence, a large mass of people consider the Gorkha community's identity crisis as one of the reasons for their marginalization.

Coincidentally, after the withdrawal of the Gorkhaland Movement and when Subash Ghisingh took charge of DGHC, the Singalila was upgraded to National Park in 1992 along with Neora Valley in Kalimpong, by the

Government of West Bengal, and in 1993 the Darjeeling Zoo was transferred to the West Bengal Forest Department. Subsequently, the management plan for the SNP was prepared. Meanwhile, the grazing pattas were stopped and hitherto mobile pastoralists were evicted either with threat or with promise of compensation (Thomas, 2011). Resistance by local communities in the SNP failed to advance into larger regional politics and promises was made to compensate those evicted with jobs in the Forest Department. Thus, Samuel Thomas writes,

> In the Singalila National Park, evictions started in earnest right after the declaration of the National Park and most of the goths (herder huts/ cattle stations) were evicted. Some held on until threatened with force. The recording and settling of rights were arbitrary, and by most accounts, unfair. Only two herders got 'jobs' with the forest department, as a form of compensation, and both remain on the casual workers list even today. Some herders who kept cattle in the lower reaches, and also worked on road maintenance, were settled in the villages on the periphery. Most nomadic pastoralists got nothing.
>
> (2011: 11)

In no time, the national park of great significance evolved and soon became a major tourist destination in West Bengal. The story of eviction and suppression soon faded away from public memories, and the Singalila became one of the important tourist destinations in Darjeeling Hills. Scholars on conservation argued that "the eviction of humans from areas of conservation interest helps sustain the myth of 'pristine' or 'natural wilderness'" (Saberwal et al., 2001: 44). In the Singalila Range, although population density is comparatively very less since the colonial era owing to its isolated location, it is not right to completely deny the historical existence of various communities in the Singalila. Thus, the notion of "pristine" is a coproduction of human and nature built in a harmonious way.

Notwithstanding, with its distinguished varieties of floras and faunas of the Eastern Himalayan origin often characterized by unique species of temperate and alpine forest, the SNP became a biodiversity conservation zone with varied theories and designs nowhere representing the voices of those evicted. While on the Nepal side of the park, the community forestry system, though initiated by some external agencies with the support of the government, fuels the argument further that "the mobile pastoralism is bad for the environment" (Thomas, 2011: 12) and restricted the movement of grazers/ herders. The state erected Sashastra Suraksha Bal (SSB) camp in the major location of the park that perhaps have eased the check on mobilities in the border. One of the local respondents while sharing his everyday life exclaimed that "there is no border for us, we can easily move around but sometime there will be restriction on transport of cattle."[10] The construction of a border for nonhuman directly affects the mobility and livelihood condition of human and reconfigured their subsistence economy. Thus, a majority of people who were earlier pastoralists now shifted toward settled cultivation and tourism as their source of income. This displacement of community and transformation of the economy not only poses a rapid change in the ecosystem of the SNP but also affects the coexistence between human and wildlife—redefined by new notion of human, economy and conservation. The reconfiguration of the relationship between the landscape, wildlife, and human after the establishment of the national park invites critical engagement with the scholarship on conservation and in the following sections we will attempt to bring the contestation over red panda conservation in the SNP.

Rearing the Red: Categorizing "Human" in the Singalila National Park

The establishment of a zoo in Darjeeling town reinforces the existing relationship between red panda and wilderness into a new imagination of "pristine" Singalila.[11] The Darjeeling Zoo adopted red panda captive breeding conservation as their major project to save the critically endangered red panda (*Ailurus fulgens*) and declared Singalila as a major wild habitat. The Wildlife Department also put an immense effort to convert the Singalila into a major destination for wildlife conservation. In 1994, the red panda conservation breeding program was started with an individual from the Cologne Zoo, the Madrid Zoo, Belgium, and the Rotterdam zoo.[12] After the recommendation of the IUCN–SSC action plan for Ailuridae and procyonids, the Darjeeling Zoo received a global network for the red panda captive breeding program. This rare red panda of the Eastern Himalayas distribution thus entered into a new discourse of modern conservation. However, before going into its conservation politics, it is essential to know the brief history of the animal, their distribution and significance.

The "red panda" first appeared in modern scientific discourse in 1821 when English general Thomas Hardwicke wrote his manuscript in Darjeeling, but published only after two years when Frédéric Cuvier gave the name "Panda," referring to the animal's preference for bamboo (Ziegler et al., 2010: 79). There are several local names used to describe panda: Lepcha call it as *sak nam*, or in some part of eastern Nepal, it is also known by the name *Bhalu Biralo* (A bearcat) and *habre*; Sherpa often calls it *ye niglva ponya* and *wah donka*. However, it is generally believed that the word "panda" is derived from the Nepalese word *ponya* meaning bamboo eater. The panda with its solitary and cute character soon became "the most handsome mammal on earth" (Cuvier, 1824) and "in its size and shape somewhat recalls a cat, although it may be distinguished by the circumstance that in walking it applies the whole sole of the foot to the ground" (Lydekker, 2018 [1907]: 365). In 1869, scientists, especially Mr. A. D. Bartlett and W. H. Flower, established a family group of a panda by associating it with the raccoon family. They discovered the panda's resemblance to American raccoons in terms of its habits and anatomical relationship with the raccoons. Though the animal shares its name with giant panda (*Ailuropoda melanoleuca*), the two are not closely related to each other and the red panda is now generally placed in the monotypic family Ailuridae (Glatson, 1994 as cited in Roka, 2011: 1). In fact, the red panda, not the giant panda, was the first of the pandas to become known in the West (Catton, 1990: 4). However, Chris Catton further argued in his book *Pandas* that

> The fossil history of the red panda can be track back to the Oligocene and early Miocene epochs, more than 25 million years ago, when the raccoon-like ancestors of the red panda were distributed through–out the northern temperate zone.
>
> (ibid.: 33)

Toward the end of the last ice age, due to drastic changes in climate a radical shift emerged in the distribution pattern of both giant and red panda, with the red panda moving westward into its present range (ibid.). Today, the red panda is found mostly in high mountain elevation of about 7,000–12,000 feet, as far westward as Nepal and extending eastward through the mountainous district of Assam into Yunnan (Lydekkar, 2018 [1907]: 367) and spend most of their time in trees. They hunt for food mostly in the evening and in the morning, they climb back to its roost and spend the day sleeping aloof. The red pandas are among the shyest mammals of the wild and hence solitary. This rare red panda mostly depends on bamboo for its diet and is endemic to the eastern Himalayas where bamboo is abundant. Among many varieties of bamboos, *Yushania Maling* (local name *Malingo*) that is easily available in the eastern Himalayas constitutes one of the important diets of the red panda. Unlike other mega faunas with high economic and commercial value such as one-horned rhinoceros, the red panda has less economic and cultural significance as argued by earlier scholars. Nevertheless, a recent study (Glatston and Gebauer, 2011) has highlighted the significance of the red panda in cultural practices among some communities in Nepal, Arunachal Pradesh, and even Bhutan where red panda is available. Following a Swiss anthropologist, they highlighted the example of some cultural practices in western Nepal where

> the ramma or shamans of the Northern Magar tribe in the Dhaulagiri Region of Western Nepal using the skin and fur of the red panda in their ritual dress. These people consider the red panda to be a protective animal which guards the wearer against the attacks of aggressive spirits; for this reason, its body is hung on the shaman's back when he undergoes a dangerous ritual in the course of his healing séances.
>
> (ibid.: 13)

Similarly, some of the tribal people in Arunachal Pradesh and Bhutan use the red panda as their mascot, and especially its tail and fur are used to make bridegroom wedding hat by the Yi people of Yunnan.

The red panda also has less economic value as their meats are rarely reported to have been consumed by human. Due to its slow and seasonal reproduction process that usually happens in the spring and summer, this rare species has been a crucial part of many conservation efforts in the Himalayan regions. Most notably, due to a rapid expansion of land for cultivation, tourism, and other activities, there has been a serious transformation in red panda habitat which perhaps resulted in the decline in its population in the wild. Thus, many institutions, agencies, and even countries have adopted this little rare red panda as a flagship of their conservation ideology. Sikkim, the small state in northeastern India, declared the red panda as a "state animal"; "thereby wants to show that conservation of its unique wildlife is ranking high on its political agenda" (Ziegler et al., 2010: 79). They also adopted captive breeding of the red panda in its state zoo. Similarly, the government of India has issued a postage stamp with the image of the red panda.

As stated above, the red panda conservation in Darjeeling received its institutional character when the Darjeeling Zoo adopted a conservation breeding program for ex situ and in situ experiment in the early 1990s. However, it was not the first time that the wild red panda was introduced to the human world; in fact, red pandas were even used as a pet by the then Prime Minister of India, Indira Gandhi. Citing from the autobiography of Ms. Gandhi, *My Truth* (pp. 72–73) Mahesh Rangarajan writes,

> More famous was a friendly and photogenic red panda, presented to them during a tour of Assam. Nobody in the entourage quite knew what the creature was and it was only identified using a book on Indian animals. Soon joined by a mate, it was housed in a special treehouse. Fed bamboo shoots by Nehru each morning, the pair spent their winters in Shimla.
>
> (2015: 164)

Even the Darjeeling Zoo has a history of keeping red panda since the beginning (Roka, 2011). However, it became a centerpiece of conservation only recently. What seems interesting in such conservation efforts is the increasing interaction of this species with human as limited natural surroundings built around the designated enclave inside the zoo often pushed this shy animal to encounter crowded zoo visitors.[13] Since "red panda like other wild animals are more active during dawn and night" (Roka, 2011: 110) and remain silent during the daytime, changing habitats and increasing visitors to the red panda exhibit might lead to significant changes in the mobility and adaptability of the red panda. However, the successful adaptation and increase in red panda population inside the zoo gave a positive dimension to the conservation breeding efforts.

After the 1990s, many Western conservation societies started taking a keen interest in red panda conservation that resulted in the increasing export of

Himalayan exotic wild animals like giant panda and red panda to the Western world.[14] Many conservationists in Darjeeling Hills believed that the Singalila would provide a suitable ecosystem for the conservation of red pandas in the wild. Hence, in 2003, the Darjeeling Zoo after successful ex situ conservation of 22 red pandas decided to release two zoo-born red pandas, namely Mini and Sweety, back into the wild (ibid.: 114). For that, Gairibas in SNP was chosen as the most appropriate location due to two main reasons (though not exclusively): (a) Based on a pre-release survey conducted by the Wildlife Wing of the Forest Department, Government of West Bengal, in collaboration with the Darjeeling Zoo, it was found that the region contains a good amount of bamboo and most probably high density of red panda population (Jha, 2011), and (b) the state presence through SSB and Forest Department made it relatively easy to monitor the mobility of human and domestic cattle that possess threat to red pandas in the wild.[15]

However, besides this effort to protect this rare red, conservation in the SNP reflects two contrasting opinions. The local community (mostly in villages like Jaubari and Gairibas in Nepal side) believed that unprecedented infrastructural developments such as road, trekkers huts, and home stay are causing threat to the red panda's habitat in the wild. While few bureaucratic argument holds the opinion that the expansion of land for cultivation and human activities on the Nepal side pose a serious threat to wildlife conservation in the SNP (see SNP Management plan). Under such contestation, conceiving the notion of "Anthropocene" requires engagement with the "kinds of people" (Hacking, 2007) defined by the state.

In the case of SNP, we use tourist inflow data to discuss how different kinds of people were categorized and how this exerts differential pressure on landscape and wildlife. Figure 2.2 shows the increasing inflow of tourist in the Singalila National Park and this alarms us to rethink our approach to human interference inside the park. As argued above, the increasing inflow of tourist, even though helps in building up local economy in many ways, does lead to unprecedented pressure on landscape and ecosystem and hence wildlife. With the construction of road in the erstwhile isolated and arduous terrain, the number of vehicles to carry tourist has increased drastically over the year. However, few conservationists tend to overlook such factors and solely see the local community as a threat to wildlife, thereby denying their historical coexistence with nature and wildlife – resulting in the creation of a contested terrain for conservation.

Notes: We were able to procure the tourist inflow data only from 2003 due to some bureaucratic protocol. Since the Singalila National Park remains closed from mid-June to mid-September; the tourist inflow data are month-wise. The data for 2019 are not included in the chart above because the chapter was first written in August 2019. However, for data till April 2019, we are putting them in numeric form here (inflow of tourist tills in 2019; 8,035 Indian and 834 foreigners).

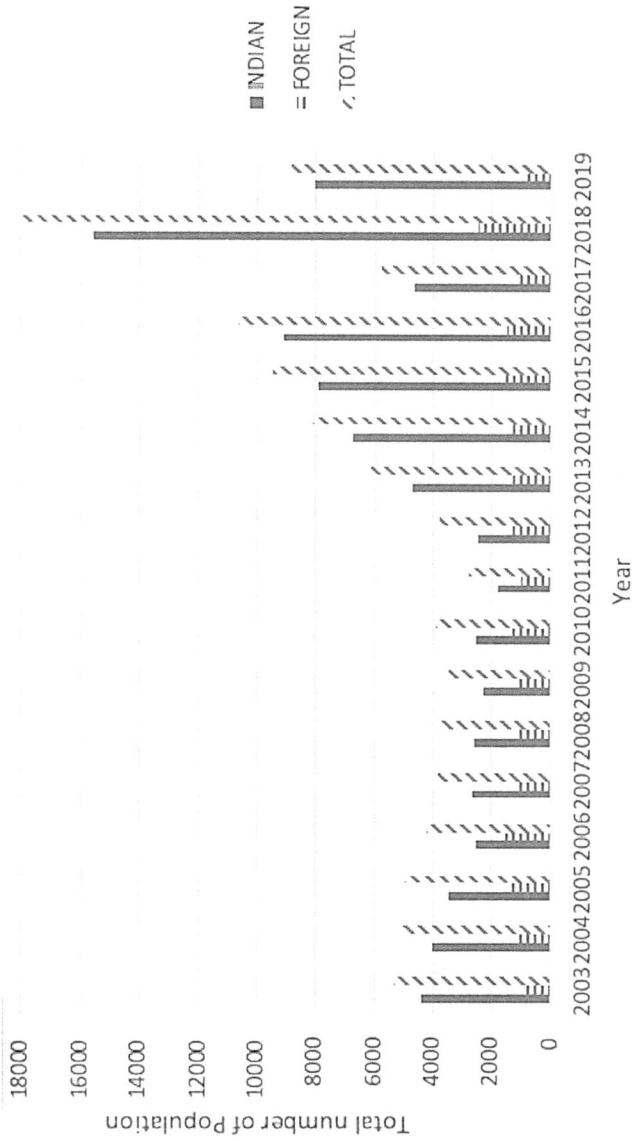

Figure 2.2 Tourist inflow in the Singalila National Park

Sources: Wildlife Office (Singalila National Park Check Post, Maney banjyang).

Disputed Bamboos: Missing Links in the Wildlife Conservation

In 2006, based on the life of Mini and Sweety in the wild monitored through non-triangulation location, a documentary titled *Red Panda: Cherub in the Mist* was published that unraveled the red panda conservation in the wild. *The Hindu*, while recounting the film, writes,

> Although the Red Panda is protected under the Wildlife Protection Act (1972) its numbers have been falling and today only about 2,500 Red Pandas survive in India, Nepal, Bhutan and Myanmar. The primary reason for the dwindling numbers is habitat loss due to prime forest area being cleared up for agriculture along the Nepali side.
>
> Besides, bamboo trees are recklessly felled by the Nepali villagers for building houses, denying them their staple diet. The film has won nearly half a dozen prestigious awards across the globe including the Best Film award in the Revelation category (sponsored by the Ministry of Environment and Forests) at the recent Vatavaran film festival in Delhi.
>
> (2nd November 2008)

In the movie, the border villages in Nepal side are presented as one of the biggest threats to red panda habitats, and even the park management plan described that "the western boundary suffers from anthropogenic pressure and requires greater protection measures" (see Management plan 2013–2014 and 2022–2023). The former Chief Minister of West Bengal, Buddhadeb Bhattacherjee, "on seeing the film has decided to fence the border of Singalila with Nepal" (The Hindu, November 2, 2008). However, the SNP share a long open border between Nepal and India with no clear boundaries. Even though the fencing of the border in the Singalila is not possible practically concerning the protection of wildlife, several efforts put to protect the red panda in the Singalila often get conflicted with livelihood issues at the local level. In the lower periphery of the Singalila National park, an adjacent reserved forest under Tonglu Range where bamboo is abundant with dense forest, the conservation project of red panda encounters local livelihood issues. Locals from the lower periphery of the park such as Group forest village, Dhotray, and 10th Mile depend directly or indirectly on bamboos (*malingo*) for their sustenance. Apart from local uses such as roofing, fencing, and various other domestic purposes, the malingo also have commercial values. Thus, a large number of people are involved in the everyday bamboo collection from the forest and for every bamboo stick (*Malingo*) (around six feet in height) they collect, they get one rupee from the local dealer who negotiates with the forest department for transit permit (TP)[16] and with dealers in Siliguri for price and quantity.[17]

The TP (transit permit) system, which has its colonial origin in the Indian Forest Act of 1878 and retained in the 1927 version, gave the state the sole power to control the uses of the forest and non-forest products. Thus, locals in those forest villages claim that "the forest department has stopped issuing TP for felling and transportation of bamboo from in and around the park for almost a year (in 2017) based on the rumors that Red Panda is vanishing from the SNP."[18] However, few forest staffs whom I contacted either refused to talk about the controversies or ignored the question. Also, we failed to get any leads on such ban for bamboo even though many local villagers claim it to be true. Assuming that the ban in bamboo collection was true, one could argue that this kind of conflict resonates with a political ecology of conservation in the region that "defines the relations of power and difference in interactions between human groups and their biophysical environments" (Gezon and Paulson, 2005). The power structure controlled by a certain kind of conservation ideology produces an adverse impact on local communities who depend on the forest for their everyday livelihood. This missing link between the idea of conservation and local involvement has created a hostile relationship between the forest department and local communities in the region.

Nevertheless, due to a rapid change in its habitat pattern, it is extremely difficult to locate the red panda habitat in the SNP but few believe that "where there is bamboo, there is red panda." Bamboo here is what multispecies scholars call it "border web of life" (Dooren et al., 2016;) that includes not only human and animal but also plant and other living factors like fungi and bacteria. It is a marker of interaction between human and nonhuman while also intersecting with multiple other lives. Hence, a ban on bamboo felling has put great challenges not only to community depending on forest for their subsistence but to the entire ecosystem that has historically coevolved through human and nonhuman interaction. It is undoubtedly true that "without the bamboo, there would be of course no pandas" (Catton, 1990: 38); however, the coexistence between bamboo for pandas and bamboo for people is essential. A study on bamboo flowering has suggested that "bamboo recovers and persists in the same sites, acting as a pervasive ecological and cultural marker of human influence in the landscape" (Raman et al., 1998 cited in Raman, 2018: 257). Also, bamboo being one of the fastest-growing plants has been one of the most important markers of eastern Himalayas since time immemorial. Thus, separating human from the use of bamboo impacts not only on species generation but also wildlife as there exists nothing called "pristine" in this earth (Cederlof and Rangarajan, 2018) and human and wildlife do not survive in distinct domain but have coevolved and coexisted within the lap of nature.

Needless to say, that bamboo ban and the red panda have been a crucial part of the political ecology of the region as its manifestation (without local knowledge) found a new discourse when endangered red panda became a prominent symbol (along with snow leopard) of the regional political depiction of "Gorkha heritage" and local NGOs' depiction of environmental protection (Besky, 2017). Similarly, to legitimize the community's attachment with the land, history, and heritage, the red panda was used as a mascot of the Tea and Tourism festival of Darjeeling. Such manifestation of community identity and wildlife in the public sphere through animated picture not only elided everyday encounters between human and wildlife (ibid) within the structural domination of the state but also brought the notion of "wilderness," "pristine," and "conservation of endangered red pandas" into the realm of community struggle for land and autonomy in Darjeeling Hills. Similar to Assamese claims of one-horned rhino as "their" animals and Bodo sub-nationalism using it as their mascot, the Gorkha sub-nationalism in Darjeeling (though not exclusively like one-horned rhino) claimed the red panda as a mascot for their regionalism. Will the representation of wildlife in the assertion of regional politics accord rights to community for conservation? Will it help to reconcile tension between conservationist and community per se in the efforts of various institutions (such as wildlife and forest department) to protect wildlife?

Conclusion

No doubt, Anthropocene has been the biggest threat to a red panda in the wild, but there is a need to rethink our approach to Anthropocene. In the case of SNP, there are two ways to (re)think our conservation policy, first, to identify the historical role of the local community who have coexisted with nature and understand their culture and incorporate them into the discourse of conservation. Second, there is a need to create more stringent measures for the short-term visitors who see nature only for their luxury and short-term comfort.

Acknowledgment

I would like to express my gratitude to the West Bengal Forest Department for granting me permission to do research in Darjeeling Hills and also various forest staffs and officials in Darjeeling Forest and Wildlife Division for sharing me their valuable times. Local communities in and around the SNP have greatly facilitated me throughout this study by allowing me to part of their everyday life in the forest villages. Interaction with few members from ATREE (Ashoka Trust for Research in Ecology and the Environment) Darjeeling and Sikkim has greatly facilitated this chapter in various ways. I would also like to thank Dr. Lakpa Tamang (University of Calcutta) for preparing a map of the SNP. Last, but not the least, I am grateful to Victor Teo for inviting me to present my paper at Hong Kong University, the result of which is reflected in this book.

Notes

1 Interview with staff at the SNP (2019, May).
2 Some older people from the neighboring villages in and around the park share this thought (Fieldwork, 2018).
3 https://www.bbc.com/news/science-environment-51632790, last accessed on July 12, 2021.
4 Forest officer B. B. Osmaston, who was posted in Darjeeling forest department during the early 19th century, writes: "big game was so scarce in the forests around Darjeeling that I never carried a rifle" (102). He further said that natives, especially Gorkhas (*sic*), have largely killed off hunting animal in the region. However, many colonial officials' travelogues regularly mentioned the tracks of wild animals and even the story of tiger jungle in the Terai Jungle or most particularly Sevoke Jungle but didn't describe in detail about the hunting game in the hills (See Donaldson, 1900).
5 Chebu Lama was believed to be one of the colonial officials, Campbell's, closet Lepchas associates who mediated East India Company and Sikkim diplomacy initially as a vakil. In return for his loyalty toward the Company, Campbell granted him a tract near Darjeeling (cited by Warner, 2014: 30). He was also a very influential man in maintaining land and labour in the eastern Himalayas even before the Company intervened.
6 Some form of wildlife protection rule was in existence since 1912, but the British forest department was less concerned about implementing it in the region. A closer reading of the forest working plan reflects that wildlife featured in the working plan only after the independence of the country. This proved the presence of minimal laws for animal protection in Darjeeling unlike in other parts where colonial forest officials like "Edward Stebbing constantly accused local of threatening wild animals through poaching" (Beinart and Hughes, 2007: 123).
7 The VIII working plan for the forest of the Darjeeling Division (1940–1941 and 1959–1960) state that "the rotation was fixed at 125 years and the possibilities of 20 acres per annum after the groups have been amalgamated by feeling annually at the rate of 300 trees of 4 feet 9 inches firth and over. For lack of communication it was not possible to work up to even half the prescription."
8 The presence of the trans-human grazing in the Singalila region has been well summarized by B. B. Osmaston in his dairy while referring to kinds of aconite in the alpine pastures on the hills overlooking Nepal. He writes, "When the villagers bring down their sheep in huge flocks from their high-level grazing grounds they are obliged to take precautions to prevent them from eating the aconite, which would prove fatal. So, they prepare basket muzzles which are attached to all the sheep when passing through this zone" (104).
9 Indira Priyadarshini Nehru Gandhi, the first woman Prime minister of India, was considered as one of those conservationists who introduced various environmental policies in Independent India. For details on her environmental and political life, see Rangarajan (2015), especially the chapter titled "Striving for a Balance: Nature Power, Science and India's Indira Gandhi 1917–1984." For more elaborated history, see Ramesh (2017).
10 Personal communication with one of the residents of the village on Nepal side, April 2019.
11 It was only in 1916 that the Darjeeling Zoo was brought under the Bengal Public Parks Act 1904 under Bengal notification NoI.87M dated 10 January.
12 For details on the Red panda Conservation project in Darjeeling Zoo, see http://ww.pnhzp.gov.in/urls/conservationredpanda:html, accessed on 24 January 2019: 2:57 pm.

13 An opinion expressed by one of my fellow researchers from India who studied animal biology in his PhD thesis (July 2019).
14 For details on the early export of giant panda from China to the Western world, see Catton (1990), especially the introduction and also see Croke (2006).
15 The second opinion has been shared by one of the researchers from ATREE (Ashoka Trust for Research in Ecology and the Environment) who had worked in the Singalila regions for many years.
16 As per the recently made amendment in the Indian Forest Act of 1927, bamboo has been removed from the category of trees and bamboo grown on agricultural land will not require a transit permit (TP), for felling and transportation, from the forest department. However, bamboo grown in the forest would continue to require a TP. This amendment has been made to encourage farmers to cultivate bamboo in hopes to raise their income and green cover of the country. For conflict regarding bamboo as a source of income and its TP policy, see https://www.financialexpress.com/opinion/there-is-little-evidence-that-permit-systemhelps-our-forests/1065163/, accessed on 17 August 2018, 3:46 pm.
17 Interview with villagers in Group Forest Village and Dhotrey Forest Village, April 2019.
18 Personal communication with villagers who wish to remain anonymous, April 2019.

Bibliography

Agrawal, Arun and Gibson, Clark C. 1999. "Enchantment and Disenchantment: The Role of Community in Natural Resource Conservation." *World Development* 27, no. 4: 629–649.

Beinart, William and Hughes, Lotte. 2007. *Environment and Empire*. New York: Oxford University Press.

Besky, Sarah. 2017. "Land in Gorkhaland: On the Edges of Belongingness in Darjeeling, India." *Environmental Humanities* 9, no. 1: 18–39. DOI: 10.1215/22011919-3829118.

Catton, Chris. 1990. *Pandas*. Bromley: Christopher Helm (Publishers) Ltd.

Cederlof, Gunnel and Rangarajan, Mahesh (eds.). 2018. *At Nature's Edge: The Global Present and Long-Term History*. New Delhi: Oxford University Press.

Cheng, Y. 2014. "Telling Stories of the City: Walking Ethnography, Affective Materialities, and Mobile Encounters." *Space and Culture* 17, no. 3: 211–223.

Chettri, Mona and McDuie-Ra, Duncan. 2018. "Delinquent Borderlands: Disorder and Exception in the Eastern Himalaya." *Journal of Borderland Studies* 35, no. 5: 709–723. DOI: 10.1080/08865655.2018.1452166.

Corbett, G. 1929. "The Founding of the Himalayan Club." *The Himalayan Journal: Records of the Himalayan Club* 1, no. 1: 1–3.

Croke, Vicki. 2006. *The Lady and the Panda: The True Adventures of the First American Explorer to Bring Back China's Most Exotic Animal*. Random House.

Cuvier, F. 1824. "Panda." In *Histoire naturelle des mammiferes, avec des figures originales, colorées, desinées d' apres des animaux vivants (1824-42)*, vol. 2, edited by G. Sainthilaire & F. Cuvier, 1–3. Paris, plate 203.

Donaldson, Florence. 1900. *"Lepcha Land" Or Six Week in the Sikhim Himalayas*. London: Sampson Low Marstono and Company Limited.

Dooren, Van Thom, Kirksey, Eben and Munster, Ursula. 2016. "Multispecies Studies: Cultivating Arts of Attentiveness." *Environmental Humanities* 8, no. 1: 1–23. DOI: 10.1215/22011919-3527695.

Gezon, L. L. and Paulson, S. 2005. "Place, Power and Difference: Mmultiscale Research at the Dawn of the Twenty-first Century." In *Political Ecology across Spaces, Scales and Social Groups*, edited by Paulson, S. and L. L. Gezon, 1–16. New Brunswick, NJ: Rutgers University Press.

Glatston, Angela R. and Gebauer, Axel. 2011. "People and Red Pandas: The Red Panda's Role in Economy and Culture." In *Red Panda: Biology and conservation of the first Panda*, edited by Angela, R. Glatston, 435–463. United Kingdom: Elsevier.

Hacking, Ian. 2007. "Kinds of People: Moving Target." *Proceedings of the British Academy* 151: 285–318. (British Academy Lecture).

Jha, Alankar K. 2011. "Release and Reintroduction of Captive-Bred Red Pandas into Singalila National Park, Darjeeling, India." In *Red Panda: Biology and Conservation of the First Panda*, edited by Angela, R. Glatston, 435–463. United Kingdom: Elsevier.

Lydekkar, Richard. 1907. *The Game Animals of India, Burma, Malaya, and Tibet: Being a New and Revised Edition of "The Great and Small Game of India, Burma and Tibet"*. London: R Ward Limited.

Mackenzie, John. 1997. *The Empire of Nature*. Manchester: Manchester University Press.

Marcus, George E. 1995. "Ethnography in/of the World System: The Emergence of Multi-Sited Ethnography." *Annual Review of Anthropology* 24: 95–117.

Norgay, Tenzing and James, Ramsey Ullman. 1955. *Tiger of the Snows: The autobiography of Tenzing of Everest*. Written in Collaboration with James Ramsey Ullman. New York: G. P. Putnam's Sons.

O'Gorman, Emily and Gaynor, Andrea. 2020. "More-Than-Human Histories." *Environmental History*: 1–25. DOI: 10.1093/envhis/emaa027.

O'Malley, L. S. S. 1985 [1907]. *Bengal District Gazetteers: Darjeeling*. New Delhi: Logos Press.

Osmaston, H. 1999. *Wildlife and Adventure in Indian Forests: From Dairies of B. B. C.E.I. 1868-1961*. Imprint Unknown: Second Revised Edition.

Raman, T. R. Shankar. 2018. "Expanding Nature Conservation: Considering Wide Landscapes and Deep Histories." In *At Nature's Edge: The Global Present and Long Term History*, edited by Cederlof, Gunnel and Rangarajan, Mahesh, 249–267. New Delhi: Oxford University Press.

Ramesh, Jai Ram. 2017. *Indra Gandhi: A Life in Nature*. New Delhi: Simon & Schuster India.

Rangarajan, Mahesh. 2015. *Nature and Nation*. New Delhi: Oriental Blackswan Private Limited.

Rasmussen, Mattias Borg, Adam, French and Conlon, Susan. 2019. "Conservation Conjunctures: Contestation and Situated Consent in Peru's Huascarán National Park." *Conservation and Society* 17, no. 1: 1–14.

Roka, Bhupen. 2011. *Study of Red Panda (Ailurus fulgens) in Ex-Situ Facilities in Co-Relation with In-Situ Facilities for Conservation Breeding*. Project Report submitted at Padmaja Naidu Himalayan Zoological Park Darjeeling.

Saberwal, Vasant, Rangarajan, Mahesh and Kothari, Ashish. 2001 [2002, 2004]. *People, Parks and Wildlife: Towards Coexistence*. New Delhi: Orient Longman Private Limited.

Kumar, Samanta Samir. 2018. *Geomorphology of the Southern Singalila Range*. Rhito Prakashan.

Thomas, Samuel. 2011. "A Tragedy on the Commons: Old Prejudices and Partial Science Combine to Destroy a Particular form of Mobile Pastoralism in the Darjeeling and Sikkim Himalayas." *Current Conservation* 5, no. 3: 8–16.

Warner, Catherine. 2014. "Flighty Subjects: Sovereignty, Shifting Cultivators, and the State in Darjeeling, 1830-1856." *Himalaya, the Journal of the Association for Nepal and Himalayan Studies* 34, no. 1: 23–35.

Ziegler, Stefan, Axel, Gebauer, Melisch, Roland, Sharma, Basant Kumar, Ghose, Partha Sarathi, Chakraborty, Rajarshi, Shrestha, Priyadarshinee, Ghose, Dipankar, Legshey, Karma, Pradhan, Hari, Bhutia, Nari Tshering, Tambe, Sandeep and Samir Sinha. 2010. "Sikkim—Under the Sign of the Red Panda." *Zeitschrift Des Kolner* 53, Jahrgang, Heft 2: 79–92.

3 A Mongolian Muzzle in the Chinese Grasslands

The Shifting Uses of the Camel in Nomadic Pastoralism and Festivities[1]

Aurore Dumont

Introduction

The camel is widely associated with the sandy deserts of Africa and the Middle East, where it occupies a fundamental role in the regional economies. Although less well known in Northern Asia, the Mongolian camel (*temee*[2]) has nonetheless been herded in this part of the world for centuries. In the Mongolian deserts and steppes, it constitutes one of the "five muzzles" domesticated by the autochthonous populations, alongside horses, sheep, goats, and cattle. The two-humped herbivore is not only appreciated for its strength and exceptional carrying capacity that make it a valuable tool in the pastoral economy; its docility and flexibility have also contributed to creating a symbiotic relationship with its owner. The traditional usages of the camel in nomadic pastoralism, combining the production of meat, milk, and wool together with transportation, lasted until major socioeconomic transformations in the mid-20th century altered the animal's basic functions. In many places where the camel was an essential tool for nomadic mobility, it was gradually replaced by motor vehicles.

In Northern Asia, under the socialist regimes of the People's Republic of China (1949–), the Mongolian's People Republic (1924–1990), and the Soviet Union (1922–1991), the implementation of collectivization policies aimed at modernizing pastoral economies redefined the composition of herds and the position of animals within the flock. It has also led to the modification of herding practices and the creation of new relationships between men and their animals and people and their environment. The belief that humans can and should control and reform the natural world so as to better humanity and nature was deeply rooted in the Soviet Union and throughout the world (Breyfogle, 2018: 5). For decades, the socialist mode of production placed the human experience at the heart of a process in which humans dominated the environment rather than adapting to it. This conception of human–nature relationships is the foundation for what we today conceive as the *Anthropocene*, a new planetary era in which humans have become the dominant force shaping the Earth's bio-geophysical composition and processes (Chua and Fair, 2019: 1).

DOI: 10.4324/9781003212089-3

After the opening-up reforms in the PRC in the late 1970s and the collapse of the Soviet Union in 1991, the liberalization of the planned economy gave a new insight to human–animals–environment relationship that has gone hand in hand with religious and ethnic revival among minority ethnic groups. In the Inner Mongolia Autonomous Region, this new interest in Mongolian culture and the growing tourist industry have placed the camel in a new light. It has been integrated into diverse festivities and has become a racing animal symbolizing Mongolian ethnicity, thus entering the new era of the Anthropocene.

In this chapter, I investigate how the camel in the Inner Mongolia Autonomous Region switched from being a beast of burden in nomadic pastoralism to a racing animal used in contemporary recreational activities actively promoted by the Chinese authorities. What does the changing function of the camel within the local domestic economy add to our understanding of human–animal and human–natural environment interactions? How can we grasp herder–state relationships in the peripheral areas of the People's Republic of China? Is the contemporary racing camel a new opportunity for minority groups to revitalize a vanishing nomadic way of life? By answering these questions, I will show how pastoralists have adapted to this major change and how camel racing, as well as other "invented traditions," has slowly reshaped domestication practices and relationships between men and their natural environment. My aim is also to rethink our understanding of human–animal experience by placing the camel and its herders within the larger context of the environment, native skills, and politics. By exploring the intersections between these realms, I wish to discuss how local societies have faced the changing paradigm of the Anthropocene over the past seven decades. More precisely, the case study I present may fit with what Hazard (2017: 1) has conceptualized as an *anthropo-scene*, that is to say a localized or territorialized specific context that is valuable for understanding global environmental change on a local scale.

In order to underscore local adaptation to these multifaceted changes, I purposely give prominence to the concept of "traditional ecological knowledge." Those who live with herds and know them best offer important insight into resilience in the Anthropocene. Many studies on pastoral nomadism in Northern Asia have highlighted the intertwined connection between herders, their animals, and the environment (Fijn, 2011; Gardelle, 2010; Humphrey and Sneath, 1999; Lavrillier, 2005; Marchina, 2019; Stammler, 2005). Much more than a simple exploitation of natural resources and domesticated animals by man, nomadic pastoralism relies upon herding skills, techniques and a deep knowledge of the surrounding environment. Indeed, a herder knows precisely the name and use of each type of grass covering the pastures, is able to find his way according to animal tracks on the ground, and so on. This autochthonous experience of the animal environment is what numerous anthropologists have called *traditional ecological knowledge* (TEK). Inglis defines TEK as "an intimate and detailed knowledge of plants, animals and natural phenomena, the development and use of appropriate technologies for hunting, fishing, trapping, agriculture and forestry, and a holistic knowledge or 'world view' which parallels the scientific discipline of ecology" (Inglis, 1993: vi). Lavrillier and Gabyshev go further by stating that "TEK is not only practical knowledge but also corresponds both to the elements of the environment that people exploit and the surroundings (insects, plants, vegetal cover, etc.) that they observe during movement" (Lavrillier and Gabyshev, 2017: 15). In the present research, I introduce Mongol camel herding as *traditional ecological knowledge* in terms of two parameters: the process of domestication (human–animal interactions) and the mobility of men and camels (usage of the land). As I will describe in the following sections, domestication and mobility are also both linked to the attachment the herders feel to their homeland (*nutag*). The exploration of TEK allows us to better highlight the human capacity for ecological reflexivity, or, in other words, how humans innovate with new forms of exploitation in relation to their environmental relationships and thus affect developments in the era of the Anthropocene (Franklin, 2015: 65).

This research is based on ethnographic fieldwork data and written sources with a special focus on two anthropo-scenes: the Hulun Buir municipality (*Hulunbei'er shi* 呼伦贝尔市) and the Alashan League (*Alashan meng* 阿拉善盟), situated in the northeastern and northwestern corners of Inner Mongolia, respectively. The two areas present different pictures of camel herding that deserve to be compared: while the grasslands of Hulun Buir have a few hundred camels mainly used in the private domestic economy, the Gobi Desert of the Alashan League is the cradle of large herds of camels belonging to private companies. I conducted fieldwork between 2011 and 2019 among the Mongol and Tungus[3] pastoralists living in the southwest of Hulun Buir. One of my goals was to investigate contemporary nomadic practices and, more precisely, the way herders have adapted their skills to a new socioeconomic context. The camel caught my attention since it is almost only seen during celebrations, in contrast to the other four muzzles that are easily visible in the pasturelands. In order to understand the atypical status of this animal and gain insights into today's camel herding, I conducted interviews with camel herders and riders. I also had discussions with different people from the autochthonous community who explained to me what the camel means for them and how they perceive its new uses. Furthermore, I practiced participant observation during various festivities where the camel is in the limelight. Written sources complemented the oral data. They comprise travel records from the beginning of the 20th century, newspapers, and ethnographic studies that look at the past and present usages of the camel in other parts of the Mongolian cultural world. In addition, I used contemporary Chinese newspapers and magazines to expose how national and local media present the animal and its socioeconomic value in China.

The chapter follows a chronological structure in order to reflect more accurately the shifting usages of the camel throughout the 20th century. The first section briefly introduces an emic conception of the homeland, mobility, and ecological knowledge. The second part describes the economic policies launched by the new communist regime in the 1950s to "modernize" the nomadic domestic economy with new mechanized vehicles. It shows how the introduction of tractors led to camel herding specialization and made the animal's mobility function obsolete. Finally, the last part shows that since the camel was integrated into races and beauty contests from the 2000s, it has become a symbol of "Mongolian traditional culture" and given Mongolian people a way to assert their ethnic belonging.

Camel Herding as *Traditional Ecological Knowledge*

Known as the Bactrian camel (*Camelus Bactrianus*), the two-humped camel[4] is well adapted to the cold and dry areas of northwestern and northeastern areas of China (Figure 3.1).

A large population of camels is distributed across the Inner Mongolia Autonomous Region from the Gobi Desert of the Alashan League (a third of China's total camel livestock) to the Bayannur Prefecture 巴彦淖尔市, the desert area of the Shilin Gol League 锡林郭勒盟 and the grasslands of Hulun Buir.[5] In Inner Mongolia, camel herders belong for the most part to diverse Mongolian-speaking groups[6] merged into the officially recognized "Mongol ethnic minority" (蒙古民族).[7] In our case study, these are, from west to east, the Oirat Mongols (卫拉特蒙古人) of the Alashan League and the Buryat Mongols (布里亚特蒙古人) of the Hulun Buir Prefecture. Although sharing a common Mongolian language and "ethnicity," both peoples, 2,700 kilometers distant from one another, are distinguished by living environment, dialect, and history.

I shall begin by briefly presenting the camel as it was used within the nomadic economy of the Mongols up to the 1960s. The description of this

Figure 3.1 Camels in Hulun Buir, 2019
(photograph by the author).

past nomadic experience is based on written materials and, above all, on people's own experience and narratives.

The Camel within the "Five Muzzles"

The camel is part of the multispecies livestock herding practiced by the Mongols. Having different types of livestock "represent[s] an asset not only for the variety of herds' uses but also in the daily management of the pastures" (Marchina, 2013: 174). For nomads, the "five muzzles" were the most valuable form of property and constituted the basis for economic survival. Families constituted their herd according to the ecological environment, their economic conditions and needs. They habitually kept a large number of sheep and goats (a third of the livestock), some horses, cattle, and a few camels, all herded on similar pastures. Although all five of the "muzzles" provided milk and meat, each offered specific resources that could be used differently. Sheep and goats were preferred for their meat and wool. Apart from being a beast of burden, cattle gave meat and milk, which could be transformed into butter, cream, yoghurt, cheese, and other "white foods"[8] (*tsagaan idee*), as well as dried dung (*argal*) which was used as a combustible. Horses were especially valued as riding animals and enjoyed great prestige among the Mongols: this was the only animal used in the Mongolian "three manly games" (*Eriin gurban naadam*), usually called *naadam* and made up of horse racing, archery, and wrestling.

Among the other four muzzles, the camel has a unique position by virtue of its size and behavior. It has a strong body, with a male weighing around 650 kilograms and a female around 480 kilograms: it is about 170 centimeters high. It has a short head, big eyes, small ears, and a long neck. The hair color varies from apricot to reddish brown and white. The Mongols use image categories to describe their domesticated animals according to their morphology, behavior, and degree of domestication. The camel belongs to the "long legs," together with horses and cattle, as opposed to the "short legs" represented by sheep and goats. The long legs go out to pastures and come back by themselves, receiving less attention while also being subject to especial care. Unlike the short legs, which are herded in great number and slaughtered quickly for economic resources, the camel is a long-term companion since it may live for up to 40 years. The camel is an herbivore that feeds on a variety of plants and grass and is resistant to hunger and thirst: "Unlike slow-moving cattle and intensively grazing goats, camels are economical feeders that never overgraze the vegetation" (Gauthier-Pilters and Dagg, 1981: 22). Many people I talked with remembered that when they were young (in the 1940s and the 1960s), every family possessed at least one camel. Although the camel has always made up the smallest percentage of all livestock, it was valued for its resources and labor.

The herders of the Alashan Gobi depict camel resources as follows: "the camel has the milk of the goat, the hair of the sheep, the meat of the cattle and the carrying function of the horse" (Wuricaihu, 2012: 16). Camel milk is known for its high nutritional value and is habitually consumed as a fermented milk drink. The lactation period lasts from 8 to 10 months, during which the she-camel can give 1–2 kilograms of milk per day. The animal possesses two layers of hair quite similar to cashmere: these shed in April. The 4–7 kilograms of wool collected annually is used to make winter clothes, wool, and felt. Although a camel can offer up to 350 kilograms of meat, the Mongols rarely consume it except in cases of emergency or when they possess large herds. Lattimore wrote that

> a caravan man may not slaughter a camel, nor eat camel flesh, nor sell the hide of a camel. If a camel becomes too weak to follow the caravan, it is left by the trail to die. The owner will not kill it, for fear that its soul might follow the caravan, haunting the other camels.
>
> (Lattimore, 1928: 502)

In contrast, the Chinese people were fond of camel humps, considered a delicacy by the nobility (Roux, 1959: 68, 73).

Camel Domestication and Attachment to the Homeland (nutag)

The pastoralists of Inner Mongolia attribute particular importance to what they call *nutag*. Commonly translated as "homeland" or "land," the Mongolian term *nutag* refers to one's birthplace or homeland but may also designate the area of seasonal migration, grazing land (Namsaraeva, 2012: 141) and living place. The *nutag* is also perceived as a sacred landscape since it is believed to host powerful deities worshipped annually at holy cairns called *oboo*. By making offerings to the local spirits of their homeland, people hope to obtain in return the fertility of the herds, good pastures, and the well-being of the community. The *nutag* is thus a complex whole, encompassing various socioeconomic, environmental, and spiritual dimensions intertwined with *traditional ecological knowledge*. In Mongolian cultural areas, the concept of *nutag* is also associated with the notion of attachment. People are attached to everything that is constitutive of their homeland: nature, animals, and spirits. Charlier (2016) has demonstrated that movement and fixity are complementary aspects of attachment to the land. It is not only established by birth but also maintained through mobility and ritual actions. Following the emic idea of *nutag*, I will describe how camel domestication practices and knowledge are also part of this intimate connection with the land.

In the same way the herder has detailed knowledge of his/her *nutag*, he/she possesses a profound understanding of his/her camel. A herder can distinguish one of his/her camels by the color of its fur or its footprints. He/she can even tell whether the camel was loaded or if it was watered recently. Raising *temee* is a delicate matter, and its domestication requires a specific traditional savoir-faire. The Mongols take especial care of their livestock, following various rules aimed at maintaining both the economic value and the well-being of the animals on their homeland. One such rule is the great care herders provide to the newborn in the herds. This is especially true for camels. The long gestation period involved with pregnancy makes the reproduction process quite difficult, with a new calf born every two years. The fragile baby camels are sometimes kept in the yurt (*ger*), the mobile felt dwelling of the Mongols, and covered with felt for protection. Herders follow the growth of the young camel and its lactating mother. The first major step in a domestic camel's life happens between the ages of one and three years: the nose is pierced and a nose-peg (*bujila*) inserted in order to guide and control the animal. It is worth noting that the *bujila* was one of the brand marks (*tamaga*) used by the Mongols on their horses to indicate the positions of owners in patrilineal kin-groups. Although small, the camel peg-nose had a powerful and sacred value because it creates control over the camel and is situated high off the ground (Humphrey, 1974: 475). The nose-peg indicated that the camel was tame enough to be trained to follow a camel caravan, carry loads, and be ridden. Through regular interactions, men and their camels have developed a close relationship based on joint adaptation in their homeland. Since the adult camel is also attached to its *nutag*, it does not require a considerable amount of surveillance compared to other livestock. Freedom of movement is essential for the well-being of camels, which are often seen roaming freely in the desert or grasslands. Since a male camel leads the rest of the herd, the animals return home by themselves. Herders need to check the location of the herd every five to eight days by looking at the prints left on the ground.

The Camel: A Indispensable Companion for Nomadic Mobility

Camels have played an important role in the transportation of goods and men in the deserts of the world, which earned it the nickname "the ship of the desert." Numerous Western travelers of the late 19th and early 20th centuries[9] described the extensive use of camel carts in Mongolian cultural areas. As the strongest beast of burden, the camel's employment as a pack animal made perfect sense to Mongol pastoralists who had to move continuously between pastures. Searching for new feeding pastures required extensive seasonal movements of herders and their animals on their familiar *nutag*. Nomadizations from one camp to another were conducted thanks to the camel, which can work a full week without food. Prior to the 1950s, a nomadic family might have moved every week, depending on the environmental conditions and the needs of the animals. The great endurance of the camel and its exceptional carrying capacity made it an indispensable pack animal within the domestic economies of nomadic families.

The *temee* could carry heavy loads of 150–200 kilograms and cover up to 50 kilometers a day. The camel is the only animal within the "five muzzles" capable of carrying on its back an entire yurt. When the nomads moved, they loaded their camels with camp items such as portable yurts, fire stacks, and small items of furniture. The camel was sometimes ridden with a felt saddle placed between the two humps. The animal was controlled with a halter and reins while a rope was attached to the nose. When the camel was used as a pack animal, the loads were put on the animal's side, held in place by two wooden poles tied on the front and back of the camel as well as underneath its belly (Atwood, 2004: 75).

Apart from loading items directly on camels, the Mongols also invented several wheeled vehicles hauled by camels. This is the case of the Buryats, who introduced into Hulun Buir two types of wheeled carts pulled by camels. The first, simply called "the camel cart" (*temegen tereg*), was a non-covered two-wheeled cart made of metal and wood. The cart was mainly used to carry items such as dry dung and water and could carry loads of 1,000 kilograms. The second, the "fire cart" (*gal-tei tereg*), was also a two-wheeled cart but topped with a cover. With one window in the front and one door on the side, the wagon was equipped with an interior stove and a fire stack, keeping the interior warm (hence its name, "fire cart"). It was used for nomadizations during the winter season (Xu et al., 2009: 157).

When Tractors Replaced Camels: Redefining Interactions between Men, Herds, and Their Environment

Camel Herding Specialization

The camel and the way it is used by herders are illustrative indicators of the changes and adaptations experienced by nomadic societies. After the foundation of the People's Republic of China in 1949, state planners wanted to modernize all aspects of production. In the country's margins where millions of nomads lived, the implementation of what can be labeled a "socialist project of modernity" led to a radical shift in people's way of life, the organization of the domestic economy, and the environment. In this respect, the transformation of camel herding in the 1950s by state policy illustrates the way pastoral people have adjusted their traditional ecological knowledge to imposed rules from the central state and to the new era of the Anthropocene. In the pastoral areas of Russia, Mongolia, and China, where collectivization was adopted as a leading directive, the composition of livestock changed significantly. The number of "short legs" grew dramatically[10] to support the intensive production of meat and wool, and herders were encouraged to give priority to sheep and goat, thus redefining their traditional herding practices suited to the ecological zones of the desert and the steppe. Collectivization thus coincided with the spread of specialized herding of a single species at the expense of traditional multispecies herding. While the "five muzzles" were still herded together, some people became specialized in one or two animal species. Sheep and goat herders, horse herders, and camel herders became full-time jobs. Camel herding became a predominantly pastoral job, with the camel herder being the only person in charge of domestication and taming. Finally, camels belonging to different owners were kept together by a specialized herder, breaking the link that used to tie a camel to its owner in their homeland. From then on, herders had to use the natural resources of the land quite differently. While multispecies herding allowed for alternating use of the pastures according to the seasonal variation, one species restricted the herder to using only a single portion of the pasture.

The camel population also underwent a rapid expansion. In Inner Mongolia, the total number of camels in 1947 was 110,000 before reaching a peak of 344,000 in 1978 (Atwood, 2004: 76). The strategy adopted by the local authorities to increase livestock was built on the use of new technologies such as hay, fodder crops, and shelters to reduce die-off and thus allow much higher growth in livestock numbers (Atwood, 2004: 16). The doubling of camel numbers also reflected the government's desire to boost the commercialization of camel meat and wool. While in the pre-socialist period the herders kept camels mainly for their carrying capacity, during the socialist era the camel was converted into a meat/wool provider kept apart from other livestock. Furthermore, since the state farms concentrated all the camels in the household of a specialist camel herder, other herders did not have camels for transportation (Humphrey and Sneath, 1999: 39). This was the first sign that the use of the camel as a pack animal was coming to an end. The transportation function of the camel became obsolete not only because the animal was privileged for meat and wool, but also (and more importantly) because motorization was one of the key features of the socialist project in the pastoral areas of China.

Roads, Trucks, and the Degradation of Pastureland

From the 1950s, the systematic construction of large-scale infrastructure throughout the country reconfigured economic, ecological, and environmental features. In the communist countries of the 20th century, the ambition to dominate nature shaped the socialist experience of industrialization. Analyzing the heavy industrialization of the Kola Peninsula in the Arctic, Bruno has shown how interactions with the natural world both enabled nature and power in industrial livelihoods of the Soviet North and curtailed socialist promises (Bruno, 2016: 6–7). In Inner Mongolia, too, nature was a participative tool in the Chinese communist project of industrialization and development. Economic, geographical, and ecological features of this peripheral territory offered opportunities for the new state to meddle and deploy its political strategies. Pastoral areas were gradually connected with roads, railway networks, modern transport facilities and factories (Dumont, 2018: 13), while the grasslands and desert were divided into villages, pastures, and haymaking zones to support the thousands of Han Chinese migrants arriving in the area.

The introduction of motorized vehicles in the 1950s, together with campaigns of sedentarization, contributed to the redrawing of patterns of nomadic movement. The policies of the Chinese government were driven by the idea that highly mechanized pastoralism was synonymous with progress and raising livestock. Considered to be of modern technological value, the tractor officially became one of the major motorized vehicles implemented in herding management from 1958. It was a multipurpose tool mainly used for grass-cutting and transportation (of grass, wool, nomadic items, etc.). The tractor pushed the herders to learn new techniques and to develop new skills, adapting seasonal tasks to the new vehicle. The tractor gradually replaced the camel, whose transportation function became obsolete in the 1960s, even though it was still used sporadically by some families. From this time onward, the camel was raised, kept, and used in a totally different way. The relationships between men and their camels, formerly based on day-to-day interactions, became ones constituted by irregular contact: in some cases, they died out altogether. Meanwhile, the camel was no longer domesticated to carry loads and lead caravans, resulting in a more distant relationship between the camel and the herder.

Another reason for the decline in the camel's mobility function was the new connection between people and their *nutag*. The construction of fixed houses in pastoral areas led to a reduction in mobility, reduced usage of the mobile yurt, and the alternating use of sedentary and nomadic spaces. The large annual nomadic cycles involving a high level of human and animal mobility, once supported by various forms of autochthonous *traditional ecological knowledge* and technologies such as pack animals, were converted into reduced movement over delimited territories: residential nomadic items were transported by tractor or truck on summer and winter pastures instead of a pasture for each season. The replacement of the camel as a means of transportation by mechanized vehicles was certainly not restricted to socialist countries. The same process occurred in numerous parts of the world, as shown by Gauthier-Pilters and Dagg (1981) in the eastern and central Sahara and by Pâpoli-Yazdi (1982) among the Kurds of Iran. After reaching a high point in the 1970s, the camel population in China suffered a serious decline in the following three decades, from a total of 222,900 in 1990 (of which 149,700 lived in the Alashan League) to only 68,000 in 2003 (Atwood, 2004: 76).

The changing role of the camel within the Inner Mongolian domestic economy reflects the gradual transformations that nomadic societies have undergone in the People's Republic of China. The new uses of the pastures (sedentarization, mechanization, and overgrazing) combined to a massive immigration and industrialization led to a growing degradation of the land, especially during the Maoist period. Examining the relationship between political repression and environmental degradation, Shapiro has demonstrated that the environmental dynamics of the period suggest a congruence between violence against human beings and violence by humans toward the nonhuman world (Shapiro, 2001: 1). At the beginning of the 1980s, environmental instability was already a serious issue within peripheral areas, leading to serious desertification and more numerous droughts. Herders and their livestock had to face reduced access to their pastures and to adapt again to these ecological changes. In the mid-2000s, new government policies promoting environmental protection, combined with ethnic revival among the Mongol people, led the camel back to center stage.

The Return of the Camel: The Invention of Racing and Beauty Contests in the 2000s

The Camel as Support for Environmentalism and Cultural Heritage

When I first met Beljir, one of the few Buryat camel herders living in Hulun Buir, he told me: "Today, no one wants to herd camels anymore. It is because they are useless and do not bring money." (Figure 3.2)

This observation points out a paradoxical situation: while the camel has lost its working function within the pastoral economy and become unnecessary for most families, it has simultaneously received a new destiny. Since the late 2000s, the animal can often be seen at every festival including the members of "ethnic minorities" of Inner Mongolia. It is exhibited during beauty contests and is promoted by the Chinese authorities as one of the main symbols of Mongolian culture during camel racing. The camel is more than an ethnic emblem. It is a profitable political element for conveying Mongol cultural heritage and environmentalism.

Figure 3.2 Beljir, a Buryat camel herder in Hulun Buir, 2019
(photograph by the author).

This new interest in the camel began with the "Open Up the West" policy (西部大开发). Launched by the Chinese central government in the early 2000s, the campaign sought to bring socioeconomic development to non-Han Chinese areas of the country by developing tourism, infrastructure (roads railways connecting villages to cities), telecommunications, and recovering natural resources after ecological damage. Over recent decades, the "environment" has become heavily institutionalized. Environmental policies have been firmly embedded across the globe in national political systems, from environmental ministries to environmental laws, planning agencies, and national environmental plans (Biermann, 2020: 61). In Inner Mongolia, this has resulted in the relocation of nomadic people to urban settlements, a new environmentalist policy for managing pastureland and the transformation of some ethnic villages into tourist resorts. An emblematic case study of what Yeh (2005) has dubbed "green governmentality" is what is locally called the "ecological migration" (生态移民) experienced by the Evenki reindeer herders.[11] In 2003, in order to protect the forest and wild animals, this minority group was resettled from its traditional grounds in the forest zone of Hulun Buir to a new ecotourist village next to a city, while the reindeer became a tourist attraction (Dumont, 2015: 83–84) and a protected species. In the grasslands, things happened quite differently because the Mongols are more numerous and scattered over a larger territory. There, the provincial authorities initiated the "Grassland Culture" (草原文化) project, a political and economic strategy devoted to promoting and preserving the material and immaterial culture of the minority groups of Inner Mongolia. This led to the creation of subcategories of "culture," including "camel culture" (骆驼文化). From the mid-2000s on (most prominently since the mid-2010s), "camel culture" was deployed throughout Inner Mongolia through a multitude of festivities, cultural centers, and governmental research institutes capturing the sounds of a "disappearing nomadic culture." Depending on the locality, one can find the "festival of camel culture" (骆驼文化节), the "camel *naadam*" (骆驼那达慕), and so on taking place in new spaces dedicated to these novel celebrations. This echoes what Litzinger witnessed in Yunnan, another peripheral region "ideologically rethought and spatially reorganized as a site not just for investment but also for cultural struggle" (Litzinger, 2004: 490). The camel of the Mongols was converted into an exotic animal and associated with an ethnic people and their environment. In the margins of China, an animal becomes exotic only when it has lost its traditional function within domestic economies. From that moment on, it gains official status with redefined cultural and aesthetic values. This is not the case for cattle, sheep, or goats, since they still provide meat and milk for the herders and are valuable products on the Chinese market.

In 2004, China ratified the UNESCO Convention for the Safeguarding of Intangible Cultural Heritage. Not surprisingly, ethnic groups' "traditions," such as dances, songs, and herding practices, have been inscribed as the precious heritage of the Chinese nation. Following the fad for cultural heritage, local authorities from Inner Mongolia have turned "camel culture" into Mongolian heritage and inscribed several practices related to it on the National List of Intangible Cultural Heritage of China. In 2007, "camel racing of the Mongol nationality" (蒙古族赛驼) was inscribed on the Inner Mongolia regional list of cultural heritage (Ding et al., 2015: 54). In 2008, the "Mongol nationality's custom of raising camel" (蒙古族养驼习俗) in the Alashan League was inscribed on the national list of cultural heritage under the category of folk customs, suggesting that camel herding is in decline and needs to be safeguarded by the state. In 2014, the "camel spring legend" (骆驼泉传说) of the Xunhua Salar autonomous county in Qinghai was integrated into folk literature, while "Mongol nationality camel polo" (蒙古族驼球) of the Urat Rear Banner was put on the list of traditional sports.

The selection of some practices supposed to represent "camel culture" raises the issue of the involvement of the Chinese state in different layers of culture. Intangible cultural heritage today plays a central role as a new concept able to organize the preservation of traditional cultures (Gros, 2012: 24). This is particularly true in areas populated by autochthonous populations whose vanished customs have been revived, redesigned, and even fully invented in order to match the "traditional culture" designed by the state (Dumont, 2016: 281). In areas where the camel was traditionally herded, the animal was first labeled as a symbol of Mongolian traditional culture in need of preservation and then converted into an aesthetic and race animal used in contests and games. Although they are presented as "traditional festivities" of the Mongols, most of these recreational activities were fully or partially invented. In this new context, how are men and their domesticated animals integrated into these new practices? How do herders perceive the transformation of the camel into a race animal? How have domestication practices been affected or adapted? And how have relationships between men, their herds, and the environment been reshaped?

A New Manly Game: Speed Performance, Mobility, and Domestication Practices

One of the most obvious changes in the uses of the camel is its integration into the *naadam*, or the three manly games. These consist of horse racing, wrestling, and archery. Very popular among the Mongols, the *naadam* are held during the religious summer festival to propitiate local deities and to celebrate the birth of livestock and the renewal of the environment. Socialist states picked up this celebration early on in order to turn it into a sports competition and a cultural celebration supporting political propaganda.[12] As Aubin (2012) has shown for Mongolia, national governments have always retained the traditional and popular character of these festivities in order to efficiently reach the population. In China, the *naadam*, cut off from its religious dimension and reshaped as an "invented tradition" (Hobsbawm and Ranger, 2014 [1983]) by the local government to meet a political agenda, remain a well-liked celebration.

Figure 3.3 Camel racing during *naadam* in Hulun Buir, 2012
(photograph by the author).

The systematic integration of the camel into the *naadam* dates back from the mid-2000s, following the development of tourism and the new interest in "camel culture."[13] In the summer *naadam*, the camels perform in different races, the common aim of which is to achieve great speed (Figure 3.3).

Although the races differ slightly depending on the locality,[14] they normally involve races on camel back and camel polo (mainly held in the Alashan League). The races are habitually held in May, July, and December in order to respect camel reproduction and grazing cycles. Camel races work in the same way as horse races. In camel-back races, the animal runs different distances depending on the sex and age categories, and also according to locality. In Hulun Buir, the camel runs 1,800, 3,600, or 5,400 meters, while in Alashan it runs 1,000, 3,000, 5,000, and 8,000 meters. Racing usually takes place during the day on a dedicated race course outside the city. There are several dozen riders, each identified with a number. Before starting, the riders circumambulate around the track three times on camel back, sometimes around a sacred fire, in order to attract luck. Presented as a "traditional custom," this circumambulation is as new as camel racing itself. Circumambulation associated with the number three is well known in Buddhist practices. It is done, for instance, during the *oboo* cairn ritual, a religious event held every summer by Mongol communities. The riders then line up at the starting line. Equipped with a crop, participants control the speed of their animal with different techniques. The winner gets awards of medals and money, financed by the local or regional government.

The summer *naadam* was joined in the late 2000s by the newly invented "winter *naadam*" (冬季那达慕), another creation of the Inner Mongolia government. Part of winter tourism development, the winter *naadam* comprises various dances, performances, and horse and camel races. In Hulun Buir, the winter camel race includes racing on camel back and the cart camel race (Figure 3.4).

The content of the winter *naadam* has been entirely artificially shaped and has nothing to do with the original *naadam* summer festival, made up only of the "three manly games." Camel races are always presented to the audience as a traditional Mongolian "manly game," even though they were invented less than ten years ago. A parallel may be drawn with the Arabian Gulf, where an analogous process of "invention of the tradition" has occurred. Khalaf (1999, 2000) has examined how camel racing became part of a vast Bedouin (Badu) cultural revival in the Arab Gulf after the marginalization of the camel during the oil revolution in the 1960s and 1970s. From the 1980s and actively supported by the Emirati state, camel racing and "the invention of camel culture [...] provide links to the historical past of the Emirates' pastoral way of life that has been swept away by oil-triggered modernization" and gives meaning to Emirati identity (Khalaf, 2000: 244, 259). This is similar to Inner Mongolia's case, where invented camel games have become a political symbol of "Mongolian nomadic culture."

Another essential aspect of invented camel racing is the institutionalization of competitions: this has generated new standardized rules for speed performance and the domestication process. When the camel was used as a

Figure 3.4 Camel racing during winter *naadam* in Hulun Buir, 2010
(photograph by the author).

pack animal, its strength and ability to carry heavy loads over long distances was the principal quality sought by herders. Now a ridden camel is expected to achieve a high speed over a short distance of a few kilometers. In other words, camels are now trained to achieve rapid movement.

How have relationships between camels and their owners been reshaped by these new recreational activities? What is obvious today is that camel racing has created a closer relationship between man and his camel, a privilege restricted until recently to the horse, known as the best companion of the Mongols. Indeed, after decades of connections becoming distant, camels and those who herd them now tend to be closer. Beljir, the Buryat camel herder, explained to me how he developed a deep attachment to his camels and how he adapted his herding skills over the years. In his youth, he herded cattle and sheep. After he acquired one camel, he rapidly became passionate for the animal and built up a flock. With the development of camel racing in the 2000s, Beljir started to herd camels as racing animals, which required specific new skills. Nonetheless, not every camel can become a race animal. A camel suitable for racing is chosen at a very young age according to strict criteria indicating that it will run fast. The camel's hump should be low, the hair thin, and the neck hair quite short. The animal's character is equally important. It should be docile, not peevish, and able to withstand a harsh training regimen. A well-proportioned and quiet camel is trained from the age of two and will take part in races at three years old. Racing helps distinguish ordinary camels from notable ones within the flock. While the entire herd receives special care, racing camels are given even more remarkable attention, reinforcing the human–animal connection. Beljir explains that he takes care of his four racing camels every day, feeding them with specific food and training them very carefully. He has developed an exclusive relationship with his racing camels: nobody but him can ride them. A few days before racing, the camel follows a special diet: it drinks less and eats high-protein specialized food. The camel then keeps its physical strength during the competition and will lose less hair after it (Lin, 2013: 23). Like Beljir, a camel rider is often a herder.

This intimate relationship also characterizes another connection linking the man and his racing camel: prestige and economic value. With the growing popularity of racing, riding animals have become valuable, with their prices reaching 30,000–40,000 RMB, depending on their weight. Beljir feels true pride for the medals and awards he won with his camels. Awards not only represent his mastery of riding and herding skills but also highlight how the herders of Inner Mongolia have adapted their traditional ecological knowledge to an invented recreational activity.

The Camel Beauty Contest: Aesthetics, Skills, and Ethnic Revival

Camel racing is not just about speed: it also emphasizes Mongol social values. As Humphrey has pointed out about the horse in Mongolia, the Mongols do not judge a man by his clothes or accent but by his riding horse (Humphrey, 1974: 485). In the same vein, since the camel became a racing animal, it has marked out the social status of its owner. During races, the camel always looks great, with a good shape, fresh haircut, and a beautiful saddle. The more presentable a camel looks, with perfect proportions and gait, the more the man who rides it gains respect from his community. As such, the camel herder trains his animal with the greatest level of personal involvement.

In the framework of the development of tourism, regional government has understood the equal importance of aesthetics and speed for the Mongols. With the growing popularity of camel racing, the authorities have initiated another "invented tradition": the camel beauty contest. This may be organized as part of the summer *naadam*, as part of the festival of camel culture or as an independent event. In the Alashan League, a camel beauty pageant requires males and females be judged separately. For male camels, the quality of their head, color, and hair is compared, as well as the beauty of the reins and halters. Female camels are judged by the color of their hair, the length of the humps, and the apparent strength of their limbs (Menghedalai, 2015). The winner and his camel are awarded with medals and money. On the other side of the border, in Mongolia, camel beauty contests were begun in 1997 by the nongovernmental Camel Protection Association to protect camels and boost their number after the decline of the 1990s (Kwong, 2019). In both countries, camel beauty contests have helped increase the economic and social value of the animal. They also support growing interest in traditional camel herding and nomadic culture. However, in Inner Mongolia, camel beauty contests also encompass an ethnic dimension. On the one hand, the camel is promoted by the government as an indispensable component of the traditional Mongolian way of life: it symbolizes a certain idea of pastoral aesthetics attributed to the herders. When a camel's beauty is rewarded, it not only honors its owner but also the Mongolian man and his *traditional ecological knowledge*. On the other hand, camel beauty contests display Mongolian herding skills and domestication techniques, thus giving herders an opportunity to express their ethnic belonging to the Mongol community. The contest can be a means for expressing an ethnic identity. Taking part in a beauty contest makes the participant a member of the Mongolian community, structured as an exclusive social group. Furthermore, the Mongols today consider recreational camel activities as a cultural distinction linked to their pastoral way of life.

Concluding Remarks

Camels have been raised for centuries to carry goods and people. Domesticated camels accompanied the Mongols into the grasslands following cycles of nomadization, and were an indispensable muzzle within the pastoral economy. In the 1950s, with the rise of mechanization and the planned economy, the camel's function as a means of transportation was yielded to motorized vehicles, and the animal fell behind the more economically productive cattle, sheep, and goat. However, in the 2000s, the development of tourism in Inner Mongolia, combined with the promotion of ethnic cultural heritage and environmental issues, explains the renewed interest in camels. Camels have become racing animals incorporated into newly created sports, beauty contests, and other recreational festivities promoted by the local government. In order to adjust to the new requirements for speed, mobility, and aesthetics, herders have adapted their domestication practices, leading to a closer relationship with their animals. By examining the shifting usages of the camel, from a pack animal in pastoralism to a saddled racing animal used in recreational activities, I have highlighted the complex interplay between politics, economy, and resilience and also how human-animal-environment relationships are constantly fluctuating. I have also underlined the important changes experienced by local Mongol societies, adapting their *traditional ecological knowledge* to new socioeconomic, environmental, and political challenges. Furthermore, the present case study was aimed at showing how autochthonous populations on the margins of China have manifested ecological resilience over the past seven decades within the various Anthropocene eras presented above.

This invented "camel culture" and its related recreational activities provide Inner Mongolia's political authorities with an ideal model to celebrate the political ideology of the multiethnic Chinese state and the "traditional culture" of the "Mongol ethnic minority." While these festivities are used to consolidate ideology, they are also adopted by local people. Indeed, over the last few years, some Mongol communities have started organizing their own camel races and beauty contests. In January 2012, for example, a Buryat family decided to organize its own winter camel race in Hulun Buir. Camel riders from all over Inner Mongolia were invited to take part in this event, which was presented by the organizers as an opportunity to celebrate an important event for the Buryats and Mongols of the area.

In the Alashan League, camel milk has emerged as an increasingly desirable product on the Chinese market. Recently, it was recognized by the Inner Mongolia Institute of Camel Research[15] as highly nutritious: it is especially recommended for strengthening immune systems. In 2014, the camel milk factory Shamo zhi shen Co. Ltd (沙漠之神) was founded and started producing various milk products such as milk, drinking powder, soap, and other skincare products. A growing number of camel herders are now selling fresh milk to the company to make extra money (Xin, 2018). It seems that the harnessing of the camel by the autochthonous community, the local companies, or the Chinese state highlights each party's ecological and economic concerns in different ways.

In Hulun Buir, Beljir became the head of the camel association of the Evenki autonomous banner created in 2019. Local people argue that the association will organize more events related to camel culture in Hulun Buir in the coming years. In the Mongolian cultural world, it seems that the camel is regaining, step by step, its past social superiority.

Notes

1 This research was funded by the European Union's Horizon 2020 research and innovation programme under the *Marie Skłodowska-Curie* grant agreement no. 893394.
2 Unless otherwise noted, Mongolian terms are given in italics and are put in parentheses.
3 The Tungus consist of multiple small groups with diverse ethnonyms who speak different languages of the Altaic language family. They are mainly scattered in the north of China. They are officially represented by four "ethnic minorities": the Manchu, the Evenki, the Oroqen, and the Hezhe.
4 Four to six distinct brides of Bactrian camel are recognized according to their ecological environment, geographical location, and morphology (size, shape, and color) (Zhao, 1998: 345).
5 Camels are also dispersed in relative low numbers in the Xinjiang Uygur Autonomous Region and the Qinghai and Gansu provinces.
6 Other Mongol groups living in Xinjiang, Qinghai, and Gansu also herd camels. Besides the Mongols, some Uygurs and Kazakhs groups of Xinjiang also engage in camel herding.
7 The Mongols represent one of the major "ethnic minorities" of China, numbering almost six million people. Despite the official picture depicting them as a homogeneous people, the Mongols are made up of several distinct groups.
8 The Mongols divide their food according to a binary system based on a dichotomy of ingredient color. Dairy products are classified as "white food" and meat products as "grey-brown food" (Ruhlmann, 2016: 86).
9 Mongolian territories became appealing sites for travelers, explorers, and adventure writers between the late 1870s and the 1940s. These observers gave detailed accounts of their journeys, mainly done by camel cart. See, for instance, the Russian geographer Nikolaj Prževal'skij (1880), the Russian explorer Pyotr Kuzmich Kozlov (1910), the Scottish missionary James Gillmour (1893), the Danish anthropologist and writer Haslund-Christensen [1935] 2000, the British travel writer Bulstrode (1920), the American writer Owen Lattimore (1941), and the American journalist Rosholt (1977).
10 The tendency to favor sheep and goat dates back to the 1920s when the strong foreign market for sheep wool resulted in a rapid growth in the number of sheep and goats (Atwood, 2004: 16).
11 The Evenki are a Tungus people. In the People's Republic of China, they are one of the 55 officially recognized "ethnic minorities" and number around 30,000 people. The Evenki are divided into three different subgroups: the Solon, the Khamnigan, and the reindeer herders.
12 In Mongolia, the religious ceremonies connected with the *naadam* were eliminated in 1923 (Atwood, 2004: 396): in China, the religious components were suppressed much later, in the mid-1950s.
13 Although most camel races appeared in the 2000s, the first camel race took place in September 1982 within the second national festival of ethnic minorities' traditional sports in Hohhot (Gao, 1999: 39). In 1985, it became an official component of *naadam* (Lin, 2013: 23).
14 For instance, among the Yugur people of Gansu province, camel racing is mostly performed with a ball (Li, 2017: 114).
15 This official institute was created in 2014. Its main goal is to study the camel in all its aspects and to promote the camel culture and economy in China.

Bibliography

Atwood, Christopher. 2004. *Encyclopedia of Mongolia and the Mongol Empire*. New York: Facts on File.

Aubin, Françoise. 2012. "Jeux, fêtes et loisirs chez les nomades: le cas de la Mongolie à l'époque communiste." *Anthropos* 107: 87–102.

Biermann, Frank. 2020. "The Future of 'Environmental' Policy in the Anthropocene: Time for a Paradigm Shift." *Environmental Politics* 30 (1–2): 61–80.

Breyfogle, Nicholas B. 2018. "Toward an Environmental History of Tsarist Russia and the Soviet Union." In *Eurasian Environments: Nature and Ecology in Imperial Russian and Soviet History*, edited by Nicholas B. Breyfogle, 3–19. Pittsburgh: University of Pittsburgh Press.

Bruno, Andy. 2016. *The Nature of Soviet Power. An Arctic Environmental History*. Cambridge: Cambridge University Press.

Bulstrode, Beatrix (Mrs. Edward Manico Gull). 1920. *A Tour in Mongolia with an Introduction Bearing on the Political Aspect of That Country, by David Fraser*. London: Methuen & Co. Ltd. https://digital.library.upenn.edu/women/bulstrode/mongolia/mongolia.html

Charlier, Bernard. 2016. "Actions rituelles, mobilité et attachement au « pays natal » parmi des éleveurs nomades de Mongolie." *Études mongoles et sibériennes, centrasiatiques et tibétaines* 47. http://journals.openedition.org/emscat/2779; https://doi.org/10.4000/emscat.2779

Chua, Liana, Fair, Hannah. 2019. "Anthropocene." In *The Cambridge Encyclopedia of Anthropology*, edited by Felix Stein, Sian Lazar, Matei Candea, Hildegard Diemberger, Joel Robbins, Andrew Sanchez Rupert Stasch. Accessed July 5, 2021. https://www.anthroencyclopedia.com/printpdf/512

Ding, Zhiying 丁志英, Shang, Wentong 商文通, Shang, Yong 尚勇. 2015. "Neimenggu feiwuzhi wenhua yichan caixie: Mengguzu saituo" 内蒙古非物质文化遗产采撷:蒙古族赛驼 [Inner Mongolia's Intangible Cultural Heritage Data Collection: The Camel Racing of the Mongol Nationality]. *Shijian (Sixiang lilun ban)* 1: 54.

Dumont, Aurore. 2015. "The Many Faces of Nomadism among the Reindeer Ewenki: Uses of Land, Mobility and Exchange Networks." In *The Ewenki of Aoluguya: Reclaiming the Forest*, edited by Ashild Kolås, Xie Yuanyuan, 77–97. Oxford: Berghahn Books.

Dumont, Aurore. 2016. "Le 'patrimoine culturel du renne': pratiques touristiques et trajectoires nomades chez les Évenk de Chine (Mongolie-Intérieure)." *Autrepart. Revue de sciences sociales au Sud* 78–79: 277–291.

Dumont, Aurore. 2018. "From Horse-Drawn Haymaking to Tractors: A Century of Technological Adaptation among the Herders of Inner Mongolia." Unpublished draft paper presented at the international workshop: *Closing the Gap. How Technology Changes Spatial Relationships between Humans and Animals*, Nordic Centre, Fudan University, Shanghai, PRC, March 2018.

Fijn, Natacha. 2011. *Living with Herds. Human-Animal Coexistence in Mongolia*. Cambridge: Cambridge University Press.

Franklin, Adrian. 2015. Ecosystem and Landscape: Strategies for the Anthropocene." In *Animals in the Anthropocene. Critical Perspectives on Non-Human Futures*, edited by the Human Animal Research Network Editorial Collective, 63–87. Sydney: Sydney University Press.

Gao, Yingjian 高迎健. 1999. "Mengguren de nadamu 蒙古人的那达慕" [The *naadam* of the Mongols]. *Tiyu bolan* 4: 39.

Gardelle, Linda. 2010. *Pasteurs nomades de Mongolie. Des sociétés nomades et des États*. Paris: Buchet/Chastel.

Gauthier-Pilters, Hilde, Dagg, Anne Innis. 1981. *The Camel, Its Evolution, Ecology, Behavior, and Relationship to Man*. Chicago and London: The University of Chicago Press.

Gillmour, James. 1893. *Among the Mongols*. London: Religious Tract Society.

Gros, Stéphane. 2012. "L'injonction à la fête. Enjeux locaux patrimoniaux d'une fête en voie de disparition." *Gradhiva* 16: 25–43.

Haslund-Christensen, Hennig. [1935] 2000. *Men and Gods in Mongolia*. Kempton, PA: Adventures Unlimited Press.

Hazard, Benoît. 2017. "Anthropocene versus Anthropo-scenes." 1–3. Accessed July 6, 2021. https://halshs.archives-ouvertes.fr/halshs-01505329/document

Hobsbawm, Eric, Ranger, Terence. 2014 [1983]. *The Invention of Tradition*. Cambridge: Cambridge University Press.

Humphrey, Caroline, Sneath, David. 1999. *The End of Nomadism? Society, State and the Environment in Inner Asia*. Durham: Duke University Press.

Humphrey, Waddington Caroline. 1974. "Horse Brands of the Mongolians: A System of Signs in a Nomadic Culture." *American Ethnologist* 1 (3): 471–488.

Inglis, Julian. 1993. *Traditional Ecological Knowledge: Concepts and Cases*. Ottawa: International Program on Traditional Ecological Knowledge, International Development Research Centre.

Khalaf, Sulayman. 1999. "Camel Racing in the Arab Gulf: Notes on the Evolution of a Traditional Cultural Sport." *Anthropos* 94: 85–106.

Khalaf, Sulayman. 2000. "Poetics and Politics of Newly Invented Traditions in the Gulf: Camel Racing in the United Arab Emirates." *Ethnology* 39 (3): 243–261.

Kozlov, Petr Kuzmich. 1910. "The Mongolia-Sze-Chuan Expedition of the Imperial Russian Geographical Society." *The Geographical Journal* 36 (3): 288–310.

Kwong, Emily. 2019. "Where Camels Become Beauty Queens: Inside Mongolia's Biggest Camel Festival." *National Public Radio*, May 11, 2019. https://www.npr.org/2019/05/11/721738620/where-camels-become-beauty-queens-inside-mongolias-biggest-camel-festival

Lattimore, Owen. 1928. "Caravan Routes of Inner Asia: The Third 'Asia Lecture'." *The Geographical Journal* 72 (6): 497–528.

Lattimore, Owen, 1941. *Mongol Journeys*. New York: Doubleday Doran.

Lavrillier, Alexandra. 2005. "Nomadisme et adaptations sédentaires chez les Évenks de Sibérie postsoviétique : 'jouer' pour vivre avec et sans chamanes." PhD diss., École Pratique des Hautes Études.

Lavrillier, Alexandra, Gabyshev, Semen. 2017. *An Arctic Indigenous Knowledge System of Landscape, Climate, and Human Interactions. Evenki Reindeer Herders and Hunters*. Fürstenberg, Havel: Kulturstiftung Sibirien.

Li, Bo 李波. 2017. "Gansu sheng Sunan Yuguzu zizhixian sai luotuo xianzhuang diaocha yanjiu" 甘肃省肃南裕固族自治县赛骆驼现状调查研究 [Research on the Current Situation of Camel Racing in the Sunan Yugu Autonomous County in Gansu Province]. *Wushu yanjiu* 2 (4): 114–116.

Lin, Sitong 林思桐. 2013. "Neimenggu caoyuan de wenhua shengshi – sai luotuo" 内蒙古草原的文化盛事——赛骆驼 [An Important Cultural Event in the Grasslands of Inner Mongolia: Camel Racing]. *Minsu tiyu* 10: 23.

Litzinger, Ralph. 2004. "The Mobilization of 'Nature': Perspectives from North-West Yunnan." *The China Quarterly* 178: 488–504.

Marchina, Charlotte. 2013. "L'élevage mongol à plusieurs espèces." In *Nomadismes d'Asie centrale et septentrionale. Sociétés, mobilités et environnement*, edited by Charles Stépanoff, Carole Ferret, Gaëlle Lacaze, Julien Thorez, 172–175. Paris: Armand Colin.

Marchina, Charlotte. 2019. *Nomad's land: Éleveurs, animaux et paysage chez les peuples mongols*. Brussels: Zones sensibles.

Menghedalai 孟和达来. 2015. "Luotuo wenhua" 骆驼文化 [The Culture of the Camel]. *Alashan youqi renmin zhengfu* 阿拉善右旗人民政府, July 27, 2015. http://www.alsyq.gov.cn/2018_show/30/3313.html

Namsaraeva, Sayana. 2012. "Ritual and Memory in the Buriat Diaspora Notion of Home." In *Frontier Encounters: Knowledge and Practice at the Russian, Chinese and Mongolian Border*, edited by Franck Billé, Grégory Delaplace and Caroline Humphrey, 137–163. Cambridge: Open Book Publishers.

Olsen, Stanley J. 1988. "The Camel in Ancient China and an Osteology of the Camel." *Proceedings of the Academy of Natural Sciences of Philadelphia* 140 (1): 18–58.

Pâpoli-Yazdi, Mohammad-Hossein. 1982. "La motorisation des moyens de transport et ses conséquences chez les nomades kurdes du Khorâssân (Iran)." *Revue Géographique de l'Est* 22 (1–2): 99–115.

Prževal'skij, Nikolaj Mihajlovič. 1880. *Mongolie et pays des Tangoutes*. Translated from Russian by G. Du Laurens. Paris: Librairie Hachette.

Rosholt, Malcolm. 1977. "To the Edsin Gol: A Wisconsinite's Journey in Inner Mongolia, 1935." *The Wisconsin Magazine of History* 60 (3): 197–227.

Roux, Jean-Paul. 1959. "Le chameau en Asie Centrale: son nom – son élevage – sa place dans la mythologie." *Central Asiatic Journal* 5 (1): 35–76.

Ruhlmann, Sandrine. 2016. "Are Buuz and Banš Traditional Mongolian Foods? Strategy of Appropriation and Identity Adjustment in Contemporary Mongolia." In *Eating Traditional Food: Politics and Identity*, edited by Brigitte Sébastia, 86–103. London and New York: Routledge, Taylor & Francis Group.

Shapiro, Judith. 2001. *Mao's War against Nature: Politics and the Environment in Revolutionary China*. Cambridge: Cambridge University Press.

Stammler, Florian. 2005. *Reindeer Nomads Meet the Market: Culture, Property and Globalisation at the 'End of the Land'*. Münster: Lit Verlag.

Wuricaihu 乌日才呼. 2012. Mengguzu Gebi mutuo wenhua duiyu shengtai wenming jianshe de yiyi—yi Bayannao'er Wulate houqi bayin gebi sumu weili 蒙古族戈壁牧驼文化对于生态文明建设的意义—以巴彦淖尔乌拉特后旗巴音戈壁苏木为例 [The Significance of the Mongols' Domesticated Gobi Camel Culture for the Construction of an Ecological Civilization: the Case of Bayin Gebi Village of the Urat Rear Banner in Bayannur Prefecture]. MA diss., Inner Mongolia Agricultural University.

Xin, Xu. 2018. "Across China: Camel Milk, Latest Entry in Milk Alternatives Race." *Xinhua Net*, December 2, 2018. http://www.xinhuanet.com/english/2018-12/02/c_137646091.htm

Xu, Zhanjiang 徐占江, Zhao, Yuxia 赵玉霞, Long, Tao 陶龙, Bi, Jinjie 毕金杰, Wang, Zhaoguo 王召国, Daxizhamusu 达喜扎木苏, Simujide 斯木吉德. 2009. *Zhongguo Buliyate Mengguren* 中国布里亚特蒙古人 [The Buryat Mongols of China]. Hulun Buir: Neimenggu wenhua chubanshe.

Yeh, Emily T. 2005. Green Governmentality and Pastoralism in Western China: 'Converting Pastures to Grasslands'." *Nomadic Peoples* 9 (1–2): 9–30.

Zhao, X. X. 1998. "Types and Breeds of the Chinese Bactrian Camel (Camelus Bactrianus)." *Revue d'élevage et de médecine vétérinaire des pays tropicaux* 51 (4): 345–352.

4 Reindeer, Taiga, Ethnic Culture

State-Forced Resettlement and the Changing Human–Animal Interactions in the Aoluguya Ewenki Community

Hang Lin

Within the taiga mountains of the Greater Khingan Range, China's largest continuous area of primitive forest, live the Manchu-Tungusic Ewenki people of Aoluguya with their reindeer. For centuries, the life of the merely 200 Ewenki remained intact: living on reindeer herding. As the only reindeer herders in China, the Ewenki inhabit the taiga and kept small herds of domesticated reindeer and used them for milking, riding, and carrying loads. As their major living source and partner, the reindeer allowed the Ewenki to maintain a mobile lifestyle dictated by their seasonal hunting and herding activities, exerting enormous symbolic as well as instrumental value (Fondahl, 1998: 3). Yet as the local government undertook to resettle them as "ecological migrants" (*shengtai yimin* 生态移民) to a new town in 2003, they suddenly became the focus of the public both in and outside China.

The state introduced the resettlement to the Ewenki of Aoluguya and promised them that they would have a new sedentary way of life, since in the new town they and their reindeer would be provided for (Wu, 2003). Only a short time after the relocation, however, the reindeer were falling ill and beginning to die due to lack of food source. Some Ewenki herders, together with their reindeer, were left with no choice but to move back to the mountains. With their homes now settled in the new town, which provides all necessary facilities for living and schools for children, they are not entirely leaving the settlement but to rotate between the town and the camps in the forest where they herd reindeer. A growing number of the Ewenki, on the other hand, begin to restrain from engaging in reindeer herding and gradually adopt a new life of sedentary residence, involving themselves in service industry branches centering on tourism.

Already prior to the 2003 movement, the Aoluguya Ewenki, as the only reindeer herders in China, have attracted scholarly attention, both domestically and internationally. In their pioneering ethnographic accounts on the reindeer Tungus in Manchuria (today's Aoluguya Ewenki), Sergei Mikhailovich Shirokogoroff (Chin.: Shi Luguo 史禄国, 1887–1939) and Ethel J. Lindgren, respectively, have produced in-depth documents on the Ewenki way of living and their trade with the Cossack farmers (Shirokogoroff,

DOI: 10.4324/9781003212089-4

1929, 1935; Lindgren, 1935, 1938, 1939). The hitherto most comprehensive study of the history and ethnography of the Aoluguya Ewenki is provided by Kong Fanzhi, who has elaborately traced their route to the old Aoluguya and charted their herding system and indigenous cultural practice (Kong, 1994). As their story became known to the world in 2003, there has been a growing body of literature focusing on the relocation and their changed life: Xie Yuanyuan has vividly described the resettlement process and cogently pointed out that the problem of housing distribution and unemployment after the move (Xie, 2005, 2010, 2015). Based on extensive fieldworks from 2008 to 2009, Huang Jianying has recorded the social development of the new settlement and living conditions of the Ewenki in detail (Huang, 2009). The relocation has urged Richard Fraser and Åshild Kolås to focus on the different perspectives of the central government, the local authorities, and the Ewenki (Fraser, 2010; Kolås, 2011). A recent volume, edited by Åshild Kolås and Xie Yuanyuan, is devoted to the efforts of the Ewenki to reclaim their forest lifestyle and develop new forest livelihoods (Kolås and Xie, 2015).

These inspiring works have laid the cornerstone for our understanding of various aspects that pertain to the relocation of the reindeer Ewenki, yet how did this resettlement impact the ethnic culture and mind setting of the Ewenki remains largely a neglected topic in the anthropological literature. The changed human–natural environment relationships and the lifeworld of the Ewenki has remained largely unventured, in particular concerning the peculiar intimate mutual relationships between human, taiga, and reindeer. What were the relationships between the Ewenki and reindeer and how did such relationships shape the economic and religious lifeworld of the Ewenki? How did the resettlement, together with the resulting spatial control and constraints, influence their physical and spiritual distance to the reindeer? In what way did this changed distance, in addition to the increasing importance of tourism, transform their indigenous way of economic, social, and religious living? How does the community attempt to cope with the crises and how is their chance to be successful?

To tackle these questions, this chapter sets out to analyze the interactive human–animal relations between the Aoluguya Ewenki and the reindeer by analyzing them through the perspective of the Anthropocene, a discourse developed by Paul J. Crutzen and Eugene F. Stoermer (2000). While much of the discussion on the Anthropocene has since then centered on anthropogenic climate change and the human response to the challenge, recent scholarship has begun to advance a relational model for human and nonhuman life (Human Animal Research Network Editorial Collective, 2015). Applying the perspective of tracing the effects of human political and economic systems on animals, this study focuses on the evolving role natural environment and animals played in the life and ethnic culture of the Ewenki, as well as how the changed environment produces new forms of life and alters the very nature of species. By tracing the state-forced resettlements in the past decades, with a particular emphasis laid on the 2003 relocation, it shall explore the immense impact of such resettlement, as well as the consequent growth in tourism, on the long-term interest of the community. Through the lens of Anthropocene (Crutzen, 2002; Zalasiewicz, 2011), it aims to explore the multidimensional influence of human behavior (e.g., the state-forced resettlement) on the environment and the animal (e.g., the reindeer), and vice versa, and by doing so probes into the complex relationship between human, animal, and environment.

Ewenki Ethic Culture: Reindeer and Taiga

As a branch of the Ewenki, which means "the people who live deep in the mountains" in their language, the Ewenki of Aoluguya originated in the region around the Lake Baikal and they moved to live in the forests north of the Armur River some 300 years ago (Kong, 1994: 35–36; Neimenggu zizhiqu bianjizu, 2009: 129). Whereas most Ewenki people in China, the Solon and the Khamnigan Ewenki, have settled down in agrarian and pastural areas, the Aoluguya Ewenki tribe, however, began to nurture reindeer-breeding skills as early as the 17th century.

Unlike the large-scale reindeer ranchers of Scandinavia and north Siberia, who reside in tundra areas and live off the meat of large herds of reindeer, the Aoluguya Ewenki, similar to other reindeer herding peoples such as the Soyot of Buryatia, the Tofalar of Irkutsk Oblast, and the Tsaatan of Mongolia, practice a peculiar form of reindeer husbandry by raising small herds in the taiga as pack and riding animals for their milk, while wild game stands as their principal source of food.[1] Used to being saddled and either

Figure 4.1 Close-up of a Reindeer

(Public Domain Photo by LTapsaH from Pixabay).

ridden or burdened with a pack and the does are used to being milked, the Ewenki reindeer are tamer than tundra reindeer, which enables a closer relationship between the herder and the reindeer. As a result, the reindeer come to depend on specialized technologies requiring intensive contact to the humans, such as firing smoke to protect against biting insects, provision of salt, and protection from predators.

Such way of herding has a long history, as the Ewenki initially found that the reindeer is a means of food reserve, which could make up for the uncertainty of prey acquisition. In the process of domestication, other characteristics of the reindeer are gradually recognized. First, the reindeer foraging migration in the forest is consistent with the lifestyle of the Ewenki hunters. In order to match the migration characteristics of the reindeer, the Ewenki people also follow a relatively fixed migration route and frequency, roughly in a one-year cycle (Lin, 2018: 6). With reindeer by their side, the hunters can easily obtain a better harvest, insomuch as the reindeer attract their natural enemies, the target of the Ewenki hunters.

Second, the reindeer has a strong load capacity and is an important tool for packing and riding. Its large hoof allows it to walk on the marsh without much difficulty, and even through bushes and mountain rocks. When reaching adulthood, a single reindeer can carry about 40 kilograms of goods and walk 5–6 kilometers per hour for 10 hours. When equipped with a sledge, a reindeer can draw up to 160 kilograms, far exceeding the capacity of cattle and horses, which enable their herders to carry their households and freely move in the vast forest (Zhao, 1975: 25). Together with the hound, the reindeer constitute the most important assistant for their herders. As an old Ewenki saying goes: "There is a reindeer and a hound. It is especially labor-saving when hunting. The reindeer can ride the game, the hound prevents the beast from escaping. It is convenient for the hunter in the mountains and forests" (Abenqian, 2015: 46).

Being a hoofed species uniquely adapted to the tundra and taiga areas, the reindeer are able to venture far during the winter to searching for lichen, their principal food. Thanks to their extraordinary olfactory sense, the reindeer can smell the lichen and even scoop their food from the snow. While they do not need to be fed or driven to pasture land, the reindeer would rely on their herders to acquire their favorite refreshment, for example, salt. Not being a local product easily accessible to the reindeer, their herders become the major source. Based on such fondness for salt, the Ewenki can easily summon their herds by tapping the salt bag, since when hearing the sound, the reindeer would return to the campsite from afar and compete for salt (Kong, 1994: 176). Before their departure for the next campsite, the herders feed the reindeer with salt in advance, so that the herds become more tame and docile and their vitality and endurance will be greatly increased.

In spring, reindeer cows give birth to calves. Throughout the summer they come back to the encampment everyday to benefit from the herders' smoky fire that protects them from insects. The Ewenki herders pile up thick logs to set up a wooden frame and ignite a layer of yellow-green wet lichen to produce smoke. As the anophelifuge smoke rises, the reindeer gather around the smoke to escape from mosquitos and midges. The herders then observe the smoke and regularly add wood and lichen to the fire, ensuring the smoke can continue to function until the reindeer get enough time for rest (Tang, 1998: 92). Typically, herding activities are shared by gender, in which men manage the herds, organize moving, cut wood, and cutting off the antlers, while women are responsible for making food, looking after calves, and milking.

In this way, the herders and herds together form a reindeer-pastoralism necessitating intimate human–reindeer relations, which Florian Stammler and Hugh Beach term "symbiotic domestication" (Stammler and Beach, 2006: 8) As herders, the Ewenki provide salt to their animals while lighting smoke to deter biting insects. In return, the reindeer rely on their herders for refreshment and in doing so seek interaction with human without coercion. With the assistance of shamans, the Ewenki herders perceive their world as a universe populated by spirits of various kinds.[2] While such view is on a par with many Tungusic peoples in North Asia, what makes the Ewenki particular is their perception of the reindeer as the proper domain, from which the complex relationships can be maintained between the sky, the earth, the taiga, the human, and other animals.

Under the influence of such understanding, the Ewenki even attribute to their reindeer characteristics typically reserved for humans. As Richard Fraser remarks, when the reindeer are led back to the campsite, the herders would recount descriptions concerning their herds' experience from the perspective of the animals themselves (Fraser, 2010: 335). In another case Joachim Otto Habeck also observes that an Ewenki sledge driver describes that his draught reindeer "listens well" and "obeys well" (Habeck, 2006: 133). Illustrating the animal's sensual perception, the herders characterized their herds as possessing their own personalities and thus redefine the human–reindeer relations. Yet as the living environment change, such relations also change in accordance.

From the Taiga to New Aoluguya

Following the repeated incursions of Tsarist Russia in northeast Asia, a few groups of Ewenki reindeer herders crossed the Armur River between the late 18th and mid-19th centuries to settle in the Chinese territory (Heyne, 2002). They remained anonymous to the Chinese authorities until the early 20th century, while they continued to pay taxes to the Tsar and marry in the Orthodox Church (Shirokogoroff, 1929: 67–68). The Ewenki herders also engaged in regular trade with Cossack farmers, borrowing many Russian words that are still in use today, including names.[3] While conducting fieldwork among the Ewenki clans in the early 1930s, the anthropologist Ethel J. Lindgren noted that the area of their nomadic pasture was about 7,000 km^2, covering the large territory along the Armur and Argun rivers on the current Sino-Russian border (Lindgren, 1938: 609).

Although the Ewenki reindeer herders chose to remain on the southern banks of the Armur after the Russian Revolution, their contact to the Chinese state was few. In 1957, the first Ewenki "ethnic township" (*minzu xiang* 民族乡) was established in Qiqian south of the Armur. Recognizing the medical effect of the antlers, the state collectivized the reindeer, yet they still remained under the care of the Ewenki herders and continued to live in the taiga encampments (Lü, 1983: 12). As the Sino-Russian conflicts intensified in the early 1960s, the Chinese authorities became anxious about the Ewenki's Russian connection and sought to relocate and sedentarize them. Finally in 1965, 35 households of Ewenki and their 900 reindeer were moved southward to a newly built village named Aoluguya, which means "flourishing aspen" in the Ewenki language (Nentwig, 2003: 36). Facilitated with wooden houses and an antler-processing factory, the settlement evolved as the principal domain of the community and some herders began to find employment in other sectors such as the forest industry.

During the following two decades, reform policies promoted livestock business in Aoluguya, with the reindeer population peaked in the 1970s at more than 1,080 animals. The reforms of the 1980s initiated the state to redistribute the 755 head of reindeer to 24 Ewenki families in 1984, whereas the antler industry remained under state control, in which the state-owned enterprise was responsible for processing and sale of the antlers in exchange for 20% of profit (Beach, 2003: 34; Huang, 2009: 62). In doing so, the state attempted to turn the rather small-scale reindeer herding into an industry of reindeer-breeding, somehow in the Soviet form of "production nomadism" (Vitebsky, 1990: 348). It was also during this time that the grazing lands of the reindeer herders became a major attraction for the rapidly growing forest industry, which resulted in appropriation of land by the forestry authorities and gradual reduction of reindeer pastures. The deterioration of pastures and forests forced the government to take actions in the late 1990s. In 1996, hunting was brought to an end when firearms were confiscated. Two years later, following the nationwide campaign to Open Up the West (*Xibu da kaifa* 西部大开发), the policy of "Converting Pasture to Forest" (*tuimu huanlin* 退牧还林) was deployed which aimed at "adopting settled residences and controlling livestock stocking rates" (Wu and Wen, 2008: 18). In 2003, eventually, as China called for "ecological migration" to better protect the forests along the Greater Khingan Range, the local government undertook to resettle the 62 Ewenki households and their reindeer 260 kilometers southward to Genhe and built a new township at its western outskirts, known as "New Aoluguya."

Coping with the New Environment: Returning to the Forest

Located only 4 kilometers to the center of Genhe, the new Aoluguya town consists of 31 chalet-style residence houses, 48 reindeer pens, an antler processing factory, a school (soon turned into a hotel), a museum, a hotel, and a government building, as well as some medical and shopping facilities, which were offered to the Ewenki herders as compensation for relocating and accepting the hunting ban (Xie, 2005: 52). Following the plan formulated by the Finnish consulting firm Pöyry, the houses have been completely refurbished in Finnish design, with each house, all equipped with modern amenities such as running water and central heating, divided into two halves, one for each family.

With all these facilities for a sedentary way of life, the resettlement was introduced to the Ewenki herders with a promise from the authorities that they would no longer need to move and hunt in the forest. According to the plan, the reindeer were to be hand-fed and kept permanently in enclosures outside the settlement. But it neglected the fact that although the new Aoluguya is still located in the taiga, since it is too close to a city, its environment is not suitable for the growth of lichen. Moreover, the reindeer are not accustomed to the enclosures in which their moving space is considerably limited. In consequence, within only several weeks after the move, the reindeer were falling ill and perishing (Xie, 2005: 54; Kolås, 2011: 398). Anger and despair spread among the herders as they realized that the government was incapable of restoring the health of their herds. Finally, they were left with no other viable options than to return to the forests, back to the areas where reindeer could find sufficient food. Thus only a short time after the relocation, many Ewenki, together with their reindeer, swarmed back to live in the mountains.

Their life in the mountains and forests, however, has changed substantively. Located in the taiga off smaller logging paths that diverge from the main road, there are now six campsite located from south to north along the railway line that connects Genhe to Mangui.[4] The size of their herds varies, with the smallest consists of 50 reindeer and the largest about 700 (Fraser, 2010: 330). Each encampment is made of two to six tents, including the traditional cone-shaped tepee-tents known as *zuoluozi*, and modern ridge ones facilitated with solar panels and wood-burning stoves (Dumont, 2015: 88–89). Altogether some 50 Ewenki herders live and work in the taiga, among which a small portion spent almost all of their time in the campsites, whereas the majority stay only temporarily and visit the settlement on regular basis, primarily to purchase daily supplies, visit family and friends, or to deliver antler harvest. No bus service is available, and only once in a month, the local government sends a car to bring some basic supplies such as vegetables to the campsites. But for other necessities such as rice, oil, medicine, and alcohol, the herders need to rely on themselves to ride motorcycles or taxi service.

A certain level of mobility continues to be an essential feature of reindeer herding, as the camps move around four times a year, yet their movement across the taiga is no longer an act of free choice. Already before 2003, the state has established special zones for natural conservation (*ziran baohuqu* 自然保护区) in the forest, which has drastically reduced the area available for the move of the herders. As the forests are state-owned, the herders are required, in accordance with environmental legislation, to report the location of their campsites to the local forestry station (*linchang* 林场) responsible for overseeing their movement and stay within its confines. In consequence, although the Ewenki herders view the forest as pasture and hunting ground, they now have to move with demarcated frontiers in mind and follow established routes allotted to them. Such boundaries, as Emily Yeh argues, expand the state power by creating greater control over both the resources and the people within this sphere, thus restricting the mobility of the herders (Yeh, 2005: 16). Whereas Siberian Ewenki reindeer and their herders travel ca. 1,000 kilometers per year, the movements of Aoluguya Ewenki are significantly limited, annually moving only 15–20 kilometers.[5] The strict spatial control has led to constraints on the rotation routes of the reindeer and their interaction with other groups in other parts of Siberia. Unlike their relatives in Mongolia and Russia, the size of the Aoluguya reindeer community is considerably smaller and their biological diversity and vitality is also much more restricted.

The spatial restriction also exerts an enormous impact on the herders. At the campsite, consistent with the observations made long ago by Shirokogoroff and Lindgren (Shirokogoroff, 1929; Lindgren, 1933), everyday life still revolves around the reindeer, with all tasks entertained focusing on the husbandry enterprise. But unlike the traditional campsites that were organized by clans, the new ones are run by one single family or several nuclear families together. After the decollectivization of the reindeer in 1984, the reindeer are now privately owned, the pastoral activities are communally organized, ranging from setting the site, sharing firewood and salt, and most notably, cropping of antlers. The organization of herding labor, however, has substantively changed from a family occupation to a male dominant model, in which men, some of them do not possess reindeer but are hired by the owners, take care of the traditional responsibilities of the women, such as cooking, tending animals, and milking. Women and children, on the other hand, stay almost permanently in the settlement or even in Genhe, where schools and other facilities are available.

With the ban of hunting and the reliance on antlers, the Ewenki ontology of the nature has accordingly shifted to a growing emphasis on herding. While previously reindeer were considered as the pack animals that facilitated their hunting movements in the taiga, as their living equipment become more modernized and weighty, motorcycles and trucks replace the reindeer as the main means for riding and transporting (Dumont, 2015: 84–86). As a result, it is the reindeer, not anymore the game, that are envisaged as the primary reason for the herders to remain in the forest, since the forest is the only place which allows them to perpetuate the human–reindeer relations. In this way, the relationship between the herder and the reindeer is again strengthened, in which the herders provided for their reindeer the necessary substance (e.g., lichen) and, in return, taking the antlers from the reindeer. This reciprocal relationship thus maintains the Ewenki personhood and lifeworld, "not only for their movements in the landscape, but also for their sustenance and reproduction, their life and death" (Stammler and Beach, 2006: 12).

New Aoluguya: Changed Environment and Changed Life

During the migration project that lasted from August to September 2003, 62 families (162 persons) of Ewenki were relocated to the new Aoluguya settlement, to each a half of the residence house was assigned. In the state-sponsored media, all the Ewenki involved in the relocation were described as "practicing" reindeer herders; however, only 24 families received reindeer in the reform of decollectivization in 1984 and were therefore herders in the true sense (Wu, 2003). Such discrepancy is resulted from the fact that prior to the movement, a number of Ewenki have already lost their connections to the reindeer lifeworld, some even before the 1984 reform. Moreover, the new Aoluguya also become a mini Ewenki community, since all the 162 residents are Ewenki, whereas their non-Ewenki neighbors, mostly Han-Chinese and Daur Mongols, remain in the old town or are required to look for new residence with certain compensation.

As aforementioned, the greatest impact of the movement has been on those who are still connected to the reindeer, especially those herders and their family members who do not reside permanently in the campsites. Since almost all financial funding and living facilities are allocated to the new Aoluguya settlement, the government support that the herders receive is very limited, except for the trucks that occasionally help their movements in the forest. Next to a minimal amount of welfare allowance provided by the local government, the herders need to rely on their settlement-based families, or in some cases employers and taxi drivers, to deliver food and daily necessities and to sell the cropped antlers. In addition, they have to cope with the new challenge of dividing their time to travel between the campsite and the settlement. As for their families, only few women reside in the campsites, typically because they maintain shops to sell antlers or their children attend schools. In this sense, the new Aoluguya becomes an indispensable space in Ewenki life, since even though the reindeer are largely herded in the campsites in the forest and not all Ewenki reside permanently in the settlement, still it is the administrative and herding headquarters, which offers living spaces for the herders' families and the sale of the antlers, thus acting as the connecting point in their changed life as "mobile reindeer herders" (Dumont, 2015: 89).

The new Aoluguya, however, was not positively perceived by all the Ewenki herders. Although facilitated with residence houses and reindeer pens, the settlement gives an impression of an artificial settlement, as it lacks essential social and economic infrastructure such as school, hospital, and food store. In fact, there are no commercial services distinct from the tourism industry and very limited means for community interaction. Such tourism-oriented structure is not a design error or a malpractice of the master plan, rather an intention of the authorities. Officially described as "ecological migration," the relocation was rationalized by the state to recover the environment and to accelerate the development of the Ewenki community (Xie, 2010: 100–110). On the regional level, however, the protection of wildlife was not the only argument for the movement. In their funding proposal to the central administration, the local government expressed explicit desire centered on need for infrastructure modernization and economic growth and ethnic culture (Huang, 2009: 22). To modernize the Ewenki community and integrate them into the development scheme of the region, the local authorities introduced tourism into the new settlement.

According to the state plan, the settlement is to be transformed into the "Aoluguya Ethnic Reindeer Resort," which intends "to preserve the Aoluguya cultural heritage and old livelihoods by turning the old skills and unique lifestyle into a tourism product" (Pöyry, 2008:7). Under such plan, the school was turned into a hotel, which, after two rounds of expansion and renovation, has a capacity of accommodating 110 guests. Featuring reproductions of cone-shaped tents made of birch bark, the forest theme park, called the "Aoluguya Reindeer Tribe" (*Aoluguya shilu buluo* 敖鲁古雅使鹿部落) welcomes the tourists to sip tea from bark cups and feed the some ten reindeer kept there, while the tour guide dressed in "traditional Ewenki costumes" tell the stories of Ewenki herders and hunters. After visiting the museum, the tourists are encouraged to stay in the hotel or home-style inns offered by the Ewenki residents to gain a sense of the ethnic uniqueness of the reindeer tribe. At the highest peak around the settlement, a wood reindeer gilded in gold has been placed, marking this the home of the reindeer herders.

Aimed at providing modern housing for the Ewenki while simultaneously creating tourism-related employment opportunities for the community, the local government envisioned new sources of income for the local residents, including hotel services, souvenir production, and ethnic show. The efforts of the local government to promote the image of China's reindeer people are relatively well received and rewarded. Though not a well-known tourist destination, Aoluguya has attracted over 15,000 tourists from its opening in June 2007 to the end of that year (Huang, 2009: 77). As of 2017, a total of 51,2000 tourists visited Aoluguya to get a sense of the reindeer herding lifestyle, among which 30,000 came from countries outside China (Aoluguya shilu buluo jingqu, 2018). In 2010, China, represented by the Aoluguya herders, was admitted to the Association of World Reindeer Herders and Aologuya hosted the 2013 World Reindeer Herders Congress.

Taiga, Reindeer, Ethnic Culture: The Dilemma

Although the Ewenki have already experienced several rounds of relocation in the preceding decades, none of them has so dramatically altered the lifeworld of the Ewenki, both herders and non-herders, as the 2003 movement. In the relocations of 1957 and 1965, they were also moved as required by the state but one has resulted in to such discrepancy as seen in 2003. The reason behind this difference is because they were relocated within the taiga so that their economic and cultural base of reindeer herding was largely maintained. While their ties to the forest were not cut by the relocation, in the settlement they were provided with more alternatives next to herding and hunting. Next to their traditional way of living with the reindeer, new opportunities became available to enjoy the material benefits of the modern life. This time, however, the relocation project was carried out with a hitherto unseen level of authority, with the state actively integrating the community into its scheme of modernization and economic development. Being assigned to a sedentary settlement so close to an urban environment, the herders were moved out of the taiga and a large share of them thus lost the opportunity to engage in subsistence hunting and reindeer herding.

The resettlement was imposed upon the Ewenki by strong state will and their life has thus irreversibly changed. In their effort to react to the change, some gradually accepted their fate and tried to perceive it in a positive light. Pu Lingsheng, former mayor of Aoluguya, is one of them. As one of the increasing numbers of Ewenki who do not own reindeer, he is among the first to appeal to the government that they be given a new settlement so as to modernize the community. For him, it is the move to the new Aoluguya, with this privileged location and modern infrastructure, that makes the antler processing industry possible and valuable. Although reindeer and their antlers lay the economic ground of the Ewenki, the reindeer-based tourism, which has been emerging rapidly, shall be prioritized, because it will "not only increase the income of the Ewenki, but also protect and revive the ethnic culture" (Huang, 2009: 193). Pu's point is shared by Wu Xuhong and Suo Ronghua. When asked about her feeling for the new settlement, Wu, holding a university degree and currently working for the local government, firmly answers that she would doubtlessly prefer the new over the old, since "it is near the city of Genhe and the living conditions are much better, which makes life more convenient than before" (Huang, 2009: 126). Running a store for souvenirs, she is pleased to see that Aoluguya is welcoming more and more tourists, who have brought significant increase in the family income. For her, the new environment is more comfortable as in comparison to the forest and the close distance to Genhe it also means better education for her children (Huang, 2009: 143).

Despite the fact that the relocation has improved the material well-being of the Ewenki, it is also to a great extent damaging for the community's long-term interest. With a large number of Ewenki being sedentarized and increasingly attracted to the lifestyle provided by the urban infrastructure, the new generation of Ewenki are now largely separated from the taiga where their ancestors lived for generations. This separation has exerted its negative impact on the Ewenki way of living and their indigenous cultural practices. Due to the hunting ban, the Ewenki stopped hunting and except for few old hunter, no one knows how to use guns, ground arrows, or birch bark skis. The reduction in the number of reindeer hide and the complexity of the traditional way of skinning have kept the young generation away from learning it, while the feather garment is quickly replaced by the more fashionable and convenient modern clothing (Wang, 2015: 957). Even the reindeer, which used to be the most important means of transport for the Ewenki, have given place to motorcycles and trucks, and the production and living activities such as moving now have been replaced by various cars and motorcycles. As a result, the traditional skills of training reindeer for riding and loading are also on the verge of loss. Today, children and youth are seldom seen in the campsites, yet most of them are students who only stay there temporarily during holidays and vacations and only few of them are willing to continue practicing reindeer herding.

In addition, the Ewenki are also experiencing growing threat to the preservation and inheritance of their language and religious belief. In the school, Ewenki children are sitting in classes with their Han-Chinese peers, with all classed held in Chinese. During their work in the tourist industry, the formal hunters and herders also speak Chinese, as their customers are overwhelmingly Han-Chinese. Thus within only one generation, most Ewenki in Aoluguya have changed from speaking predominantly their own language to conversing in Chinese and now only about 40 people can still speak the traditional Ewenki language (Fraser, 2010: 317). Although believed in by the Ewenki for generations, Shamanism, which has played an irreplaceable role in their spiritual beliefs and cultural practices, is gradually faded out of their religious lifeworld, especially with the death of Niula, the last Shaman, in 1997 (Lin, 2018: 10). Since then, there is no new shaman in Aoluguya and the Ewenki have never held any shaman activities.

Certainly, such negative experiential implications of the relocation are shared by many Ewenki. Although some of them are still directly or indirectly involved in the reindeer herding, a growing number feel themselves dragooned into the relocation and lament that they are forced to distance themselves from hunting and herding economies. The opinion of Anta, one of the few Ewenki who are still well versed in traditional skinning skills, is representative of the former herders and hunters, as she expressly notes that although she gradually accepted the settlement, she still prefers the forest, since there "it is free and unstrained, which is irreplaceable by the material comfort of the sedentary life" (Huang, 2009: 124). While Anta remains in the settlement, others even sternly choose to return to their campsite in the taiga. For instance, the sisters Maria Bu and Hasha, two eldest in the community, finds the divide between the taiga and the settlement insurmountable. To cope with this, they abandoned their places in the pension facility to live in the forest, as they feel themselves sentimentally attached to the taiga and their reindeer. After living almost the whole life in the forest, they "are accustomed to everything there" and only with the reindeer that they can meet the spirits of the nature and thus "feel good" (Xie, 2010: 66).

The displeasure with the changed life and the spiritual attachment to the forest are also shared by the younger generation. Wang Ying, an Ewenki woman in her early 40s, is among those who stay in the settlement while her husband remains the forest to take care of their reindeer, since her daughter visits school in Genhe. Complaining that her family income is restricted despite hard work, she misses the old days in the forest, when the whole family could live on hunted game and stay with their reindeer (Huang, 2009: 132). Wu Xusheng, aged 37, herds his reindeer seasonally while residing in the settlement, expressed explicitly his dismay at the new

Aoluguya, for "it is too close to the city and the temperature is higher, which is unfavorable for the growth of lichen." Admitting that life is much harder in the forest, he would rather choose to bear the hardships at the campsites than stay at the settlement, where there are too many people but no animals. For him, the living conditions of his family and himself are secondary to that of the reindeer, and only in the taiga, where lichen is abundant, that the reindeer can survive and that the Ewenki community can survive (Huang, 2009: 154–155).

The Ewenki are aware of the crisis and they have attempted to actively deal with the issue, such as those moving back to the taiga with their reindeer. For those who cannot bear the sedentary life at the new settlement but are not able to return to the forest, they looked for passive alternatives. Some found alcohol to cope with the grief at their changed life. Asuo, who used to a renowned hunter in the community, lost his source of income and rhythm of life after the hunting ban. Since he is among those who do not own reindeer, he had nothing to do but to live on welfare allowance provided by the government. Although he had been assigned a job as tour guide in the forest theme park, he was soon dismissed for he was drunk all the time. Asked why he cannot stop drinking, he replies: "I am from the forest, but here there is nothing for me. All I can do now is drink everyday" (Xie, 2010: 172–174). While alcohol is not foreign to the Ewenki as it is an essential drink to keep them warm in the winter, but after the relocation, their drinking manner has considerably changed. Traditionally, the Ewenki, similar to many other reindeer herders across North and Inner Asia, make ritualized offerings before drinking, indicated by tipping the liquor three times, as showing respect to the sky, earth, and the hearth.[6] Through such offerings, the hunters and herders are seeking to appease the nonhuman agents of the forest. In the new Aoluguya, however, such act is now seldom seen, replaced by scenes of people drinking directly from their own bottles and often reaching a degree of intoxication, as Asuo frequently does.

In the face of the state's policies that have been constricting and strangulating the herd, little can be observed among the Aoluguya Ewenkis as how they collectively contest to such policies. Many herders residing other parts of the world have confronted similar challenges of strong state intervention and they have organized initiatives, though in different ways and to different extents, to fight back against the state, including the Nenets and Sami in Northern Europe (Nuttall, 2009), the Wales people in Alaska (Ongtowasruk, 2014), or the Tozhu and the and the Ienisseysk Evenki (Vlassova, 2006; Arakchaa, 2014), as well as among the Even and Itelmen of southern Kamchatka (Sharakhmatova, 2011). Yet given the disadvantaged situation of the Aoluguya Ewenkis, in particular the extremely small number of the community and that of their reindeer, they are not in the place to claim agency in order to negotiate with the powerful state, in contrast to their counterparts in Europe or North America. China's entry into the Association of World Reindeer Herders, which has been collectively initiated and supported by the Aoluguya Ewenki and other Ewenki peoples outside China, has been characterized as a positive step toward a more active role of the Aoluguya Ewenki. After the 2013 World Reindeer Herders Congress in Aoluguya, however, their connections to other Ewenkis and reindeer herders out of China have been strictly restricted. According to the informants, the state welcomes such networking but it should be kept in a very confined sphere and only under official surveillance.

Concluding Remarks

For many Ewenki, from their ancestors on their life has always been tightly connected with the taiga and the reindeer, maintaining particular relationships to the domain that contains all natural resources and creatures, which together consist of their lifeworld. In this lifeworld, the reindeer are doubtlessly playing the central role. In contrast to the large-scale tundra reindeer herders in other parts of the world, the Ewenki enjoy a much intimate connection to their reindeer, developing a particular model of "symbiotic domestication" in which the herders and the reindeer are mutually dependent and meanwhile also beneficial to each other. As pack and riding animals that facilitate their movement in the forest, the reindeer is a principal source of income, storytelling, cultural practice, empowerment, and identity.

While literature on Anthropocene informs us about the immense impact of human imprint on the global environment, in the case of the Aoluguya Ewenki we also see another side of such impact, namely that of the changed environment on human. In face of the state authority that forcefully integrates the community into its project of economic development and socialist modernization, however, the Ewenki herders and hunters are left with little space to negotiate their preference for way of life. Being relocated to a new settlement distant to the forest but close to the urban center, they are now far from the environment that once nurtured their ancestors and shaped their economy and ethnic culture. The Ewenki looked for ways to cope with the risks and crises brought up by the changed environment and living space. Some of them welcome the material benefits provided by the sedentarized way of life, residing permanently in the modern houses and transforming themselves to tourist guides and souvenir salespersons. Yet many found difficult to adjust to the drastic transition, choosing either to return to the campsites to stay with their reindeer or to indulge themselves in alcohol. No matter how the relocation is experienced by different individuals, the Ewenki are now in the middle of a "confrontation with the natural as well as the social environment" (Bird-David, 1999: 84). In other words, the whole Ewenki community is somehow caught in the dilemma of coping with life between the taiga and the forest, reindeer herding and tourism, ethnic culture and modernization.

What is hidden behind such dilemma is indeed the seemingly unresolvable ambivalence of how to deal in the changed environment with the human–reindeer relations that has structured the social organization of the Ewenki and defined their cultural identity. Although different members of the community tried different ways to cope with the situation, their chance of success seems rather pessimistic. While resisting the modern urban life to a certain extent, it is evident that the Ewenki, as a mini society of no more than 250 members, are too powerless to act against the state force. They can choose to go back to the taiga, as some have already done, but given the spatial restrictions and hunting ban, they are no longer able to return to their traditional life as that of their ancestors. For the majority, however, they can only passively react in hope of receiving the care and help from the government. As one Ewenki herder expresses:

> We hope that the government can care about the development of the reindeer in the campsites, but the government believes that it is better for us to leave the forest. Isn't it good for both the government and us that assistance is offered according to our will?
>
> (Xie, 2010: 204)

Acknowledgments

I am grateful to Victor Teo for his comments on earlier drafts of this chapter. I also owe my gratitude to Bai Lan, Qu Feng, and Qi Jinyu for their invaluable inputs into Ewenki culture and tradition. The research is funded by a grant of Zhejiang Provincial Social Science Fund (23QNYC18ZD).

Notes

1 On their way of living and use of reindeer, see Donahoe and Kyrgyz (2003: 12).
2 For a discussion of the Ewenki worldview and Shamanism among the community, see Heyne (1999).
3 For examples of Russian words used by the Ewenki, see Dumont (2017: 524).
4 The 24 herding families, to which the reindeer were assigned to in the decollectivization reform in 1984, formed five campsites. That of Maria So, also the farthest to Aoluguya, was divided into two in 2009.
5 On the movement of Siberian Ewenki, see Lavrillier (2011: 217).
6 On the drinking rituals among reindeer herders in North and Inner Asia, see Humphrey and Onon (1996); Vitebsky (1990).

Bibliography

Abenqian 阿本千. *Ewenke lishi wenhua fazhan shi* 鄂温克历史文化发展史 (History of Historical Cultural Development of the Ewenki). Beijing: Zhongguo shehui kexue chubanshe, 2015.

Aoluguya shilu buluo jingqu 敖鲁古雅使鹿部落景区. "Aoluguya jingqu AAAA jingqu baogao shenqingshu" 敖鲁古雅景区AAAA景区报告申请书 (Aoluguya Application for AAAA Scenic Spot). Unpublished material, 2018.

Arakchaa, Tayana. "The Effects of Climate Change on Hunting and Reindeer Herding Practices among the Tozhu of Southern Siberia." *Alaska Journal of Anthropology* 12, no. 2 (2014): 61–73.

Beach, Hugh. "Milk and Antlers: Chinese Dual-Ownership System Remains a Hopeful Model Despite Forced Relocation from Olguya (Inner Mongolia)." *Cultural Survival Quarterly* 27, no. 1 (2003): 33–35.

Bird-David, Nurit. "'Animism' Revisited: Personhood, Environment, and Relational Epistemology." *Current Anthropology* 40, no. S1 (1999): 67–91.

Crutzen, Paul J. "Geology of Mankind." *Nature*, no. 415 (2002): 23.

Crutzen, Paul J. and Eugene F. Stoermer. "The 'Anthropocene'." *Global Change Newsletter* 41 (2000): 17–18.

Donahoe, Brian and Chaizu Kyrgyz. "The Troubled Taiga: Survival on the Move for the Last Nomadic Reindeer Herders of South Siberia, Mongolia, and China." *Cultural Survival Quarterly* 27, no. 1 (2003): 12–18.

Dumont, Aurore. "The Many Faces of Nomadism among the Reindeer Ewenki: Uses of Land, Mobility, and Exchange Networks." In: *Reclaiming the Forest: The Ewenki Reindeer Herders of Aoluguya*, ed. by Åshild Kolås and Yuanyuan Xie. Oxford: Berghahn Books, 2015, 77–97.

———. "Declining Ewenki 'Identities': Playing with Loyalty in modern and Contemporary China." *History and Anthropology* 28, no. 4 (2017): 515–530.

Fondahl, Gail A. *Gaining Ground: Ewenkis, Land, and Reform in Southeastern Siberia*. Boston, MA: Allyn and Bacon, 1998.

Fraser, Richard. "Forced Relocation amongst the Reindeer-Evenki of Inner Mongolia." *Inner Asia* 12 (2010): 317–345.

Habeck, Joachim Otto. "Experience, Movement, and Mobility: Komi Reindeer Herders' Perception of the Environment." *Nomadic Peoples* 10, no. 2 (2006): 123–144.

Heyne, F. Georg. "The Social Significance of the Shaman amongst the Chinese Reindeer-Evenki." *Asian Folklore Studies* 58, no. 2 (1999): 377–395.

———. "Between the Hinggan Taiga and a Moscow Public Library." *Asian Folklore Studies* 61 (2002): 149–154.

Huang, Jianying 黄健英, ed. *Aoluguya: Ewenke zu liemin xincun diaocha* 敖鲁古雅: 鄂温克族猎民新村调查 (Aoluguya: Survey at the New Settlement of the Ewenki). Beijing: Zhongguo jingji chubanshe, 2009.

Human Animal Research Network Editorial Collective, eds. *Animals in the Anthropocene: Critical Perspectives on Non-Human Futures*. Sydney: Sydney University Press, 2015.

Humphrey, Caroline and Urgunge Onon. *Shamans and Elders: Experience, Knowledge, and Power among the Daur Mongols*. Oxford: Clarendon Press, 1996.

Kolås, Åshild. "Reclaiming the Forest: Ewenki Reindeer Herding as Exception." *Human Organization* 70, no. 4 (2011): 397–404.

Kolås, Åshild and Yuanyuan Xie, eds. *Reclaiming the Forest: The Ewenki Reindeer Herders of Aoluguya*. Oxford: Berghahn Books, 2015.

Kong, Fanzhi 孔繁志. *Aoluguya de Ewenke ren* 敖鲁古雅的鄂温克人 (The Aoluguya Ewenki). Tianjin: Tianjin guji chubanshe, 1994.

Lavrillier, Alexandra. "The Creation and Persistence of Cultural Land-Scapes among the Siberian Evenkis: Two Conceptions of 'Sacred' Space." In: *Landscape and Culture in Northern Eurasia*, ed. by P. Jordan. Walnut Creek, CA: Left Coast Press, 2011, 215–231.

Lin, Hang 林航. "Ewenke zu xunyang xunyu de bentu zhishi" 鄂温克族驯养养驯鹿的本土知识 (The Ewenki Local Knowledge of Methods of Domesticating Reindeer). *Yuanshengtai minzu wenhua xuekan* 原生态民族文化学刊 10, no. 4 (2018): 3–11.

Lindgren, Ethel J. "Northwest Manchuria and the Reindeer-Tungus." *The Geographical Journal* 75, no. 6 (1933): 518–534.

———. "The Reindeer Tungus of Manchuria." *Man* 35 (1935): 44–45.

———. "An Example of Culture Contact without Conflict: Reindeer Tungus and Cossacks of North-Western Manchuria." *American Anthropologist*, New Series 40, no. 4 (1938): 605–621.

———. "The Khingan Tungus (Numinchen)." *Man* 39 (1939): 19.

Liu, Yunshan 刘云山. *Aoluguya Ewenke fengqing* 敖鲁古雅鄂温克风情 (Landscape and Feelings of Aologuya Ewenki). Hohhot: Neimenggu renmin chubanshe, 2010.

Lü, Guangtian 吕光天. *Ewenke zu* 鄂温克族 (The Ewenki). Beijing: Minzu chubanshe, 1983.

Neimenggu zizhiqu bianjizu 内蒙古自治区编辑组. *Ewenke zu shehui lishi diaocha* 鄂温克族社会历史调查 (A Survey of the Social History of the Ewenki). Beijing: Minzu chubanshe, 2009.

Nentwig, Ingo. "Reminiscences about the Reindeer Herders of China." *Cultural Survival Quarterly* 27, no. 1 (2003): 36–38.

Nuttall, Mark. "Living in a World of Movement: Human Resilience to Environment Instability in Greenland." In: *Anthropology and Climate Change: From Encounters to Actions*, ed. by Susan Crate and Mark Nuttall. Walnut Creek, CA: Left Coast Press, 2009, 292–310.

Ongtowasruk, Davis. "The Ongtowasruk Herd of Wales, Alaska." *Alaska Journal of Anthropology* 12, no. 2 (2014): 52–60.

Pöyry. *Aoluguya Ethnic Reindeer Resort Master Plan*. Beijing: China Pöyry [Beijing] Consulting Company Limited, 2008.

Sharakhmatova, Viktoria Nikolaevna. *Nabliudeniia korennykh narodov Severa Kamchatki za izmeneniiami klimata. Otchet* (The Observation of Northern Kamchatkan Indigenous Peoples on Climate Change. Report). Petropavlovsk-Kamchatskii: Etno-ekologicheskii informatsionnyi tsentr, 2011.

Shirokogoroff, Sergei Mikhailovich [Shi Luguo]. *Psychomental Complex of the Tungus*. London: Kegan Paul, Trench, Trubner & Co., 1935.

———. *Social Organization of the Northern Tungus. With Introductory Chapters Concerning Geographical Distribution and History of These Groups*. New York: Garland Publishing, 1979 [1929].

Stammler, Florian and Hugh Beach. "Human-Animal Relations in Pastoralism." *Nomadic Peoples* 10, no. 2 (2006): 6–25.

Tang, Ge 唐戈. "Ewenke zu de xunlu wenhua" 鄂温克族驯鹿文化 (Reindeer Culture of the Ewenki). *Heilongjiang minzu congkan* 黑龙江民族丛刊 14, no. 2 (1998): 90–94.

Vitebsky, Piers. "Centralized Decentralization: The Ethnography of Remote Reindeer Herders under Perestroika." *Cahiers du monde russe et soviétique* 31, no. 2/3 (1990): 345–358.

Vlassova, Tatiana K. "Arctic Residents' Observations and Human Impact Assessments in Understanding Environmental Changes in Boreal Forests: Russian Experience and Circumpolar Perspectives." *Mitigation and Adaptation Strategies for Global Change* 11 (2006): 897–909.

Wang, Jing, etc. "Reindeer Ewenki's Fading Culture." *Science* 347, no. 6225 (2015): 957.

Wu, Nanlan. "Last Hunting Tribe Gives Up Virgin Forest." China.org.cn, August 11, 2003. http://www.china.org.cn/english/2003/Aug/72126.htm. Accessed January 25, 2021.

Wu, Zhizhong and Du Wen. "Pastoral Nomad Rights in Inner Mongolia." *Nomadic Peoples* 12, no. 2 (2008): 13–33.

Xie, Yuanyuan 谢媛媛. "Aoluguya Ewenke liemin shengtai yimin hou de zhuangkuang diaocha: bianyuan shaoshu zuqun de fazhan daolu tansuo" 敖鲁古雅鄂温克猎民生态移民后的状况调查: 边缘少数民族族群的发展道路探索 (A Survey of Aoluguya Ewenki Hunters after the Ecological Migration: Probing the Developing Routes of Marginal Ethnic Minorities), *Minsu yanjiu* 民俗研究, 2005, no. 2 (2005): 50–60.

———. *Shengtai yimin zhengce yu difang zhengfu shijian: yi Aoluguya Ewenke shengtai yimin weili* 生态移民政策与地方政府实践: 以敖鲁古雅鄂温克生态移民为例 (Policy of Ecological Migration and Practice of the Local Government: The Ecological Migration of the Aoluguya Ewenki). Beijing: Beijing daxue chubanshe, 2010.

———. *Ecological Migrants: The Relocation of China's Ewenki Reindeer Herders*. Oxford: Berghahn Books, 2015.

Yeh, Emily T. "Green Governmentality and Pastoralism in Western China: Converting Pastures to Grasslands." *Nomadic Peoples* 9, no. 1/2 (2005): 9–29.

Zalasiewicz, Jan, etc. "The Anthropocene: A New Epoch of Geological Time?" *Philosophical Transactions of the Royal Society A* 369, no. 1938 (2011): 835–841.

Zhao, Y. J. "Ecology and Use of Reindeers in Greater Higgnan Mountains." *Chinese Journal of Zoology* 10, no. 2 (1975): 25–26.

5 Yak *Dzongs* in Sikkim Himalayas

Spaces of Conflict in the Making

Bhim Subba

Introduction

In South Asia, countries sharing the Himalayan mountain range – India, Nepal, and Bhutan – and some highlands of high central Asia, such as Afghanistan, and the Hindukush Himalayan region of Pakistan are yak habitats (Kreutzmann 2003). Besides sharing climatic similarities andcultural and geo-civilizational linkages, the region shares similar ways of life from the Qinghai-Tibet Plateau to the southern slope of the mountainous Himalayas, becoming a prominent yak habitat (Weiner et al. 2003). India's alpine highlands of Jammu & Kashmir, Ladakh, Himachal Pradesh, and Uttarakhand in western and middle Himalayas, and Sikkim and Arunachal Pradesh in eastern India, have supported native communities in yak-based social, cultural, and economic activities. The region is part of a "yak-cultural area" that has defined human–animal relationships since immemorial. Yaks and other ungulates have shaped the human–animal interactions much before the formation of the modern nation-state systems with defined and controlled territorial spaces in these alpine transboundary landscapes and dominated the indigenous knowledge systems, livelihoods, and social and cultural ethos. At the same time, yak rearing in these highlands has seen a remarkable transformation from ecological changes such as climate change, changes in the herder outlook for other employment opportunities, availability of yak fodder, and state-induced restrictions on other animals vis-à-vis yak population.

Yak-rearing highlands in Sikkim Himalaya form a part of the Kanchenjunga Landscape (KL), which constitutes the transboundary regions shared by Sikkim (India), Nepal, and Bhutan. It is one of the richest cultural and biological diversities among the 34 Global Biodiversity Hotspots (Mittermeier et al. 2004; Wu et al. 2016). Kanchenjunga National Park (KNP), in Sikkim, was established in 1977 and covers nearly 25% (1,784 square kilometers approximately) of the total geographical area of the state and is a part of the KL and is located in the north and west districts (Tambe and Rawat 2011: 443). With changing seasons, the local communities practice transhumance, which is moving their livestock to higher pastures during summers and to lower valleys during dry and snowy winters (Chettri 2008; Wu et al. 2016) (Figure 5.1).

DOI: 10.4324/9781003212089-5

Figure 5.1 Milking yak, Yambong, West Sikkim, at an average altitude of above 3,300 meters

(photo provided by Dawa Tsangpo).

Although animal husbandry is the main source of livelihood, it is yak rearing that dominates these highland pastures than other pastoral animals such as sheep and domestic cows, *hybrids* of yaks and domestic cows rightfully becoming yak fortress (*dzongs*) which is also reinforced by state's imposition of restrictions via grazing and conservation laws. However, with changing life priorities, the younger generation with formal school education has been seeking alternate forms of employment opportunities in the non-farm sector. This has resulted in the declining yak-rearing families in Sikkim (Avaste 1999), and it is compounded by the increasing stress of climate change, especially in the higher mountains where the impacts are felt much more directly than in the lower terrain. The environmental stress has led to changes in traditional family professions and their much-cherished nomadic lifestyle and opting for sedentary lifestyles and adopting a cash economy by farming medicinal plants, oil seeds, and potatoes (Sharma and Rai 2012).

Thus, this age-old human–animal interaction is being reshaped in the Anthropocene Sikkim Himalayas and has to be seen from changing priorities of the humans accentuated by the state. One of the issues emerging is the nature of differential implementation of conservation laws and policies and the lack of agency to negotiate with the state for some traditional herders in Sikkim. If the local laws have given differential prerogatives to keep yak-sheds, the increased securitization of the borders has challenged these traditional mandates of the local communities at the same time. These multilayered relationships between the local herder stakeholders, despite their social and cultural affinities with the ungulate and their tussle with the country's security and development priorities and looming challenges of climate change, have been shaping these alpine yak *dzongs* as potential spaces of conflicts and equity issues, especially among the herder and border communities. This study, therefore, shows that yak–human interactions have undergone a multilayered transformation in Sikkim Himalayas. The aim here, therefore, is an attempt to understand these multilayered interactions engendered by communities' ecological concerns and state's development policies and security priorities vis-à-vis yak abode in the alpine region.

Yak Pastoralism, Traditional Governance, and Equity

Sikkim, one of the Indian states in the Northeastern Himalayas of the Indian subcontinent, is sandwiched between Bhutan in the east, Nepal in the west, and Tibetan Autonomous Region and China in the north. Sikkim's topographic landscape, with just 7,096 square kilometers, has varied elevations from low 300 to 7,000 meters with different climate and vegetation from tropical moist to high alpine forests above 3,600 meters (State of Forest Report, 2021). Apart from being a mountainous and hilly terrain, Sikkim is the least populous state in the country with 610,577 persons as per the last census (Census 2011). Like in other alpine regions, Sikkimese Himalaya has ample pastures in the north, west, and eastern highlands, which have been prominent areas of yak rearing and farming.

Traditionally, yaks (*Bos Grunniens*) and Himalaya Blue Sheep (*baral*) have been owned and herded by communities such as *Bhutia* and the Tibetan *Drokpa* trans-Himalayan pastoralists. These communities, with their ungulates, have endured subzero temperatures while grazing and foraging resources in a sustainable manner in the highlands of North Sikkim for generations (Chanchani et al. 2011: 359). With the *Drokpas* of Muguthang valley and Tso Lhamo plateau, a southward continuation of the Tibetan plateau, the *Bhutia* communities in Lachen and Lachung, Thangu, Yumthang and other adjoining areas are yak-rearing natives in the region. Similarly, Yambong, Sangkhola, Dzongri, Tshoka in West Sikkim and Kupup, Nathang, and Tshomgo Lake (yak safari) in East Sikkim do have traditional yak farming families assisted by professional herders from other communities. Likewise, *Gurung* and *Magar* are shepherds. *Limbu* and *Lepcha* communities' primary livelihood was shifting cultivation and hunting-gathering. Besides, other communities such as *Newars* and *Marwaris* are engaged in trade and business, and *Chettris* and *Bahuns* as agriculturists in Sikkim (Tambe and Rawat 2006). These communities with specific livelihood specializations have shaped the economic, social, and cultural imagination of the region for several centuries.

As indicated in this study, Lachen and Lachung, the two most prominent township villages in North Sikkim, have been governed by a traditional political institution – the *Dzumsa*, which has continued to survive until today.

Although the local self-government in the form of Panchayati Raj institutions has been established in rural India and Sikkim, the *Dzumsa*, as a traditional system of governance, has been sanctioned and recognized in the state since 1985. This traditional administrative system provides structure and cohesion for the societies and activities with the majority of the inhabitants, as agro-pastoralists and yak herders (Sabatier 2004). Moreover, as an agency, this traditional system helps the local community to engage and negotiate with the state as a homogenous unit regarding other infrastructure and developmental activities within the jurisdiction.

Being a yak herding zone, many migrate during the grazing season, foraging pastures for their yak herds and sheep. *Dzumsa*, the village council, regulates the grazing activities and natural resource management in the zone (Chakrabarti 2011). It has a mandate from maintaining fodder to lumbering, regulating *goths*, collecting taxes and revenues and even imposing fines for violating grazing laws. This institutional mechanism has helped the inhabitants to negotiate terms with regard to developmental activities such as hydropower projects that can bring unwarranted surrender of land and construction activities. In the east and west districts, such traditional local governance institutions are missing. The state forest department and national defense ministry have sole writs in regulating grazing laws and foraging areas in the region. However, unlike in the west district, the yak owners of the Gangtok subdivision of Gnathang, Kupup and Tshomgo have successfully resisted the state's limitation in accessing some border areas. In West Sikkim, the least developed dominance of non-pastoralists and non-influential families was weaker in negotiating with the state forest bureaucracy and politics. Although the abolition of the old *patta*-system, which mandated paying a certain grazing tax in 1986, led to an increased and uncontrolled number of yak and its hybrids at the cost of other animals in the subalpine and alpine regions, its system was deemed to be unsustainable to the ecology of the region.

Sikkim has 3,379 square kilometers (47.62%) of the total geographical area of the state (Sikkim Forest Report 2017). Forestry, being one of the richest natural resources, has been under the administrative control of the state forest department with three main categories of forests: reserve forests, *khasmal and goucharan. Khasmal* forest is set aside by the government for bonafide exploitation for timber and firewood, and *goucharan* were forest areas reserved for fodder and grazing (Forest Department, Govt. of Sikkim). After Sikkim became a part of Indian union in 1975, several old forest laws were repealed, including the *patta system* which had permitted the herders to pay rent to the state for grazing and foraging. However, from 1998, the state forest law led to the banning of grazing and *goths* in reserved forests, plantations, and water source areas. With such prudent policies and stringent implementation, many dying and depleted forests due to rampant cattle grazing and deforestation regenerated with natural succession, including the wildlife (Forest Dept. Sikkim).

On the other hand, there were arguments that nomadic herders in West Sikkim were "outsider," Tibetan, and preferably from Nepal (Chettri 2015).

Such accusations, compounded with subtle weaponization of state policies, seem to discourage herding as a vocation, especially for the existing herding families in western Sikkim. This ban imposed in Sikkim was detrimental to many agro-pastoralists. The *goth* eviction was mostly concentrated in alpine areas of the west district and caused livelihood concerns for the state without adequate redressal mechanisms and employment opportunities. The impact was more felt among the non-Bhutia families who had learnt yakherding as a livelihood strategy for economic prosperity. The ban led to serious exclusion and equity issues among the highland herdsmen, unlike the cases in north and east districts where the herders could wield strong political influence in the corridors of power in the state.

In 2004, the state enacted a law on removing *goths* (sheds) from the reserved forests; many herders were implicitly coerced to sell their mixed herds of yaks, *dzo/dzomo*, sheep, and goats and cattle. This eviction was carried out in the sub-alpine zone of conifer and dwarf bamboo and other herbs becoming extinct because of overgrazing and denudation, and washing away of the topsoil. The eviction was strictly implemented in Barsey Rhododendron Sanctuary and Hee-patal and Yambong-Sindrabong in West Sikkim. The ban and eviction were a positive development in ameliorating and regenerating flora and fauna, but they also led to the exclusion of many pastoralists in the zone, thereby changing their existing livelihood practices (Bhagwat et al. 2012). The lack of agency for these pastoralists, unlike the *Dzumsa* in North Sikkim, has had an adverse impact on these agro-pastoralists and later herders to give up their nomadic lifestyle. The demand for equity in treatment with respect to state grazing laws vis-à-vis other traditional communities led to bitter-sweet relations among the communities.

These yak *dzongs* (fortresses), a zone of sacred landscape among the practitioners of the indigenous beliefs, animists, Buddhists, and Hindus alike, have also become an area of competing claims. In recent years, Sikhs also claim to believe that their founder guru or saint had blessed these places during his Himalayan sojourn, raising much consternation among the natives in the region (Singh 2019). Similarly, in Sikkim's northern plateau, most of the pastureland has been appropriated by the security forces, causing much conflict with the herder families. In West Sikkim, yak herding among the private families was introduced only in the middle of the 20th century, and until 1975, herding was the sole prerogative of the royal herds (Tambe and Rawat 2006).

Today many of these herders have their yak herds with the diverse alpine economy moving away from barter and transborder trade to ecotourism compounded with state regulation on grazing and conservation laws. To these nontraditional yak herding families, the consolidation of ownership since changes in the forest laws cost dearly as they had to give up their acquired herding cultural practices. The author's interaction with some local natives shows that there are only less than seven families and a private school trust engaged in yak rearing. Other indigenous livestock like yak-hybrid, cattle, and sheep (*banpala*), which are averse to extremely cold conditions, especially in lower alpine reaches, became victims of grazing laws when they were forced to vacate their herds from the forests leading to equity concerns in the state (Interview, July 2019). Besides, in East and West Sikkim, although there is no such direct conflict from government agencies, the concern emanates more so from the economic viability and desertion or giving up yak herding practices for other employment such as tourist guides and porters. If this transition helps in bringing in cash flows with more tourist footfalls, it also results in ecological stress in the region.

Yak in Sociocultural Imagination

If the Tibetan *Dokpas* (*Drokpas*) in North Sikkim were nomadic transboundary herders sharing strong bonds with their northern neighbors, the Bhutia, one of the early Tibetan settlers in Sikkim in the state, identify yak with a sense of pride and prestige in the social and cultural milieu. This can be heard in many folk tales associated with the animal narrated by grand nannies to younger generations (Interview, October 2019). Even some stories or narratives relating to Phunstog Namgyal, the founder of the Namgyal dynasty in 1640–1642 CE, a righteous ruler (*Chogyal*), who was crowned at Yuksam in West Sikkim was a herder in the earlier avatar (Mullard 2005). Being a herder in East Sikkim, the narratives do not show what he (Phunstog) was herding! However, it reinforces a strong association with herding, and many prominent former social and political elites, including the members of the erstwhile aristocracy and landlords, own large yak herds compared to nontraditional yak-rearing families. Most of the yak-herds in East Sikkim, especially in the India–China border region of Nathula, Gnathang, Kupup, and Tshomgo are owned by the former ruling aristocratic families, which for them is not just a source of revenue but more of a vocation of identity and status in the fast-changing sociocultural and economic landscape. Besides, with the increasing non-Bhutia population in the state, owning yak, even in single digits, has become an important symbol of cultural assertion. The census data show that yak herding is popular among Tibetan Buddhist followers in Sikkim: Nyingmapa and Gelugpa. Apart from the south district, all the other three districts are majority Buddhist areas with yak-owning families (see Table 5.1).

Table 5.1 Sikkim: Human Population (2011) and Yaks (2018) (in Numbers)

Districts	State Population	Hindus	Buddhists	Yaks
North	43,709	14,883	23,318	3,710
East	283,583	172,910	72,455	2,369
West	136,435	75,286	36,390	141
South	146,850	84,583	35,053	0
Total	**610,577**	**347,662**	**167,216**	**5,220**

Source: Compiled by the author from Sikkim Animal Census, Report 2.

North Sikkim, with the largest area, is the least populous district, and Chungthang subdivision, with a Buddhist majority population alone, had 3,384 yaks in 2012, similar to West District's Gyalshing subdivision had all 141 yaks and a Buddhist majority area. Likewise, among the three subdivisions in East District, Gangtok subdivision, a Buddhist-dominated area, had all the yaks and yak-owning families.

Even in cultural and religious events among some Buddhist communities, festivals like *Pang Lhabsol* and *Drukpa Tshechi* (*Dukpa Tshezhi*) have become important events for yaks in the region. Drukpa Tshechi (*Choekhor Duichen*), the first day of Buddha's sermon to his four disciples in Sarnath, marks an important day for yak herders in Muguthang and Thangu in North Sikkim. The local inhabitants conduct prayers, rituals, and yak races among the herdsmen (Nugo 2018). These events indicate the role of Buddhism in the lives of yaks and pastoralists inhabiting the region. It also creates awareness among the people, the visitors, and state administration to bring about prudent strategies in conserving age-old cultural practices surrounding yak pastoralism.

Unlike *Drupa Tshezhi*, *Pang Lhabsol* is a Buddhist festival unique to Sikkim when Mount Kanchendzonga, a scared peak regarded as a guardian deity, is commemorated to protect the land, that is, *Demajong* (Govt. of Sikkim). Apart from various monastic dances and the most important *pangtoed chaam* (warrior dance) performed at Tsuklakhang (royal chapel), this festival also marks the beginning of yak migration and special prayers performed by invoking mountain gods by herder families. This festival is usually held in August–September every year, usually on the 15th day of the seventh month of the Tibetan calendar, which is also observed as yak day with the organization of yak races and awareness for preserving indigenous yak culture for posterity. Prayers at special sites at hillocks are offered, wishing good health and protection of their yaks in their annual migration to the upper alpine reaches for grazing. Among non-Buddhist herder families, annual *goth pujas* (ritual ceremonies) invoking local spirits such as *devis* and *sampmangs* of *deoralis* and mountains in their prayers are initiated by shamans, *bijwas* and *damis* in *goths* (make-shifts tents) individually or collectively by the community.

Among the folk dances, Yak *chaam*, or yak dance, is one of the categories of the folk performances that have been popular among the Sikkimese Bhutia and other yak-herding communities such as Monpas and Brokpas in Arunachal Pradesh. However, Sikkim's Yak *Chaam* has become the state's main folk performance in cultural and dance presentation in the country. Every year, this *chaam* is presented to honor yak and the life of herdsmen and the ungulate's contribution to livelihood and sociocultural landscape among in Sikkim's alpine region (Singh 1995). Every year, on May 16, which is observed as the state day, yak dance is performed at Mahatma Gandhi Road, the state capital city's main boulevard becomes the center of attraction for the locals and tourists alike at Gangtok.

Yak Statistics at a Glance

Nevertheless, the few yak-owning families have consolidated their herds and have reshaped the yak-human-state relationship in recent decades. In 2012, the 19th Livestock Census was conducted in the country and Sikkim; there were 520 yak-owning households in Sikkim (Dept. of Animal Husbandry & Veterinary Services, Govt. of Sikkim 2012). The yak population was 1.38% of the total livestock in the state (Dept. of Animal Husbandry & Dairying, Govt. of India). But the share of *Drokpa* families and nontraditional herding communities in western Sikkim, however, has been declining ever since. The recent 20th Livestock Census provisional report released in October 2019 has shown an increase of 29% in the state, whereas there is an overall decrease in the number of yak population in the country (see Table 5.2).

Although Sikkim accounts for 9% of the total yak population in the country as per the recent census, this trend indicates a positive development in the overall conservation of yak culture in the alpine Himalayas of India. With the exception of Arunachal Pradesh, Sikkim is the other state that performed much better in increasing yak numbers. Jammu and Kashmir, Himachal Pradesh, Uttarakhand, and the Darjeeling region of West Bengal had a remarkable decline in the number of yaks. Jammu & Kashmir, Himachal Pradesh, and West Bengal had a decline in yak population of −51%, −33%, and −80%, respectively (Table 5.3). The yak data for West Bengal and Uttarakhand were unavailable for the 1997 census period.

However, the overall population in the country shows a marked decline from the 1977 census, which stood at 132,000 (Pal 1993), especially in the

Table 5.2 Yak Population in India and Sikkim (in Total Numbers)

Census/Year	16th (1997)	18th (2007)	19th (2012)	20th (2018)
National	58,781	84,410	76,662	57,722
Sikkim	4,781	5,225	4,036	5,219

Source: Compiled from various livestock census, Dept. of Animal Husbandry & Dairying, Ministry of Fisheries, Animal Husbandry and Dairying, Government of India.

Table 5.3 State wise Yak Population Census (India – in Total Numbers)

Census/Year	16th (1997)	18th (2007)	19th (2012)	20th (2018)
Jammu & Kashmir	33,000	61,910	54,493	26,221
Arunachal Pradesh	14,000	14,251	14,061	24,075
Sikkim	**4,781**	**5,225**	**4,036**	**5,219**
Himachal Pradesh	7,000	1,705	2,921	1,940
West Bengal	NA	26	1,089	213
Uttarakhand	NA	50	62	54
National	**58,781**	**84,410**	**76,662**	**57,722**

Source: Compiled from various livestock census, Dept. of Animal Husbandry & Dairying, Ministry of Fisheries, Animal Husbandry and Dairying, Government of India.

Jammu & Kashmir and Ladakh regions, where there has been a sharp decline in the yak population recorded from 2007 onward, and nationally there was a decline of approximately 25% for the 2012–2018 livestock census period. One of the reasons for the positive trend in Sikkim is the consolidation of small yak herds into larger groups, especially after the restrictions of grazing laws in the early 2000s that have given more carrying capacity for yaks in the alpine grasslands. To confirm this, I spoke to a yak owner from Lachung; the owner said the increase in the state's yak numbers might be due to importing yak from Kashmir and Ladakh and the ban on yak slaughter in Lachen in the recent past (Interview, 2019). To disaggregate district-wise, North Sikkim with Mangan and Chungtang subdivision has more than 90%, and the remaining 10% is shared almost equally by east and west Districts (19th Livestock Census 2012) (Figure 5.2).

Figure 5.2 Churpi, a hard dried cheese consumed as nut, Gangtok Shopping Mart, Sikkim

(photograph by author).

Yak Economy in Alpine Habitat

Apart from yak's cultural and social significance, yak herding is the fulcrum of the alpine economy in Sikkim. Yak is not only an exotic animal to some populations, but the lack of adequate knowledge due to the apathy of younger generations and migration to urban spaces seeking alternative employment has led to increased stress on the existing herder families to maintain the herd for the future. As lesser families are owning herds, but with increased consolidation, the situation does not look all positive. Besides, the reproductive capacity and milk yield of the Sikkimese yaks have declined from two to three decades earlier. The reasons can be plenty. In an interaction with a herder, the inbreeding among a limited gene pool has led to stunting and less productivity of the existing breed in Sikkim (Interview, September 2019). This can be rectified through genetic engineering and research and the mixing of different breeds.

Nevertheless, yak farming has been a profitable business despite hardships. From horn to tail, it has cultural, religious, and economic value. An adult yak of 5–6 years can fetch between INR 80,000 and 90,000 (USD 1,100 and 1,250). Besides, these animals are reared for dairy and meat, commanding premium prices in the market. Most of the milk is processed for butter and cheese. The cheese that is hard and soft can be consumed differently. The hard cheese, called *churpi*, made from yak milk, is much more expensive than processed from other domestic cattle like cow and buffalo, reaching up to USD 10–12 per kilogram, not to mention the butter that is known for its nutritional value and also used as oil for lamps in monastic and religious ceremonies. The most important items of economic value are the meat and tail. With a voluntary moratorium imposed by communities, meat has become rare. The meat commands a conservative amount of USD 10–12 a kilogram, and a black yak tail can fetch USD 60–70 per one, and the popular white ones can be approximately USD 90 for one, as informed by a respondent. Besides, many yaks are traded on the India–Nepal border in West Sikkim by a few existing herder families (Figure 5.3).

However, among other yak products, yak tail, or *yak jyukma*, has the most significant value across cultures and religions. With a simple function as a fly-wisk, it has become a status symbol and of religious importance. The "white ones were more expensive and were considered high-value items," as they were used in all South Asian religions such as "Hindus, Jains, Sikhs and Buddhists" and "the ruling and religious elites to drive away flies, insects and bees when addressing the audience" and exported to the West for wigs and beards (Harris 2014). Even the border trade that was restarted in 2006 between Sikkim, India, and Tibetan Autonomous Region, China, via Nathula border in East Sikkim has yak tail, wool, and other items such as sheep wool and hides in the trade list, as per the Ministry of Commerce, Government of India (Subba 2013; Sherpa 2017). These items also appeared in the trade list hammered by the British and Tibet in the Calcutta Convention in 1890–1893 and subsequent treaties during the colonial period. The list is

Figure 5.3 Yak tail, (*jyukma*), Darap, West Sikkim
(photo provided by author).

still not comprehensive and has undergone slight modifications and the inclusion of some new products in recent years. Today, with diverse uses from dusting tables and windshield glasses, yak *jyukma* is sold with mixed-breed *dzo* and *dzomo* (male yak and female cattle) are popular among tourists and locals alike as souvenirs for household decoration. However, unlike the trade volume undertaken by Harris (2014), the quantity today is quite insignificant, especially the imports from TAR, and Sikkim's contribution is even less, hence very expensive.

In recent times, Sikkim, emerging as one of the sought leisure destinations, there has been a rapid increase in tourist footfalls. With just a 0.64 million population, the state had more than 1.425 million tourists visiting the state in 2017, much more than 0.86 million in 2016 (*Financial Express* 2019). The visa relaxation for the Bangladeshis to visit Sikkim, Ladakh, and Arunachal Pradesh in 2018 has led to increased tourist traffic in Sikkim (Bagchi 2018). These numbers are welcome to the cash-strapped hotel and hospitality industry but can challenge the state's goal of sustainable tourism leading to stress on existing resources in the state. The state's ecotourism, a sustainable model, propagated especially in North Sikkim areas of Lachen, Lachung, and

Figure 5.4 Yaks features prominently in Sikkim Tourism
(Taken by author).

Yumthang, can bring employment to the erstwhile yak herder families in earning hard currencies from the visitors working as guides, porters, and renting yak rides. Tsomgo Lake, on the way to the India–China border at Nathula in East Sikkim, is a popular tourist spot for yak rides. Many yak caretakers who rent these animals from owners eke out a living catering to tourists from the mainland. Although the tourist seasons are relatively short in upper alpine regions with heavy showers during the long monsoon season compounded with roadblocks and landslides, it is this premium short duration that makes yak rides and safaris a must adventure for the travellers (Figure 5.4).

Yak Dzongs and India–China Border Tensions

Sikkim, absorbed into Indian Union as the 22nd state in 1975, has shaped the current social, cultural, economic, and political landscape vis-à-vis the other countries in its periphery. With a diverse population but similar cultures and livelihood practices, Sikkim's profile resembles that of its neighbors. Bhutan, Nepal, and the present TAR, China, have influenced and shaped the erstwhile

Sikkimese kingdom as an independent country before the British colonial rule in India and thereafter as a protectorate under India until 1974–1975. Within the Indian constitutional framework, Sikkim's unique identity and old laws have been preserved and guaranteed under Article 371 (F) of the Constitution of India (Basu 1982). This law has been the harbinger of Sikkim's integration into India. However, with the territorial modern Indian state, the role of physical boundary and territories has also reshaped Sikkim's interaction with these neighbors and borderland communities. Being a peripheral state, Sikkim has a strong presence of the central security forces to man the once inhabited transboundary landscape, and, many times, have seen border skirmishes with the northern neighbor – China. The increasing securitization of boundaries in the north, east, and west districts at a high-altitude alpine zone has restricted movements of yaks and herding families especially post India–China conflict in 1962, even when Sikkim was just a protectorate state of India. As a protectorate country since 1950, Sikkim's defense and external relations were under the tutelage of New Delhi, and the border conflict was a death knell to the socioeconomic lifestyle of the alpine communities, including both humans and ungulates.

The nomadic migration pattern of these communities was restricted, especially during the summer and winter months. Once free highlanders, *Drokpas* were trapped and immobile when the open borders were manned by the military-security apparatus of the two countries, India and China. From Doklam, near Bhutan in 2017, to Naku La in North Sikkim in 2020, India–China relations have undergone serious tests in these alpine regions. Even the Galwan clash of June 2020, where 20 Indian soldiers lost their lives, occurred in these high-altitude boundaries (Krishnan 2021). In 1965 in the midst of the Indo-Pak conflict, China alleged that Indian troops had stolen 59 yaks and 800 sheep from the Tibetan herdsmen at the border (Dasgupta 2020). Although denied by the Indian government, the border crossing of the herdsmen was not unusual, as many of the nomadic herdsmen reared their yaks toward the south, which is the north of the Sikkimese territory. Likewise, there are reports of yak sightings helped by Chinese soldiers in Arunachal Pradesh, India's eastern state when border skirmishes keep happening at the Line of Actual Control (LAC) between India and China at the western sector, and some suspect that ungulates might be involved in spying and espionage activities. However, it is not proven, and not so in Sikkim yet!

The geopolitics of the region affected the yak-herding patterns in Sikkim. This alpine region has been sanitized to establish security and military infrastructure across both sides of the border. As part of the so-called strategic chicken's neck, the region has been a perennial concern for policymakers and security establishments. My discussion with some respondents in North Sikkim does indicate that, in the process, many alpine grazing grounds have been appropriated by the military and causing stress on the availability of fodder for foraging animals like yaks and resulting in consternations between the security establishment and the herder families. Despite many family stakeholders, and the *Dzumsa* council raising concerns, no adequate steps have been taken to mitigate these issues, but a token guarantee from the state and central bureaucracy seems to work out for now. The herder families alleged that the army and security forces had occupied grassland depriving their livestock to graze and forage freely. The animosity with the locals also results from quarrying and excavating the pastureland leading to denudation and barren topsoil. Although yaks are being utilized for ferrying goods and logistics and generating income, the problem of controlling these grasslands persists between the state and the local communities.

Shrinking Yak *dzongs* and Climate Change Issues

With restrictions of movement across the KL and toward the northern trans-boundary across Tso Lhamo plateau into the TAR, the yak gene pool in Sikkim has been adversely affected. This has led to severe inbreeding among the species, resulting in low productivity and a high mortal rate in the state (Interview, October 2019). The yak-herding communities have been most affected because of a limited mating and a confined gene pool with no viable alternatives until a decade ago. The lack of movement among the yaks and the nomadic *Drokpas*, especially, has led to the latter's livelihood changes in the last three to four decades, leading them to sedentary lifestyles. On the contrary, both the Indian government and its Chinese counterpart could have permitted that Himalayas' bordering states of managing yak migration in a more meaningful way in transboundary regions to enhance crossbreed with their cousins on either side of the border.

Taking cognizance of the issue, the state government, under the auspices of the Animal Husbandry, Livestock, Fisheries and Veterinary Services (AH, LF&VS) department, has initiated the Yak Development Programme. This program seeks to improve indigenous yak breed through gene upgradation in terms of quality and quantity of yak body size, inter-calving period, and milk yield in two breeding centers at Zeema and Chopta valley (AH, LF&VS, Govt. of Sikkim). A similar kind of research has been undertaken at ICAR-National Research Centre on Yak in Arunachal Pradesh, the easternmost Indian state. With such institutional collaboration, some positive steps have been initiated in gene upgradation of yaks and fodder research to make yak herding profitable and more sustainable and better economic returns. Many enterprising yak-owning families are also taking individual initiatives to part-ner with other stakeholders from Nepal, Bhutan, and Ladakh to diversify the gene pool for better alternatives. Even research on fodder, an important aspect of the yak dairy development program in the alpine Himalayas, has been initiated as climate change, denudation of the pasture grassland by the military apparatus, and sudden changes in rain and snowfall patterns have inhibited the growth of nutritious yak vegetation.

In spite of this positive development in recent years, the overall scenario of yak herding as a primary occupation has undergone profound changes. Many of the younger generation *Drokpas* and herder families have given up their traditional vocations and have sought other forms of occupations such as adventure tours and travels and hospitality business such as homestays, hotels, and restaurants at their villages and in Gangtok, the state capital where they also work as sightseeing taxi drivers during peak tourist season. Seeing their parents' hardships and tough life, many of them never want to go back to the mountains (Interview, December 2019).

Sikkim's policy of infrastructure development and protection of the environment became the twin goals of the state government. The neoliberal modernization path of natural resources exploitation and harnessing of water resources for hydropower for economic prosperity, on the one hand, and the protection of forests and wildlife habitat through grassroots joint forest management committee mechanisms and Eco-Development Committees (EDCs) became the concern for the Himalayan state. The state's climate change mitigation efforts through various policies, such as State Green Mission in 2006, were welcome policy initiatives, and the establishment of national parks and wildlife sanctuaries has brought a positive impact on the state biodiversity. The formation of the Glacial Commission, the first state in the country, has put Sikkim's serious concerns vis-à-vis climate change effects on the fragile Himalayas. Although these led to substantial regeneration of flora and fauna in the evicted areas, the ecological impacts of these huge dams in the hilly mountainous state must be seen in the future. The effects have already been felt in the high alpine regions in the foothills of the glaciers in Sikkim. Therefore, the alpine yak habitats may also face a ripple effect and become zones of conflict and survival.

Receding glaciers and unpredictable weather phenomena have caused the loss of precious livestock this summer in North Sikkim. The formation of glacial lakes and the dangers of Glacial Lake Outbursts Floods (GLOFs) are subject of serious concerns to the alpine habitat and downstream human settlements (Singh 2018). For this lake to be drained, the water was siphoned downstream in 2016. Yaks, therefore, played an important role in transporting siphon pipes toward Lhonak Lake, situated at 5,200 meters (17,000 feet) situated at the India–China border in Northwest Sikkim (Wangchuk 2016). Many times, they not only help the military but also help human brethren to help secure their habitats as beasts of burden to transport relief and construction materials. If yaks and their cousins *dzo*/*dzomo* were reasons for grazing in some parts of lower alpine regions and state's strict eviction laws formulated to control them, then during a crisis, these animals have been human's best counterpart in mitigating disaster.

Conclusion

With increasing human activities in the Anthropocene, the spaces of interaction between humans and animals have become narrower. The yak *dzongs* in alpine regions are not bereft of these changes and interactions. The yak habitat in Sikkim, which occupies the transboundary region of Kanchenjunga Landscape and Tibetan Plateau, has undergone a social-cultural and economic transformation. These changes have also shaped the existing relationship between the human–animal and state. The state agency, as an actor through regulatory laws and conventions, has tried to shape the age-old human–yak relations from yak herding to pasture laws, and also has adverse ramifications for non-yak herders such as cattle and sheep farming.

Compounded with this, the vagaries of climate change are felt in the fragile regions creating serious survival challenges; without prudent policies and protecting indigenous knowledge practices, these conflicts cannot be ignored in the future. Almost 300 yaks perished to the unseasonal long snowfall leading to scaring fodder in Muguthang (PTI 2019). These are little stray incidents, but with a short season and unpredictable weather conditions, the burden will be even more in the coming future without proper regulations and alternative opportunities for the herding communities in these alpine highlands.

Once an animal of culture and identity, yak today has become a new ruler of the alpine geo-economic landscape: a transition from a cultural and nomadic lifestyle to changing social and economic landscape in Sikkim Himalayas. These *dzongs* have been witnessing layered differences vis-à-vis access to forest produce and herding rights shaped by political patronage and bargaining with the state administration and influencing traditional institutions of governance in some districts leading to marginalization of small herders in other parts of the alpine regions. At the same time, the erstwhile herder communities giving up yak rearing can be relooked from the state's environment and development policies narrative of the social and cultural politics, not as encroachers but as equal stakeholders. The stress on the alpine landscapes also emanates from the increasing geopolitical concerns between India and China, and bilateral tensions would further make these regions potential conflict spaces that are both internal and external and show that yak–human interactions have undergone a transformation in Sikkim Himalayas at multiple levels.

My gratitude to respondents who agreed to share their stories and experiences as yak owners, herders, and evictees.

Bibliography

19th Livestock Census, "State/District wise Report." Department of Animal Husbandry & Dairying, Ministry of Fisheries, Animal Husbandry and Dairying, Government of India, 2012. http://www.dahd.nic.in/about-us/divisions/statistics (Accessed on 22 October 2019).

Avaste, R. K. "Inter-Institutional Collaborative Program on Sustainable Yak Husbandry in Sikkim through Agri-Horti Technology Dissemination and Input Support' for the Dokpas (Yak Herders) of Muguthang, Sikkim." *Inaugural Speech*. Organised by ICAR-National Organic Farming Research Institute, Tadong, Gangtok; ICAR-National Research Centre on Yak, Dirang, Arunachal Pradesh and Department of Animal Husbandry, Livestock, Fisheries & Veterinary Services, Govt. of Sikkim, Gangtok, Tadong, 1999. http://www.kiran.nic.in/IICP_SY.html (Accessed on 26 October 2019).

Bagchi, Suvojit. "Now Bangladesh Citizens Can Visit Sikkim, Ladakh." *The Hindu*, 27 November 2018. https://www.thehindu.com/news/national/other-states/now-bangladesh-citizens-can-visit-sikkim-ladakh/article25608625.ece (Accessed on 23 April 2020).

Basu, Durga Das. *Introduction to the Constitution of India*. New Delhi: Prentice Hall of India, 1982.

Bhagwat, Shweta, Manasi Pathak, and Vivek Venkataraman. "Study of Ecological, Socio-Economic and Livelihood Dimensions of Grazing Exclusion in Protected Forests of West Sikkim." *Institute for Financial Management and Research—Centre for Development Finance (IFMR-CDF) in Collaboration with Department of Forest Environment and Wildlife Management, Government of Sikkim*. Gangtok: Department of Forest Environment and Wildlife Management Publications, Government of Sikkim, June 2012.

Census 2011. "Sikkim Population 2011." https://www.census2011.co.in/census/state/sikkim.html (Accessed on 3 November 2022).

Chakrabarti, Anjan. "Transhumance, Livelihood and Sustainable Development and Conflict between Formal Institution and Communal Governance: An Evaluative Note on East Himalayan State of Sikkim, India." *International Conference on Social Science and Humanity IPEDR*, vol. 5, pp. 1–7, 2011.

Chanchani, P., G. S. Rawat, and S. P. Goyal. "Ecology and Conservation of Ungulates in Tso Lhamo, North Sikkim." In *Biodiversity of Sikkim: Exploring and Conserving a Global Hotspot*, edited by M.L. Arrawatia and Sandeep Tambe, 443–462. Gangtok: IPR, Government of Sikkim, 2011.

Chettri, Nakul. "Local and Indigenous Practices on Adaptation: An Experience from Herder's Life of Western Bhutan." *Mountain Forum Bulletin*, vol. 8, no. 1, pp. 19–21, 2008.

Chettri, Simanta. "Politics of Pastoralism and Social Exclusion: A Case Study of Sikkim." Department of Peace and Conflict Studies and Management, Sikkim University, (Unpublished MPhil Dissertation), 2015.

Dasgupta, Probal. "1965—The Year China Accused Indian Troops of Stealing 800 Sheep and 59 Yaks." *The Print*, 17 February 2020. https://theprint.in/pageturner/excerpt/1965-china-accused-indian-troops-of-stealing-800-sheep-59-yaks/366554/ (Accessed on 5 May 2020).

DeMello, Margo. *Animals and Society: An Introduction to Human-Animal Studies*. New York: Columbia University Press, 2012.

"Fairs and Festivals, Tourism and Civil Aviation Department, Government of Sikkim." 2022. https://www.sikkimtourism.gov.in/Public/ExperienceSikkim/Fairs AndFestivalDetails/FF20A071?type=Festival (Accessed on 3 November 2022).

Harris, Tina. "Yak Tails, Santa Claus, and Transnational Trade in the Himalayas." *The Tibet Journal*, vol. 39, no. 1, Special Issue—Trade, Travel and the Tibetan Border Worlds: Essays in Honour of Wim van Spengen (1949–2013) (Spring–Summer 2014), pp. 145–155.

Institute for Financial Management and Research—Centre for Development Finance (IFMR-CDF) in collaboration with Department of Forest Environment and Wildlife Management, Government of Sikkim. Gangtok: Department of Forest Environment and Wildlife Management Publications, Government of Sikkim, n.d.

Kreutzmann, Hermann. "Yak Keeping in Western High Asia: Tajikistan, Afghanistan, Southern Xinjiang Pakistan." In *The Yak*, Second Edition, edited by Gerald Wiener, Han Jianlin, and Long Ruijun. Bangkok: Regional Office, FAO, June2003. pp. 323–336.

Krishnan, Ananth. "A Year on, Unanswered Questions, Sparse Details in China's Accounts of Galwan clash." *The Hindu*, 14 June 2021. https://www.thehindu.com/news/international (Accessed on 4 July 2021).

Mittermeier, R. A., P. R. Gil, M. Hoffmann, J. Pilgrim, T. Brooks, C. G. Mittermeier, J. Lamoreaux, and G. A. B. da Fonseca. *Hotspots Revisited: Earth's Biologically Richest and Most Endangered Terrestrial Ecoregions*. CEMEX, 2004.

Mullard, Saul. "The 'Tibetan' Formation of Sikkim: Religion, Politics and the Construction of a Coronation Myth." *Bulletin of Tibetology*, vol. 41, no. 2, pp. 31–48, 2005.

Nugo, Samir. "Drukpa Tshechi, a Yak Race in Lhasar Valley." *Sikkim Express*, 18 July 2018.

Pal, R.N. "Halting the Decline of the Yak Population in India." *World Animal Review*, vol. 76, pp. 56–57, 1993.

Pang Lhabsol. "Events and Festivals, Tourism and Civil Aviation Department, Government of Sikkim." 2022. http://sikkimtourism.gov.in/Webforms/General/EventandFestival/Pang_Lhabsol.aspx (Accessed on 26 September 2019).

Press Trust of India (PTI). "300 Yaks Starve to Death in North Sikkim." *Economic Times*, 12 May 2019a. https://economictimes.indiatimes.com/news/environment/flora-fauna/300-yaks-starve-to-death-in-north-sikkim/articleshow/69292179.cms?from=mdr (Accessed on 11 August 2019).

Press Trust of India (PTI). "Sikkim Calling! Record 14.25 Lakh Tourists Visited Himalayan State in 2017." *Financial Express*, 25 January 2019b. https://www.financialexpress.com/lifestyle/sikkim-calling-record-14-25-lakh-tourists-visited-himalayan-state-in-2017/1165207/ (Accessed on 11 October 2019).

Sabatier, Sophie Bourdet. "The Dzumsa of Lachen: An Example of a Sikkimese Political Institution." Translated by Anna Ballicki Denjongpa. *Bulletin of Tibetology*, vol. 40, no. 1, pp. 93–104, 2004.

Sharma, Ghanashyam and Lalit K. Rai. "Climate Change and Sustainability of Agrodiversity in Traditional Farming of the Sikkim Himalaya." In *Climate Change in Sikkim: Patterns, Impacts and Initiatives*, edited by M.L. Arrawatia and Sandeep Tambe, 443–462. Gangtok: IPR, Government of Sikkim, 2012.

Sherpa, Diki. "Sino-Indian Border Trade: The Promise of Jelepla." *ICS Analysis*, 45 May 2017.

Sikkim Forest Report, 2017. http://fsi.nic.in/isfr2017/sikkim-isfr-2017.pdf (Accessed on 24 November 2019).

Singh, Jyoti. "Glacial Lake Keeps Disaster Managers on Toes in Sikkim." *DownToEarth*, 27 September 2018. https://www.downtoearth.org.in/news/climate-change/glacial-lake-keeps-disaster-managers-on-toes-in-sikkim-61735 (Accessed on 6 May 2020).

Singh, Kripal. "Folk Songs and Dances of Sikkim." *Bulletin of Tibetology*, vol. 23, no. 1, 1995. http://himalaya.socanth.cam.ac.uk/collections/journals/bot/pdf/bot_1995_01_23.pdf (Accessed on 16 August 2016).

Singh, Sarbpreet. "The Curious Connection between Sikkim's Lake Guru Dongmar & Guru Nank." *Madras Courier*, 20 August 2019. https://madrascourier.com/opinion/the-curious-connection-between-sikkims-lake-guru-dongmar-guru-nanak/ (Accessed on 23 November 2019).

State of Forest Report 2021—Sikkim. Forest Survey of India, Dehradun. http://sikenvis.nic.in/Database/ForestResource_786.aspx (Accessed on 3 November 2022).

Subba, Bhim B. "India, China and the Nathu La: Realizing the Potential of a Border Trade." In *Issue Brief*, 205. Delhi: Institute of Peace and Conflict Studies, 2013. http://www.ipcs.org/issue_select.php?recNo=492 (Accessed on 3 November 2022).

Tambe, Sandeep and G. S. Rawat. *An Ecological Study of Pastoralism in the Khangchendzonga National Park, West Sikkim*, 53. Sikkim: The Mountain Institute – India, 2006.

Tambe, Sandeep and G. S. Rawat. "Traditional Livelihood Based on Sheep Grazing in the Khangchendzonga National Park, Sikkim." *Indian Journal of Traditional Knowledge*, vol.8, no. 1, pp. 75–80, January 2009.

Tambe, Sandeep and G. S. Rawat. "Ecology, Economics, and Equity of the Pastoral Systems in the Khangchendzonga National Park, Sikkim Himalaya, India." In *Biodiversity of Sikkim: Exploring and Conserving a Global Hotspot*, edited by M.L. Arrawatia and Sandeep Tambe, 443–462. Gangtok: IPR, Government of Sikkim, 2011.

Wangchuk, Sonam. "Photo Story: Draining a Glacial Lake in Sikkim to Prevent Climate Disaster." *Alternatives, Ecologise.in*, 5October2016.https://www.ecologise. in/2016/10/05/photo-essay-draining-a-glacial-lake-in-sikkim-to-prevent-climate-disaster/ (Accessed on 3 May 2020).

Wiener, G., Han Jianlin, and Long Ruijun. "Origins, Domestication and Distribution of Yak." In *The Yak*, Second Edition. Bangkok: Regional Office, FAO, June 2003. https://www.fao.org/3/ad347e/ad347e05.htm#bm05 (Accessed on 15 Nomember 2019).

Wu, Ning, Muhammad Ismail, Yi Shaoliang, Srijana Joshi, Faisal Mueen Qamer, and Neha Bisht. *Yak on the Move: Yak Raising in Transboundary Landscapes of the Hindu Kush Himalayan Region*. Kathmandu: ICIMOD, 2016.

Yak Development Programme, Department of Animal Husbandry, Livestock, Fisheries and Veterinary Services, Government of Sikkim. 2015–16. http://www. sikkim-ahvs.gov.in/yak_development.html (Accessed on 26 November 2019).

6 Turtles amid Healing and Extinction

International Relations and Question of Animal Agency in the South China Sea Disputes

Carmina Yu Untalan

Animals, especially sea turtles, are a hot button in Philippine-China relations. Since the Chinese government dispatched large fishery patrol vessels to the Spratly Islands in 2010, the Philippine government has reported several cases of Chinese fishers poaching sea turtles illegally. Between 2011 and 2012, Philippine authorities seized around 140 turtles from Chinese vessels around the Coral Triangle. In 2014, Philippine marine police officers caught several Chinese fishers harboring 555 endangered sea turtles near the Half Moon Shoal, considered a buying station where locals sell turtles for a high price to Chinese buyers within the Philippines' exclusive economic zone (EEZ). A regional trial court found the fishers guilty of violating the local fisheries code on illegal fishing in Philippine territorial waters. However, China's Ministry of Foreign Affairs stated that these arrests infringed Chinese "historical rights" and cautioned the Philippines against further provocative actions (*Rappler* 2014).

Endangered turtle species also figured prominently in the South China Sea Arbitration, a case the Philippines filed in The Hague against China's claim to jurisdiction within the so-called nine-dash line. One of its crucial components was China's violation of its obligation to protect and preserve the fragile marine ecosystem. The Tribunal ruled in favor of the Philippines, stating that Chinese activities have "caused severe harm to the coral reef system environment" (PCA 2016). Furthermore, it maintained that the illegal harvesting of sea turtles (along with endangered giant clams) constituted a violation of the United Nations Convention on the Law of the Sea (UNCLOS). Beijing refused to participate and rejected the ruling, arguing against the Tribunal's jurisdiction over the case and for China's sovereignty over the contested areas. Nevertheless, the arbitration demonstrated the importance of environmental protection in international politics. As scholars observe, it was a landmark case because it urged a reinterpretation of the UNCLOS, including its association with other environmental regimes (Kojima 2015; Robles 2020).

Apparent from these territorial contestations is the link between geopolitics and animals. In a time where the traditional separation between state sovereignty and environmental concerns is becoming less tenable, activities

DOI: 10.4324/9781003212089-6

that endanger wildlife constitute grounds for international action. However, it has yet to impact the typical realpolitik of international affairs significantly. It was indeed unsurprising that several analysts consider the South China Arbitration a diplomatic setback for Philippine-China relations, which both parties could turn into an opportunity to reconsider their priorities to maintain regional stability (de Castro 2020; Hsu 2015; Jimenez 2015). Both countries pursued reaffirming bilateral ties, especially after the Philippine administration shifted from the Aquino to the Duterte administration. Conservation of endangered species and marine ecosystem, arguably the Philippines' most vital point of contention against the big power China, suffered an eclipse. National interests, protection of state sovereignty, and political gains prevailed.

Such thwarted attempts to put environmental concerns at the forefront of foreign policy raise questions pertinent to the discipline of international relations (IR): How do animals influence foreign affairs? Why should IR acknowledge animals as agents of international politics? There is no simple and glib answer to these questions. One option would be to fall back into conventional explanations that preserve the primacy of nation-state and human agents, where animals matter in so far as they intervene with the Westphalian system's operation or pose a managerial problem for states and international institutions (see Youatt 2014). This perspective, however, depreciates animals' role in geopolitics. It gives a tacit nod to the anthropocentric assumptions that condone the human refusal to assume ethical responsibility toward animals and the planet. In perceiving animals as outsiders bereft of agency to affect the global, the prevailing perspective masks the contribution of state-sanctioned global degradation of nonhuman life-forms. It risks an endangered world, a future where territorial spats may not even occur because there are no resources to fight for anymore.

This chapter opts for a second, admittedly more tedious, option, that is, to seriously consider the agency of animals in international politics. Departing from the burgeoning IR scholarship on human–animal relations (Duffy 2013; Fougner 2020; Leep 2018), this chapter suggests that an Anthropocene-inspired outlook could help explore possible ways of integrating nonhuman animal life in IR. By this, I mean reflecting upon how emphasizing animals as an agent of international politics might alter our conventional conception of IR as a domain of human state actors and start thinking about the Earth as a "full-fledged actor" (Latour 2014). The main goal of this exercise is to show how IR neglects animal agency and why an IR that nurtures humanity's common bonds with the planet through an Anthropocene-inspired worldist approach is necessary (see Miyoshi 2001).

To do so, I examine the South China Sea dispute from the vantage point of human relations with animals, particularly sea turtles. The territorial dispute is an excellent case to reflect IR's anthropocentric view of global politics for three reasons: first, China's reaction is a 21st-century representation, if not rehearsal, of centuries-long conduct of international politics predicated upon great power and state-centric politics. Second, even though IR is attentive to China's unprecedented rise to global power, the planetary aspect of this rapid ascent remains underexplored. Third, and most importantly, this case elucidates how Chinese territorial expansion aids the extinction of its own people's source of healing. While other nonhuman animals are involved in territorial disputes, the choice of turtles is purposeful because of their symbolic and therapeutic role. Turtles in Chinese mythology represent the Earth and are integral components of healing. However, in what I call the "turtle paradox," the turtle's extinction becomes a condition of its life-giving function when hard politics prevail over planetary life.

In what follows, I briefly discuss the three key intersection points of human–animal relations in IR: metaphors, subjects of territorial expansion, and diplomacy. These intersections demonstrate IR's utilitarian treatment of animals, particularly as a forsaken Other. I then narrow it down to a discussion of the China-Philippine South China Sea dispute to illustrate how IR discourse has marginalized animal agency in the disputes, leaving the "turtle paradox" unresolved. Lastly, I suggest that an Anthropocene-inspired perspective could help reorient our understanding of world politics, from the politics of destruction to a politics of planetary survival, where humans recognize the agencies of animals and nature at large in foreign affairs.

It is important to note that this chapter is first and foremost about IR. It is neither an attack on China nor a celebration of the Philippines' efforts to protect marine wildlife. Moreover, since this chapter is about relations among "non-Western" actors, it is not an outright critique of the "West" or support for "non-West" ways of looking for viable solutions. I think such an approach is unhelpful in convincing a case for assuming collective responsibility toward the planet. As Horn and Bergthaller argue, "it is useless to tally up historical debts" because activities in the past that led to the present global environmental crisis did not only take place in Europe (Horn and Bergthaller 2019: 173). Studies had shown that China had conducted a mass-scale environmental intervention way before Western modernity began. China, in this regard, serves both as a foil and as an inspiration in this chapter. While China's struggle to become a global power has gravely affected the planet, recalling its traditions can nevertheless become a source for rethinking IR's relationship with animals. This chapter ends with opening a possibility for an IR recognizing the agency of its nonhuman others, where the main task is to forge relations that refuse to equate the age of the Anthropocene with a period of planetary extinction.

The Use and Abuse of Animals in International Relations

Anthropocene is making headway in global politics. Coined by Nobel laureate and chemist Paul J. Crutzen and his marine scientist collaborator, Eugene J. Stoermer, it marks a new geological epoch where human activities impact

Figure 6.1 A Sea Turtle Found in Celebes Sea off the Disputed Pulau Sipadan
(Photograph provided by Victor Teo).

the global environment (Crutzen and Stoermer 2000). Unlike previous geological periods, humans have now become geological agents and coauthors of "geostory" (Latour 2014). Notwithstanding the contested nature of the concept, the Anthropocene affords social sciences and humanities a lexicon to think about the Earth as a system that changes its "entirety" (Clark and Szerszynski 2021). In IR, scholars have broadly used the concept to reveal the discipline's anthropocentric bias and contribution to the conduct of the (Western-centric) global order to environmental crises (Simangan 2020). Moreover, its entry into the discipline marked a turning point for IR scholars and practitioners to assume ethical responsibility toward the planet (Harrington 2016).

Despite these developments, the challenge of integrating nature, which was traditionally considered outside the social field of IR, persists. In Lövbrand et al.'s (2020) structured analysis of published works on critical geopolitics and the Anthropocene, they identified three discourses: first, the endangered world, which sees human-induced environmental threats as nontraditional security threats that strong global institutions need to address. Second is the entangled world, where Anthropocene presents a "new reality where humans, non-humans and non-living things co-exist." This discourse calls for a transition from the current state-centric framework of international politics that securitizes the "Other" to an inclusive, grassroots-inspired transformative politics. Third, the extractivist world considers the global capitalist system the culprit of the environmental crisis attributed to the Anthropocene (Lövbrand, Mobjörk, and Söder 2020). While interpretations of Anthropocene differ, the three discourses share a similar dissatisfaction with the current geopolitical landscape. They, however, remain at the margins of the discipline.

Mainstream IR's relationship with animals demonstrates the discipline's aloofness from the Anthropocene. On the one hand, it treats animals as significant partners in conveying the bedrock of dominant, predominantly Eurocentric worldviews. Scholars and policymakers derive their linguistic significance from animal metaphors to "better" make sense of inter-state politics. On the other hand, animals' significance appears tenable when they are subjugated to anthropocentric needs. Human–animal relations preserve the human dominion, or stewardship over the environment, by viewing animals as food and enslaved people. Both support claims that humans, especially state agents from powerful states, have the authority over the planet's representation and activities.

The first form of nonhuman animal subjugation is linguistic. Following the "linguistic turn" in IR, scholars began to examine the function of language in representing the world and guiding actions (Fierke 2002; Neumann 2002). Metaphors are among the significant subjects of inquiry because they help cement worldviews and influence the focus of academic discourse (Marks 2011). They are also crucial in global issues' "strategic framing" (Lakoff 1999). However, while the metaphorical use of concepts appears innocuous, they misrepresent to bring cognitive familiarity and set parameters to what is knowable and otherwise (See Onuf 2013). As Goatly argues, "conventional metaphors do not unsettle our modes of perceptions" because they have "achieved currency as an acceptable way of constructing, conceptualising, and interacting with reality" (Goatly 2006: 16).

Animal representations in and of world politics attest to this. State characteristics are often illustrated through animal depictions – Russia as a bear, the United States as an eagle, China as a dragon – just as jungles and animal behavior portray the Hobbesian state of nature. This especially applies to realism, which, according to Lakoff (1999), is "rife with faux Darwinist metaphor" that characterizes the predatory nature of international politics. Analysts and commentators often use metaphors such as jungle politics, dog-eat-dog, and hawkish to connote the anarchy in world politics and the aggressive behavior of state agents. However, these metaphors also enforce epistemological power relations, particularly in determining what actions are justified or not. Animal metaphors are helpful for the powerful because by "conceptualising the animal world by projecting the state of current society onto it ... maintain the status quo because they can then argue that it is natural for humans to behave in the [ultra-competitive and war-like] way" (Goatly 2006: 21).

The use of animal metaphors to dehumanize humans works on similar logic. While acting aggressively toward the enemy is deemed normal, looking at people as animals also facilitates discrimination and violence. This is particularly evident in colonialism, where colonizers used animal metaphors to demonize the colonized. The colonial fabrication of "whiteness" as a signifier of superiority also benefited from animal metaphors. For instance, a crouching tiger visualized colonial India in colonial cartoons, while monkeys and snakes represent the "uncivilised" and "barbaric" colonial societies in Africa and Southeast Asia. The "monkey" commonly represents underdevelopment (almost human but not yet) and the position they assume when cleaning the house (almost naked in all fours). Some also use animal metaphors to dehumanize corrupt leaders and governments, primarily through satires (Eko 2007). To put shortly, IR confirms Lakoff and Turner's "great chain metaphor" of linguistically treating animals as inferior to humans (Lakoff and Turner 1989: 166–81).

Things get gloomier when these animal metaphors are set into motion. Anthropocentrism, suggesting the fixed belief in the inherent superiority of humans over animals, is central to the *raison d'état* of international politics (Youatt 2014). Colonialist representations of indigenous peoples and natives as animals entwined with the subjugation of nature. Settler colonialism, for instance, requires the destruction of indigenous land to make way for their housing and to create spaces where settlers could transplant their culture (Tuck and Yang 2012). Animal colonialism occurred, which Cohen (2017) characterized as a "dual phenomenon" of animal use for carrying out the tasks of colonization and universalizing European colonial laws and practices of human–animal relations. This applies to treating animals, such as dogs and horses, as partners in colonization and the "biological invasion" of modifying indigenous flora and fauna for human consumption and experiments. In other words, dodging concerns related to nonhuman animals were not only unimportant, but they were also necessary to impose authority outside territorial boundaries.

Globalization and independence barely altered these relations. Ecotourism has become a life source for many countries, so much so that state leaders appropriated it as a tool for attraction. Former colonies and present-day impoverished areas in Africa and Southeast Asia are considered safe spaces for experiencing the wild and the exotic. Solutions to illicit trade and killing of animals for trophies and medicine, such as the case of the ivory trade, remain elusive in world politics. More alarmingly, international agreements on climate change and environmental conservation have been suspect to scientists and nongovernmental organizations' criticism of lack of commitment. Recent proofs include the US President Donald Trump's decision to withdraw from the Paris agreement in 2017 and Brazil President Jair Bolsonaro's approval of selling hectares of the Amazon rainforest to pump up an otherwise failing political and economic system.

To be fair, there are efforts to underscore the role of animals in international politics. Some scholars and policymakers try to rectify this by integrating environmental concerns into foreign policy. For example, Richard Ullman (1983) drew a link between the environment and international security when he criticized the myopic scope of security during the Cold War era. Since the end of the Cold War, the environment has increasingly gained relevance in scholarly and policy circles (Welzer 2015). For instance, the UN's environmental security initiative aims to broaden the scope of human development goals by upholding environment protection and ameliorating environment-related destruction resulting from conflict as international security issues. The UN also mentions, albeit lightly, the role of species protection in sustaining healthy diets across cultures since the extinction of some plant and animal species also led to the extinction of nutritious food (IPCC 2019, sec. B). However, as Harrington argues, discourse on environmental security remains a managerial problem within the standard rubric of managing state interests in competitive and anarchic world order (Harrington 2016).

There are also studies about the involvement of animals outside security and managerial paradigms. Leira and Neumann (2017) draw attention to what they call "beastly diplomacy," where animals are considered objects and subjects of diplomacy. Their take on animal-as-metaphors is optimistic: animals are symbolic gifts, where state leaders exchange beasts for international status, alliance-building, and, sometimes, intimidation. However, focusing on the helpful metaphor of the present overlooks the broader consequences of geopolitics in nature. China's panda diplomacy, for instance, may have been relatively successful in softening China's image abroad, yet does not translate to good animal practices in panda mills. In a more normative analysis, Kavalski and Zolkos (2016) argue for recognizing nature as an actor in global politics. Acknowledging the persistence of wildlife in IR, they observed that its place had been detached from nonhuman systems; that is, the character matters only when it affects human activity. They advocate for a kind of recognition that speaks of symbiosis based on shared loss and precarity and of resilience and adaptability among all living things, a view that Kim (2020) somewhat shares in her study of how the environment intersects with anti–military base movements through the symbolic use of the rock formation, Gureombi in Jeju Islands and the endangered animal dugong in Okinawa.

Some of these works set important steppingstones toward collapsing IR's anthropocentrism. Animals, as much as humans, are agents of global affairs. The question is, agents of whom and for what? A conventional IR perspective's answer, pointed out by some critical geopolitical scholarships mentioned above, is that animals are agents of global human politics tasked with fulfilling *human* objectives. The puzzle is that, while the presence of animals in foreign policy and diplomacy animals attests those international human actors are aware of how influential animals are in global affairs, human–animal relations barely get the attention of IR scholars (Fougner 2020). Perhaps, it is not enough to focus on human culpability and ethical responsibility toward the planet to make the discipline more in tune with the planet. Suppose the Anthropocene provides a lexicon to think about the Earth as a system, where humans are now considered new geological agents. In that case, it is necessary to remember that natural and nonhuman animals are also agents and that they had existed and affected the workings of the planet even before humans did. What happens when animals try to exercise their agency in international affairs? The turtles in the South China Sea dispute may offer somewhat of an answer.

When Animals "Intervene": The Case of Philippine-China Territorial Disputes

As mentioned in the introduction, the Philippine-China territorial disputes present a good case for reflecting human–animal relations in IR. It shows that in situations of power asymmetry between two states, there is a tendency for the more powerful state to stick to conventional narratives of security and sovereign territoriality, while smaller states which lack power in the traditional sense act as "entrepreneurs of good practice" (see Ingebritsen 2002). Small states may promote the protection of nonhuman animals as a form of empowerment in highly unequal relations. More importantly, it demonstrates the frailty of the West/non-West dichotomy. China, whose officials have carved an international identity of differentiating itself from the powerful Western countries, has arguably used similar exceptionalism tactics to defend its stance. At the same time, the Philippines sought the help of international institutions to carry out their advocacy of conserving endangered species, especially turtles. From an Anthropocene point of view, this perhaps illustrates the initial impact of animals in the foreign policy decision-making of small state actors. However, as the following discussion will show, state-centric geopolitics prevailed.

Human-centric IR: An Overview of Asymmetrical Battle for Sovereign Rights

Before delving into the agency of turtles in the territorial disputes, a background may be helpful to illustrate geopolitical fixation with sovereignty and its anthropocentrism. The question of sovereignty over the Spratly Islands has been a permanent feature of Philippine-China relations since the end of the Second World War. In 1956, Filipino lawyer and businessman Tomas Cloma claimed the Kalayaan Islands (local name for Spratlys) based on the principle of *res nullus* after Japan relinquished its right over the Spratly and Paracel Islands in the 1952 San Francisco Treaty (Buszynski 2010). Before this, China had already asserted sovereignty over these territories under modern international law, with Chinese Premier Zhou Enlai declaring in 1951 "the inviolable sovereignty of the People's Republic of China over Spratly and Paracel Archipelago" (Samuels 2005). In 1958, the Standing Committee of the People's Congress signed the Declaration of Territorial Seas, indicating the scope of its territories (Gao and Jia 2013). Despite this, the Philippines consistently defended its juridical right despite legal arguments against it.

In February 1992, China signed a domestic Law on the Territorial Sea and the Contiguous Zone to further confirm its claim over the South China Sea. The law unilaterally *and* domestically declared the PRC's right to exercise its sovereignty over its territorial sea and safeguard its national security and maritime rights and interests ("Law of the People's Republic of China on the Territorial Sea and the Contiguous Zone" 1992). At the heart of the contention lies the ambiguity of the law. For some scholars and legal experts, it was inconsistent with international law because UNCLOS does not include historic rights, nor does it sanction states to employ baseline methods, while for others, it only reveals weaknesses of the Charter (Wang 2015). Five months later, during the 1992 ASEAN Foreign Ministers' Meeting in Manila, the ASEAN, led by the Philippines, signed the ASEAN Declaration on the South China Sea, urging all parties involved to "resolve all sovereignty and jurisdictional issues about the South China Seas by peaceful means without resort to force." While China agreed with the principles, it does not consider them binding since China is not an ASEAN country (Severino 2010).

Skirmishes between the two countries occurred in the early 1990s, including the Philippine arrests of Chinese fishers in March 1992 and September 1994 and the Chinese capture of 35 Filipino fishers in January 1995. However, the 1995 Mischief Reef incident stirred Philippine-China relations (Storey 1999). Satellite images from a reconnaissance aircraft showed Chinese structures and vessels within the reef, repudiating the Philippine jurisdiction within the 200-mile EZZ. According to Chinese Foreign Minister Qian Qichen, the facilities were not for military use. Instead, they served as protective shelters for local fishers. Nevertheless, it stirred critical reactions from neighboring countries such as Vietnam, Taiwan, Malaysia, and Brunei. Then Philippine President Fidel Ramos declared the act violated the 1992 Manila-ASEAN Declaration and that the Philippine government's only option was through diplomatic means.

The Philippines dealt with it by the ASEAN foreign ministers' recommendation to avoid actions that could destabilize the region and threaten regional security. However, between 1995 and 1999, China continued building new structures, including a helipad, communications equipment, and wharves around the area, despite Philippine attempts to settle the territorial disputes (Robles 2020). In 1998, the Philippine government caught footage of Chinese vessels unloading materials to construct additional installations. Chinese officials thwarted Philippine efforts to forge a bilateral agreement by enforcing their jurisdiction over the reef by adding telecommunications equipment and maritime elements. They also refused to ratify the 1999 ASEAN Regional Code of Conduct draft and successfully blocked the Philippine government's proposal to establish a new one three years later (Thayer 2012). Since then, the Philippines have stationed patrol boats in the Spratlys to signal its sovereignty over the islands.

The Philippine-China territorial dispute reached its crescendo in 2013. After decades of failed bilateral and multilateral attempts, the Philippine government filed a case with the Permanent Court of Arbitration in The Hague against China's UNCLOS violations. Philippine Foreign Secretary Albert del Rosario stated that the purpose of the South China Sea Arbitration was to "defend what is legitimately ours [the Philippines]" (*Time* 2014). The 4,000-page dossier enumerates China's violations as (1) China's sovereign claim to the nine-dash line is inconsistent with UNCLOS; (2) the Scarborough Shoal and several maritime features in the Spratlys "can only generate entitlement," and (3) violation of Philippine sovereign rights to freedom of navigation and breach of the Convention on Biological Diversity (CBD) (Talmon 2016). Beijing made its nonparticipation clear in a position paper released on 7 December, stating that the Tribunal has no jurisdiction over the matter, with China having an "indisputable sovereignty over the South China Sea."

Tensions between the Philippines and China mounted. Barely half a year after the Philippine representatives filed the case, the Chinese government published a "new 10-dash line map" that, policy analyst and professor Carl Thayer stated in an interview, was a move to lay sovereignty on "every feature of the island" (*GMA News* 2013). The Chinese government built a military aircraft runway, submarine harbor, and artificial islands from 2013 to 2016. Responding to official complaints from the Philippine government and Filipino fishermen, the Chinese Defence Ministry Spokesperson Hong Lei said that deploying "defence facilities in our territory is appropriate" (*CNN* 2016). Reacting to suspicions of militarizing the area, the Chinese Foreign Ministry stated that "China's sovereign right" allows it to construct defense facilities and refused to abide by any third-party ruling (*The Guardian* 2016).

After three years of grueling legal and foreign relations work, the Tribunal ruled in favor of the Philippines in July 2016. However, China has remained intransigent, building more fortifications and deploying long-range anti-ship and anti-air missiles in the Spratlys. Meanwhile, the Philippines were transitioning from the Aquino administration, which led the arbitration, to the Duterte administration. In contrast to his predecessor, President Rodrigo Duterte undermined the ruling and refused to uphold the award, something that he would "throw in the waste basket" (*Inquirer* 2021). The Chinese government continued to assert its sovereign rights over the Spratlys, and members of the Philippine government involved in the case were left to pick up the threads to uphold The Hague's ruling.

Historical Rights and the Question of Whose History?

While scholars and commentators concentrated on the geopolitical and legal aspects of the territorial dispute, the ecological part, which comprises the main thrust of the Philippine appeal, was sidelined from the mainstream discussions. The ruling was not only about the Philippines' right to the EEZ. It was also about the marine wildlife that China's activities had harmed. However, the discussions tapered into a diplomatic spat between the two countries. Animals, initially one of the main concerns of the territorial disputes, assumed a metaphorical role in ascribing the characteristics of the states involved. Political commentaries were rife with animal metaphors to depict asymmetrical power relations. Local and international news outfits published political cartoons denoting China's cunning aggressiveness (The Economist 2018; Philippine Star 2019). China was also depicted as a wicked, giant dragon wickedly embracing rock formations in the South China Sea

(Strifeblog 2017; Sydney Morning Herald 2015), a boorish dragon bullying other claimants (The Economist 2015), and a giant panda sprawled across the South China Sea (Herald Sun 2016; Pitch 2018). Article titles are sprinkled with dragon metaphors based on what China is doing or what could and should be done (Blain 2015; de Castro 2017).

Geopolitics and military-strategic significance loomed large in IR discourse. Major concerns include possible escalation, militarization and, in the context of the so-called rise of China, ways involved parties could manage China's "creeping assertiveness" (see Storey 1999). The discourse is framed in terms of conflict and power politics. China and less powerful state claimants find themselves consistently embroiled where "irresolution" is likelier in this territorial dispute (Scott 2012; Yahuda 2013). Some analyses focus on strategic incentives and disincentives for China, arguing that it has been China's interest to adjust assertive approaches with specific constraints (Fangyin 2016). Because of its interest in the Indo-Pacific and its alliance with the Philippines, the dispute also raised concerns in Washington which sees itself as a necessary force that could prevent conflict escalation (Glaser 2015). What appears as a condition ripe for big power rivalry also raised the question of whether China and the United States should deal with the situation as big powers (Cronin 2016).

While these viewpoints are valid as the situation could have severe national and regional security consequences, they may have inadvertently followed China's lead in obfuscating environmental concerns. It could be argued that the anthropocentric IR thinking enabled China to obscure the Philippines' complaint about environmental damages, thus inadvertently allowing Beijing to play its "historic rights" card, despite the Tribunal's decision. It is worth noting that in the position paper, Beijing did not mention its responsibility to protect wildlife. Its claim to historic rights over the South China Sea is based on its belief that it was the "Chinese activities in the South China Sea date back 2,000 years ago" and that the "first country to discover, name, explore and *exploit* the resources of the South China Sea Islands, and the first to continuously exercise sovereign powers over them" (People's Republic of China 2014). It could also be argued that had marine wildlife protection clauses been given the limelight, it would have been more difficult for China to press for its claims, especially of sovereignty over the whole territory, including its nonmaterial, living resources.

From an Anthropocene perspective, China's claim to "historic rights" begs the question, *whose history*? This is an important question to reflect upon because the history invoked, in this case, refers to human history. Historian Dipesh Chakrabarty argues that conceptualizing the Anthropocene entails thinking about two different time scales of Earth and world history (Chakrabarty 2018). When social scientists talk about the Anthropocene, discussions usually revolve around humanity's impact and responsibility toward the planet. Although geological time recedes in the background as world (human) history, that is, human history, takes, it is difficult to deny that geological time stretches farther than humanity's history and that the common conception of history is anthropocentric. As China's "historic rights" argument indicates, this means that any human claim to nature is circumscribed in world history. Thus, China's "historic rights" argument is controversial not only because historical maps tell otherwise, as the former Philippine Supreme Court Justice Antonio Carpio and others argue. It is controversial also because it does not and cannot integrate other nonhuman agents whose history belongs to geological time.

The way scholars, analysts, and state actors handled the dispute left no room for a planetary outlook. Even defense on the Philippine side developed into the rhetoric of the country's pride, triumph, and affirmation of the international legal order (*Rappler* 2021). However, the dispute meant more than that. As Robles argues from an international law perspective, the South China Sea Arbitration Award was notable because it was based on a meticulous analysis of the Philippine claims, unlike previous environment-related claims where the Tribunal focused on the responsibility of the parties concerned to negotiate (Robles 2020). According to him, the Tribunal spent significant efforts to underscore the importance of conserving endangered species and their habitat. This, to a great extent, signaled the possibility and necessity of rethinking international law when the preservation of wildlife is concerned.

In other words, the arbitration ruling was a step toward an Anthropocene-inspired IR. The discourse on Philippine-China South China Sea disputes exposed anthropocentric IR's limits in comprehending the global constitution, where the tedious ecological interpretation of international law appeared vulnerable to the obstinacy of the sovereignty principle among state leaders. Existing institutions are helpful, but they may not be enough to accommodate the Anthropocene's call to acknowledge human and nonhuman agencies. The two have become so entangled that it compels humans to see themselves as part of an organization of an unimaginable scale. In the long run, hardened assumptions about territorial sovereignty and the primacy of states and human actors may stand no chance against the species that existed millions of years ago. If humans do not do anything about their extinction, it may be inevitable, but the imprint of their demise could be deleterious. Following Chakrabarty, humanity has reached a point in history where it is perhaps time to integrate the "outside" into the human world (Chakrabarty 2018). It may be the time to hear them speak.

The Turtle Paradox

Letting animals speak means accepting that they can produce meaning and occupy the same metamorphic zone as humans (Latour 2014). It differs from a one-way anthropomorphism characterized above, where humans assign negative metaphorical values to animals. Instead, letting animals speak through metaphors means making them comprehensible to humans. The point is, following Latour, to "distribute agency as far and in as differentiated as way as possible" so that different types of agents, humans and otherwise, interact *with* the world (Latour 2014: 15). Put differently, acknowledging animal agency means that the purpose of telling stories about animals is not only to make humans understand their role in the planet. It is also about comprehending what the planet does and intends to do, what they feel and how they respond to human activities.

In the case of the South China Sea, the presence of turtles is undeniable. The question is, how can they speak? Articulations could take many forms. Take, for instance, a political cartoon about the South China Sea disputes where a Chinese fisherman was installing "China" placards around the islands. A turtle told a dugong, both stabbed with the placards: "the territorial claims are going too far!" (Cagle 2016). This contrasts with another cartoon that used a panda who appeared to be having a relaxing time "bathing" in the South China Sea while whistling and playing with his "toy," a Chinese aircraft carrier (Cagle 2018). While the latter uses animals to criticize Chinese activities in the disputed islands, the former is about animals expressing what they think. The turtle and the dugong were protagonists in the South China Sea dispute story.

Animals can also express themselves through their activities and the consequences of their impending extinction. Turtles particularly stood out in the territorial dispute because illegal poaching and harvesting of sea turtles had been a contentious issue long before the Philippine government filed the arbitration case. In addition, the waters surrounding the Philippine archipelago are home to five vital, yet threatened, turtle species. The International Union for the Conservation of Nature (IUCN) Red List of Threatened Species labeled critically endangered hawksbill turtles critically endangered; green turtles endangered; leatherback, loggerhead, and olive ridley sea turtles vulnerable.

The increasingly endangered status of these turtle species is alarming because they play a significant role in balancing the marine ecosystem (Lovich et al. 2018). Turtles contribute to the Earth's biomass, an essential ecological factor in measuring the amount of energy in the ecosystem. As both predators and prey they help balance the ecosystem. Green sea turtles clean the seagrass beds to prevent molds from growing, thus maintaining healthy life cycles and predator–prey relations for plant and animal species. Hawksbill turtles feed on sponges, keep reefs suitable for coral growth, and ensure food found on reefs is accessible to other animals. With their powerful jaws, loggerhead turtles break hard-shell prey, which multiplies the nutrient-recycling rate. Leatherback turtles' diet of jellyfish prevents jellyfish overgrowth, threatening fish development. Olive ridleys offer seabirds a "roosting area," protecting them from predators. There is mutual coexistence between turtles and fishes, as all sea turtle species help provide other marine species access to nutrients, while fish feed on the scrap they get from cleaning the turtle shells. In other words, turtles play a significant role in balancing the already fragile ecosystem in the South China Sea.

Aware of turtles' role in maintaining biodiversity, the Philippine government, with the help of locals, has been consistent in its policy of protecting turtle marine life environment. The Pawikan Conservation Project (PCP) was established in 1979 to develop and implement conservation and protection policies and information dissemination programs to promote marine turtle conservation. Its activities include tagging, monitoring nesting movements, habitat surveys, conducting various local awareness campaigns, and forging international agreements with ASEAN countries (Sagun 2002). In response to the massive trading of turtle by-products between the Philippines and Japan in the 1970s and 1980s, the Philippines helped ratify the 1981 Convention on International Trade (CITES). The Philippine Wildlife Act of 2001 prohibits harvesting wildlife species, including turtles. However, it is essential to note that, as Robles (2020) argues, turtle smuggling continued due to persistent high demand in neighboring countries, such as Brunei, Hong Kong, and China, despite these legal prohibitions.

Aside from domestic and international efforts, the local communities are engaged in protecting sea turtles. Locations for hatching eggs have been suffering from natural and artificial interventions, such as rising sea temperatures and the construction of beach resorts that block the coastlines. For example, in Candiis, Misamis Oriental, locals have pursued preservation and patrolling activities to protect the nesting culture of hawksbill turtles on their shores. Because of the long history of relations between their ancestors and the turtles, the locals treat the endangered animal as "neighbors." In turn, the turtles chose the area to perform their ancient nesting ritual by imprinting their "memory" with a magnetic map of the beach sand (Mascariñas 2020). There is also legal protection for indigenous communities with cultural bonds with turtles. For example, the Tagbanua tribe are legally allowed to hunt and consume turtles according to traditional spiritual practices, including serving the animal meat during special occasions such as weddings and birthdays. Because of their sound ecological practices, there are calls to integrate indigenous knowledge into ongoing scientific investigations regarding turtle conservation and management (Poonian et al. 2016).

The Chinese also have long historical and cultural bonds with turtles. Chinese cosmology forges a correlation between the shape of the tortoise and the five elements of the cosmos (Chong 2018). The tortoise's body, upper shell, and plastron correspond to heaven, while the lower plastron, which is flat, is the Earth's flatness. The turtle symbolism in the creation and healing of the world is reflected in Chinese mythology and traditional medicine. The popular story of *Ao* is a good example. *Ao* was a giant marine turtle who lived in the South China Sea. The creation goddess *Nuwa* cut his legs off to support the Earth as she was repairing it from damages another god caused. Moreover, according to a 14th-century Chinese herbology volume, *Compendium of Material Medica*, turtles are good sources of healing, especially in balancing the body's *yin* and *yang* (Li and Cheung 2001). At present, turtle shells and fluid are still widely sought-after for their therapeutic functions, and the demand is growing to serve the increasing demands of the Chinese, who believe in the turtles as symbols of healing, prosperity, and good life.

The problem of turtle conservation in Philippine-China relations is related to this demand. Filipino locals find it lucrative to sell turtles to the Chinese, willing to pay vast sums to obtain the healing creature. One of the reasons for such high demand despite the high price is customer preference. The wildlife turtle supply could not simply meet the demand of millions of Chinese people to the extent that it became necessary to commercially *farm* turtles. However, customers prefer "non-modified" and "chemical-free" turtles because they might alter their healing effects, thus increasing wildlife poaching (Lee and Liao 2014). The Chinese government has been aware of this problem. China is a signatory to the Convention on the International Trade of Endangered Species (CITES). In CITE's 2013 meeting, China, which has gained a reputation for illegal animal trade, came to the forefront of protecting freshwater tortoises and turtles. Director of Wildlife at the Humane Society International, Teresa Telecky, sums up the contradiction of this effort, "Firstly, China has all these turtle and tortoise proposals. However, it is also a consumer of illegal trade" (*Wall Street Journal* 2013). In other words, the Chinese government needs to address not its image problem abroad but the turtle paradox of deriving healing from killing the medicine.

The difference between Filipino locals and the Chinese government's response to the problem of turtles is insightful. One is based on concern for the planet; the other appears to be about international politicking. The irony is that while China seeks to be at the forefront of protecting endangered animals, its behavior toward the South China Sea Arbitration proves otherwise. The Chinese Society of International law defended China by stating that, among others, states can only perform due diligence within their capacity and that protecting their fishers and their property is part of safeguarding Chinese sovereignty (Robles 2020). Chinese officials insisted that the negotiations were primarily about conflict prevention and ensuring the region's security. These statements avoided tackling the totality of the problem, which involves degrading marine biodiversity.

The Filipinos celebrated their "victory" in the Tribunal, while the Chinese did not recognize it. Whether the ruling is legally legitimate or not may be the most critical issue at hand. As Robles concludes, the Chinese defense of their case could not reverse or prevent further damages that the Chinese nationals have done to the environment. Even with the help of the Tribunal, a small state like the Philippines may not be able to convince a considerable power like China. Nevertheless, it could have deterred environmental degradation and future criticisms toward the Chinese government had the conduct of international politics taken a radical turn. The turtle has spoken, and it might be too late for humanity to undermine its voice.

Toward a Worldist IR in the Anthropocene

Mainstream IR privileges the Westphalian state system because it believes that states and state actors play a crucial role in dealing with world issues.

However, this thinking begs the question: can the world survive with this system of thinking and doing international politics? China's response to the South China Sea arbitration tells us that the answer is no. To be precise, if the conduct of international politics continues to ignore nature, and prioritize sovereignty of states, then humanity might one day wake up with a new type of sovereign more powerful than the state, asserting itself over humans, or what eminent scientist James Lovelock calls "revenge of Gaia" (Lovelock 2007). What is necessary to avert such catastrophe somehow is a recalibration of what humanity is *and* to the planet. Specifically, to IR, this first and foremost means revising its hardened assumptions about the 'state'.

The ancient Persian animal folklore *The Scorpion and the Tortoise* provides an insightful metaphorical departure point. Not knowing how to swim, the scorpion asked the turtle if he could carry him across the river. Aware of the scorpion's poisonous sting, the turtle initially refused, yet was eventually convinced by the scorpion's assurance that if he stings the turtle, they will both die. The scorpion, unable to resist the urge, stung the turtle. As both began sinking, the turtle asked, "you said there would be no logic in your stinging me. Why did you do it?" The scorpion sadly replied, "It has nothing to do with logic," the drowning scorpion sadly replied. "It is just my character." This folklore is commonly attributed to humans' tendency to betrayal and irrationality. However, following Deleuzian interpretation of the scorpion (he comments on a version with a frog) leads us away from degrading animals toward understanding the severe consequences of staying in one's internalized character. According to him, the scorpion is neither irrational nor evil: the scorpion (as a metaphor) "is a type of force that no longer knows how to metamorphose itself according to the variation ... it can only destroy and kill before it destroys itself."

The fascinating thing about this animal folklore is that, unlike the use of animal metaphor in IR scholarship discussed above, it does not reinforce hardcore assumptions about state behavior in international politics. It unsettles them. Imagine China as the scorpion and the turtle representing the endangered sea turtles. China is endowed with and is acting upon the modern characterization of nation-states. Its goal, to "cross the river," could be interpreted as an assertion of sovereignty and power across the globe, which is "expected" of states striving for extraordinary power status. States cannot do this alone, as the folklore suggests; it requires the help and, eventually, the sacrifice of others. However, as the turtle paradox shows, China cannot keep on stinging lest it suffers the fatal consequences to its international prestige and the calamitous effects of its actions on the environment. This raises the necessity for China to revisit a worldview outside the Westphalian lens, not to compete with it, but live with it, a worldview based on Daoist, *yin* and *yang* dialectics.

Although IR scholar L.H.M. Ling was not strictly pertaining to the Anthropocene when she came up with her "worldist" approach to international politics, it may serve as a springboard for rethinking humanity's relation to animals and the planet. This is because it shares with Anthropocene-thinking

a common call for inclusivity of the "other" (Ling 2014). The core philosophy
of a worldist approach is based on Daoist dialectics' notion of yin and yang
where "each part contributes to the whole, just as no whole could form without
individual parts" (Ling 2014: 17). Indeed, the critical lesson that Daoist dialectics imparts is the possibility to address imbalances through mutual interaction
between polarities. As such, worldist international relations emerge as a syn-
thesis of an array of Western and Eastern views of international relations, such
as social constructivism, postcolonial studies, and dialectics of world order.

Since worldism works through open-minded borrowing, communications,
and negotiations, it requires four tenets. First is relationality, which pertains
to the interrelated sources that inform a holistic understanding of the world.
It draws upon Edward Said's notion of core-periphery contrapuntal relations. This calls for looking at world issues from the eyes of the "exile" to
challenge existing power relations that imprison both the Self and the Other.
In this case, this Other is not only the subaltern human other but the Other
of world history: nature and its human interlocutors. In the case of China,
this entails the abandonment of a worldview that protects the security and
territorial integrity of the state and the human appetite for consuming endangered species solely. Human–animal relations become a relational recognition of each other's agency genuinely. Such relationality, once again invoking
Latour, means sharing "agency with other subjects that have also lost their
autonomy" (Latour 2014: 5).

Second, a worldist approach involves resonance, which allows other discourse with "resonating" principles to transform the State's perspective to
what also applies to "Others." It helps reveal the concealed unity among differences by detecting vibrations in unison with the state actors and the environment. Through resonance, IR can finally hear what the animals and
nature are saying. Animal agency would matter as much as a human agency
in dealing with the potentially catastrophic impact of the Anthropocene
because this period is a period where the two had to confront *and* work with
each other. I suggest it would save China and other states from the illusion
that they could still take control of the force of late modernity and globalization. While one or two great powers could not reverse the damage that
humanity had done to the environment, at least they could listen to the
"vibrations" sent by small states like the Philippines and local actors and to
the alarms signaled by the environment.

Lastly, and perhaps the most demanding requirement of worldism: interbeing. Interbeing, or in the context of Anthropocene, interplanetary being,
urges individuals and societies to act ethically and compassionately toward
the Other. This is not merely a call toward becoming more environmentally
responsible foreign policy or pursuing "climate justice." Interbeing means
acknowledging that the human-Self is in the animal/nature-Other, the way
humans depend on nature for their healing in as much as nature depends on
humans for its healing. Acting unethically without compassion harms both
the Self and the Other. China acting based on sovereignty and territorial
expansion does not only harm the Other but counterintuitively harms itself.

Interbeing compels thinking outside the state-system box so that protection is not confined to the securitization of the Other. What could protect China and its other's ethical and compassionate judgments toward the very source of its lifeline?

Admittedly, these are visions for the future of international politics. They are visions because they have yet to materialize. Even so, they are more realistic than the utopian dream of thinking that the world could go on with prioritizing national interests and state sovereignty over preserving the planet. The world currently lives in a utopia where it can continue with its old ways of Othering without being challenged and antagonized. The realism of a worldist view of international politics shatters this utopia. The alarm would wake mainstream IR and its loyal followers to face the reality that environmental degradation is as detrimental, if not more, to humanity as wars. If the Anthropocene has a message to IR, it is that: states and their human drivers are no longer masters or custodians of the Earth; they are with the Earth.

Bibliography

Blain, Jason. 2015. 'The Dragon and the Eagle in the South China Sea: Is Conflict between China and the US Inevitable?' *Australian Defense Force Journal* 73: 73–76.

Buszynski, Leszek. 2010. 'Rising Tensions in the South China Sea: Prospects for a Resolution of the Issue'. *Security Challenges* 6 (2): 85–104.

Chakrabarty, Dipesh. 2018. 'Anthropocene Time'. *History and Theory* 57 (1): 5–32. https://doi.org/10.1111/hith.12044

Chong, Alan. 2018. 'The Metaphysical Symbolism of the Chinese Tortoise'. Nanyang Technological University Singapore. http://hdl.handle.net/10356/73204

Clark, Nigel, and Bronislaw Szerszynski. 2021. *Planetary Social Thought: The Anthropocene Challenge to the Social Sciences*. Cambridge: Polity.

CNN. 2016. 'US Says China Deploys Fighter Jets to Disputed South China Sea Island', 24 February 2016. https://edition.cnn.com/2016/02/23/asia/china-missiles-south-china-sea/index.html

Cohen, Mathilde. 2017. 'Animal Colonialism: The Case of Milk'. *AJIL Unbound* 111: 267–71. https://doi.org/10.1017/aju.2017.66

Cronin, Patrick M. 2016. 'Power and Order in the South China Sea'. Asia-Pacific Security. Center for a New American Security. https://www.cnas.org/publications/reports/power-and-order-in-the-south-china-sea

Crutzen, P. J., and E. F. Stoermer. 2000. 'The 'Anthropocene''. *Global Change Newsletter* 41 (May): 17–18.

de Castro, Renato Cruz. 2017. 'Navigating between the Dragon and the Sun: The Philippines' Gambit of Pitting Japan against China in the South China Sea Dispute'. In *Chinese–Japanese Competition and the East Asian Security Complex*, edited by Jeffrey Reeves, Jeffrey Hornung, and Kerry Lynn Nankivell, 178–97. London: Routledge.

de Castro, Renato Cruz. 2020. 'From Appeasement to Soft Balancing: The Duterte Administration's Shifting Policy on the South China Sea Imbroglio'. *Asian Affairs: An American Review*: 1–27. https://doi.org/10.1080/00927678.2020.1818910

Deleuze, Gilles. 1989. *Cinema 2: The Time-Image*. Translated by H. Tomlinson and R. Galeta. Minneapolis: Minnesota University Press.

Duffy, Rosaleen. 2013. 'Interactive Elephants: Nature, Tourism and Neoliberalism'. *Annals of Tourism Research*: 44: 88–101. https://doi.org/10.1016/j.annals.2013.09.003

Eko, Lyombe. 2007. 'It's a Political Jungle Out There: How Four African Newspaper Cartoons Dehumanized and 'Deterritorialized' African Political Leaders in the Post-Cold War Era'. *International Communication Gazette* 69 (3): 219–38. https://doi.org/10.1177/1748048507076577

Fangyin, Zhou. 2016. 'Between Assertiveness and Self-Restraint: Understanding China's South China Sea Policy'. *International Affairs* 92: 869–90.

Fierke, K. M. 2002. 'Links across the Abyss: Language and Logic in International Relations'. *International Studies Quarterly* 46 (3): 331–54.

Fougner, Tore. 2020. 'Engaging the 'Animal Question' in International Relations'. *International Studies Review*. https://doi.org/10.1093/isr/viaa082

Gao, Zhiguo, and Bing Bing Jia. 2013. 'The Nine-Dash Line in the South China Sea: History, Status, and Implications'. *The American Journal of International Law* 107 (1): 98–124. https://doi.org/10.5305/amerjintelaw.107.1.0098

Glaser, Bonnie S. 2015. 'Conflict in the South China Sea Contingency Planning Memorandum Update'. Contingency Planning Memorandum Update. Council on Foreign Relations. https://www.cfr.org/report/conflict-south-china-sea

GMA News. 2013. 'China's New '10-Dash Line Map' Eats into Philippine Territory', 26 July 2013. https://www.gmanetwork.com/news/news/nation/319303/china-s-new-10-dash-line-map-eats-into-philippine-territory/story/

Goatly, Andrew. 2006. 'Humans, Animals, and Metaphors'. *Society & Animals* 14 (1): 15–37. https://doi.org/10.1163/156853006776137131

Harrington, Cameron. 2016. 'The Ends of the World: International Relations and the Anthropocene'. *Millennium* 44 (3): 478–98. https://doi.org/10.1177/0305829816638745

Horn, Eva, and Hannes Bergthaller. 2019. *The Anthropocene: Key Issues for the Humanities*. London: Routledge. https://doi.org/10.4324/9780429439735

Hsu, Hsiao-Chi. 2015. 'The Political Implications of the South China Sea Ruling on Sino-Philippine Relations and Regional Stability'. In *Asian Yearbook of International Law*, 21: 16–33. Leiden: Brill. https://brill.com/view/book/edcoll/9789004344556/B9789004344556_003.xml

Ingebritsen, Christine. 2002. 'Norm Entrepreneurs: Scandinavia's Role in World Politics'. *Cooperation and Conflict* 37 (1): 11–23.

Inquirer. 2021. 'Duterte on PH Court Win over China: 'That's Just Paper; I'll Throw That in the Wastebasket'. Read More: https://Newsinfo.Inquirer.Net/1427860/Duterte-on-Ph-Arbitral-Win-over-China-Papel-Lang-Yan-Itatapon-Ko-Yan-Sa-Waste-Basket#ixzz6zhiDgfRH Follow Us: @inquirerdotnet on Twitter | Inquirerdotnet on Facebook', 6 May 2021. https://newsinfo.inquirer.net/1427860/duterte-on-ph-arbitral-win-over-china-papel-lang-yan-itatapon-ko-yan-sa-waste-basket

IPCC. 2019. 'Climate Change and Land: An IPCC Special Report on Climate Change, Desertification, Land Degradation, Sustainable Land Management, Food Security, and Greenhouse Gas Fluxes in Terrestrial Ecosystems'. Intergovernmental Panel on Climate Change (IPCC). https://www.ipcc.ch/srccl/chapter/summary-for-policymakers/

Jimenez, Angelo A. 2015. 'Philippines' Approaches to the South China Sea Disputes: International Arbitration and the Challenges of a Rule-Based Regime'. In

Territorial Disputes in the South China Sea: Navigating Rough Waters, 99–127. London: Palgrave Macmillan. https://doi.org/10.1057/9781137463685_7

Kavalski, Emilian, and Magdalena Zolkos. 2016. 'The Recognition of Nature in International Relations'. In *Recognition and Global Politics: Critical Encounters between State and World*, 139–56. Manchester: Manchester University Press. https://www.manchesteropenhive.com/view/9781526101037/9781526101037.xml

Kim, Claudia. 2020. 'Dugong vs. Rumsfeld: Social Movements and the Construction of Ecological Security'. *European Journal of International Relations* 27 (1): 258–80. https://doi.org/10.1177/1354066120950013

Kojima, Chie. 2015. 'South China Sea Arbitration and the Protection of the Marine Environment: Evolution of UNCLOS Part Xii Through Interpretation and the Duty to Cooperate'. In *Asian Yearbook of International Law*, 21: 166–80. Leiden: Brill. https://brill.com/view/book/edcoll/9789004344556/B9789004344556_010.xml

Lakoff, George. 1999. 'Metaphorical Thought in Foreign Policy: Why Strategic Framing Matters To the Global Interdependence Initiative'. UC Berkeley. https://escholarship.org/uc/item/4r82c6x9

Lakoff, George, and Mark Turner. 1989. *More Than Cool Reason: A Field Guide to Poetic Metaphor*. Chicago: University of Chicago Press.

Latour, Bruno. 2014. 'Agency at the Time of the Anthropocene'. *New Literary History* 45 (1): 1–18. https://doi.org/10.1353/nlh.2014.0003

'Law of the People's Republic of China on the Territorial Sea and the Contiguous Zone'. 1992. Order of the President of the People's Republic Of China. http://www.asianlii.org/cn/legis/cen/laws/lotprocottsatcz739/

Leep, Matthew. 2018. 'Stray Dogs, Post-Humanism and Cosmopolitan Belongingness: Interspecies Hospitality in Times of War'. *Millennium* 47 (1): 45–66. https://doi.org/10.1177/0305829818778365

Leira, Halvard, and Iver B. Neumann. 2017. 'Beastly Diplomacy'. *The Hague Journal of Diplomacy* 12 (4): 337–59. https://doi.org/10.1163/1871191X-12341355

Ling, L. H. M. 2014. *The Dao of World Politics: Towards a Post-Westphalian, Worldist International Relations*. New York: Routledge.

Lövbrand, Eva, Malin Mobjörk, and Rickard Söder. 2020. 'The Anthropocene and the Geopolitical Imagination: Re-Writing Earth as Political Space'. *Earth System Governance* 4: 1–8. https://doi.org/10.1016/j.esg.2020.100051

Lovelock, James. 2007. *The Revenge of Gaia: Why the Earth Is Fighting Back*. London: Penguin.

Lovich, Jeffrey E., Joshua R. Ennen, Mickey Agha, and J. Whitfield Gibbons. 2018. 'Where Have All the Turtles Gone, and Why Does It Matter?' *BioScience* 68 (10): 771–81. https://doi.org/10.1093/biosci/biy095

Marks, Michael P. 2011. *Metaphors in International Relations Theory*. New York: Palgrave Macmillan.

Mascariñas, Erwin. 2020. 'For Nesting Hawksbill Turtles, This Philippine Community Is a Sanctuary'. *Mongabay*, 31 March 2020. https://news.mongabay.com/2020/03/for-nesting-hawksbill-turtles-this-philippine-community-is-a-sanctuary/

Miyoshi, Masao. 2001. 'Turn to the Planet: Literature, Diversity, and Totality'. *Comparative Literature* 53 (4): 283–97. https://doi.org/10.2307/3593520

Neumann, Iver B. 2002. 'Returning Practice to the Linguistic Turn: The Case of Diplomacy'. *Millennium* 31 (3): 627–51. https://doi.org/10.1177/03058298020310031201

Onuf, Nicholas Greenwood. 2013. 'Fitting Methafors'. In *Making Sense, Making Worlds*, edited by Onuf, 40–50. New York: Routledge.

PCA, Permanent Court of Arbitration. 2016. 'Press Release: The South China Sea Arbitration (The Republic of the Philippines v. The People's Republic of China)'. *The Hague*. https://docs.pca-cpa.org/2016/07/PH-CN-20160712-Press-Release-No-11-English.pdf

Poonian, Christopher N. S., Reynante V. Ramilo, and Danica D. Lopez. 2016. 'Diversity, Habitat Distribution, and Indigenous Hunting of Marine Turtles in the Calamian Islands, Palawan, Republic of the Philippines'. *Journal of Asia-Pacific Biodiversity* 9 (1): 69–73. https://doi.org/10.1016/j.japb.2015.12.006

Rappler. 2014. 'China Demands 'immediate' Release of PH-Detained Fishermen', 8 May 2014. https://www.rappler.com/nation/china-demand-release-fishermen-ph

———. 2021. '5 Years on, Philippines Hails 'Legacy' of Hague Ruling', 23 June 2021. https://www.rappler.com/nation/philippines-hails-legacy-hague-ruling-2021

Robles Jr., Alfredo C. 2020. *Endangered Species and Fragile Ecosystems in the South China Sea: The Philippines v. China Arbitration*. Singapore: Palgrave Macmillan. https://doi.org/10.1007/978-981-13-9813-1

Sagun, Virgilio G. 2002. 'Updates on Marine Turtle Conservation in the Philippines'. 87–93. https://repository.kulib.kyoto-u.ac.jp/dspace/handle/2433/44165

Samuels, Marwyn. 2005. *Contest for the South China Sea*. Abingdon: Routledge.

Scott, David. 2012. 'Conflict Irresolution in the South China Sea'. *Asian Survey* 52 (6): 1019–42.

Severino, Rodolfo C. 2010. 'ASEAN and the South China Sea'. *Security Challenges* 6 (2): 37–47.

Simangan, Dahlia. 2020. 'Where Is the Anthropocene? IR in a New Geological Epoch'. *International Affairs* 96 (1): 211–24. https://doi.org/10.1093/ia/iiz248

Storey, Ian James. 1999. 'Creeping Assertiveness: China, the Philippines and the South China Sea Dispute'. *Contemporary Southeast Asia* 21 (1): 95–118.

Talmon, Stefan. 2016. 'The South China Sea Arbitration: Observations on the Award on Jurisdiction and Admissibility'. *Chinese Journal of International Law* 15 (2). https://doi.org/10.1093/chinesejil/jmw025.

Thayer, Carlyle A. 2012. 'ASEAN'S Code of Conduct in the South China Sea: A Litmus Test for Community-Building?' *The Asia-Pacific Journal* 10 (34). https://apjjf.org/2012/10/34/Carlyle-A.-Thayer/3813/article.html

The Guardian. 2016. 'Beijing Rejects Tribunal's Ruling in South China Sea Case', 12 July 2016. https://www.theguardian.com/world/2016/jul/12/philippines-wins-south-china-sea-case-against-china

The people's Republic of China. 2014. 'Position Paper of the Government of the People's Republic of China on the Matter of Jurisdiction in the South China Sea Arbitration Initiated by the Republic of the Philippines'. Position Paper. Ministry of Foreign Affairs of the People's Republic of China. https://www.fmprc.gov.cn/mfa_eng/zxxx_662805/t1217147.shtml

Time. 2014. 'The Philippines Wants the UN to Step in On Its Territorial Disputes with China', 31 March 2014. https://time.com/43362/philippines-china-un-intervention-south-china-sea/

Tuck, Eve, and K. Wayne Yang. 2012. 'Decolonisation Is Not a Metaphor'. *Decolonization: Indigeneity, Education & Society* 1 (1): 1–40.

Ullman, Richard H. 1983. 'Redefining Security'. *International Security* 8 (1): 129–53. https://doi.org/10.2307/2538489

Wall Street Journal. 2013. 'China Backs Tortoise in Race to Protect Endangered Species', 4 March 2013. https://www.wsj.com/articles/BL-CJB-17338

Wang, Zeng. 2015. 'Chinese Discourse on the 'Nine-Dashed Line''. *Asian Survey* 55 (3): 502–24.

Welzer, Harald. 2015. *Climate Wars: What People Will Be Killed For in the 21st Century*. Cambridge: Polity.

Yahuda, Michael. 2013. 'China's New Assertiveness in the South China Sea'. *Journal of Contemporary China* 22 (81): 446–59. https://doi.org/10.1080/10670564.2012.748964

Youatt, Rafi. 2014. 'Interspecies Relations, International Relations: Rethinking Anthropocentric Politics'. *Millennium* 43 (1): 207–23. https://doi.org/10.1177/0305829814536946

7 From Colonial Science to the Genome Age

The Politics of Asian Giant Salamander Conservation

Lisa Yoshikawa

Introduction

On 17 September 2019, Professor Lew's identity came under scrutiny. "The Prof.," who currently lives in London, arrived in the UK three years earlier in a cereal box with four others. The Border Force apprehended them in Coventry as they entered the country illegally; one was already dead at the scene. The Prof. was lucky and had survived the journey, was taken into custody, and given the highbrow nickname and a state-of-the-art abode in Regent's Park (ZSL 2019). For all intents and purposes, The Prof. was granted asylum, escaping the fate that her compatriots likely faced back home: extermination. By April 2019, The Prof. had become a celebrity, with the media reporting her saga and her life in her newfound home (BBC News 2019; ZSL 2021).[1]

The story changed five months later, when investigations revealed that Professor Lew might not be who Londoners thought she was. The Prof. might not be the namesake of David Armand, the 19th-century French missionary, but instead of the Marquess of Sligo, a 20th-century Irish peer. Even worse, and more likely, Professor Lew was a bastard. On 17 September, the Chinese giant salamander, hitherto considered to be one species, although suspected to be cryptic, was named to be two species, with the third designation pending. The same study reaffirmed the animals' dire state, due to habitat destruction and overhunting, and anthropogenic translocation that led to hybridization among the species taxa. Professor Lew (Figure 7.1), officially known until then as *Andrias davidianus*, was likely no longer so (Turvey *et al.* 2019).

Asian giant salamanders are the largest extant amphibians that today inhabit parts of south and southwestern China (*A. sligoi*, *A. davidianus*, and *A. jiangxiensis*) and central and western Japan (*A. japonicus*). They belong in the Cryptobranchidae family and are carnivorous apex predators in rocky, fast-flowing streams and rivers in the cooler temperate zone. They are fully aquatic and mature in about five (*A. davidianus*) to eight (*A. japonicus*) years, spawning in the summer months. Their strings of eggs are externally fertilized, often by multiple males, and brooded by the dominant male.

DOI: 10.4324/9781003212089-7

Figure 7.1 Professor Lew, ZSL, July 2022
(photograph by the author).

Asian salamanders grow to be from 150 centimeters (*A. japonicus*) to 180 centimeters (*A. sligoi*) and are estimated to live over a century (Browne *et al.* 2012a, 2012b; Chai *et al.* 2022).[2]

The Cryptobranchidae extend back to the mid-Jurassic period and likely originated in the region that became Asia, expanding into the rest of Eurasia and to North America. The genus *Andrias* emerged about 60–70 million years ago and was historically cosmopolitan. They now only exist in eastern Asia. Scientists date the split between the continental *Andrias* and those on the Japanese archipelago to 3–4 million years ago, when the two landmasses were reconnected and the East Sea (Sea of Japan) was a large freshwater lake (Browne *et al.* 2012a: 19–21; Turvey *et al.* 2019: 9–11).

In May 2018, scientists revealed that the continental *Andrias* was cryptic: multiple species that are morphologically indistinguishable but genetically distinct (Turvey *et al.* 2018; Yan *et al.* 2018). Seventeen months later, they identified three discrete clades: *A. davidianus* (Chinese giant salamander), *A. sligoi* (South China giant salamander), and an unnamed third (Turvey *et al.* 2019). Most recently in May 2022, researchers confirmed a fourth discrete species, *A. jiangxiensis*, discussed at the end of this chapter (Chai *et al.* 2022). Among those on the archipelago, scientists have identified two genetic varieties that are yet to be declared as distinct species (Matsui *et al.* 2008). The Cryptobranchidae in general have shown relatively low genetic variability as well as morphological stasis, and hence is often dubbed a "living fossil" (Browne *et al.* 2012: 20; Fujimoto and Kamishima 2004).

Today, these living fossils are endangered. The Chinese giant salamander is listed as Critically Endangered by the International Union for Conservation of Nature (IUCN), and the Japanese giant salamander as Near Threatened (IUCN 2019). Since its enforcement in 1975, the Convention on International Trade in Endangered Species of Wild Flora and Fauna (CITES) has listed the *Andrias* genus in Appendix I, prohibiting their international trade. This regulation became more meaningful in theory when the habitat nations Japan (1980) and People's Republic of China (PRC 1981) joined CITES. Domestically, PRC since 1988 has the creature listed as a Class II Protected Species, whose hunting or use requires regional authorization (Cai 2000: 117–138; Coggins 2003: 11). The government has also established 47 Nature Reserves in giant salamander habitats (Chai *et al.* 2022: 470). Japan had designated some of the habitats as protected Natural Monuments in the 1920s and the species itself in 1951. In the following year, the government upgraded its conservation status to Special Natural Monument, as one of the only 21 animal species designated today and the only amphibian. Any contact with these creatures requires the Agency for Cultural Affairs (Bunkachō) authorization.

Despite these conservation measures, the *Andrias* are declining in number. Recent investigations in China revealed that giant salamanders might be extirpated or functionally extinct in areas such as in Guizhou's Mayanghe reserve. Scientists often date the decline back to the 1950s and to anthropogenic factors: climate change, habitat degradation, and overharvesting for the luxury food and Traditional Chinese Medicine (TCM) markets. They also identify as culprits the farming industry that emerged to supply these markets, lax conservation law enforcement, and misguided conservation efforts. Professor Lew's case suggests that illegal trade adds to these problems. Since the discovery of multiple clades, scientists have highlighted anthropogenic hybridization, in addition to the general decline in populations, as exacerbating the crisis (Cunningham *et al.* 2016; Pan 2015; Pan *et al.* 2016; Pierson *et al.* 2015; Turvey *et al.* 2018; Wang *et al.* 2004; Yan *et al.* 2018).[3] The laws protecting *A. japonicus* (Figure 7.2) have been deemed more effective, but this species has not escaped climate change and habitat degradation. Certain populations of *A. japonicus* share the hybridization conundrum, not among the domestic varieties but with the continental species (Matsui 2005: 6–7).[4]

As seen in Professor Lew's case, the Chinese giant salamanders have received international media coverage for numerous interrelated reasons: the attention generated by the general Chinese world hegemony, the critical conservation status of the continental *Andrias*, the activities of an international research team, and more. The Japanese creatures remain under the global and even domestic radar by design, due to the controversial nature of the situation. The cross-border hybridization on the archipelago, like earlier cases involving other species, has become politicized to the point that summaries of the hybrid-taskforce meetings have been hidden from the public.

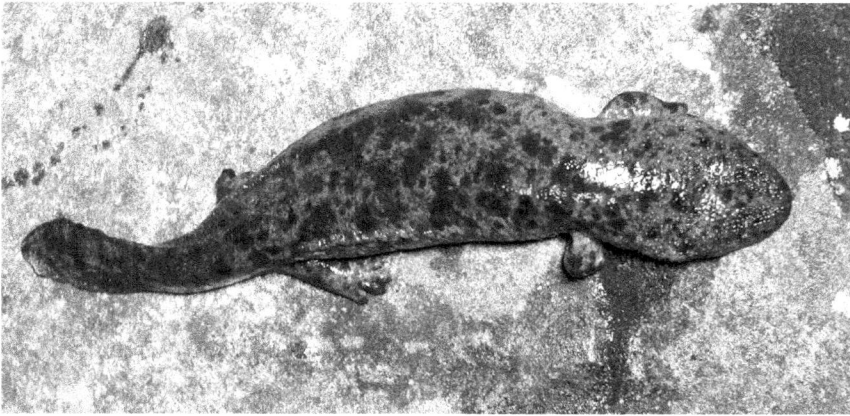

Figure 7.2 Japanese giant salamander, Okayama, August 2016
(photograph by author).

The 2001 discovery of hybridization between Taiwanese and Japanese macaques in Wakayama had led to euthanasia of 370 Taiwanese and hybrid primates. Public supporters were motivated by pest control or to protect the supposed purity of the beloved Japanese cultural icon; critics accused such sentiments as being fueled by xenophobic nationalism and fascism (Groening and Wolschke-Bulmahn 1992; Pauly 1996; Setoguchi 2003: 125). The hope in the giant salamander's case seems to be to settle the issue of hybridization quietly and quickly before internal and external opponents intervene (Kyōto-shi 2012a).

This chapter examines the conservation status of the Asian giant sala-manders from a historical perspective. It argues that the current exclusive finger-pointing at the Chinese government and people distorts our under-standing of this conservation crisis by obfuscating the context in which this impasse emerged. The *Andrias* conundrum illustrates the complexity of animal conservation today that results from a disjuncture between the latest science and long-standing politics of nationalism and imperialism. Animals discovered to modern science before the 20th century, like the giant sala-manders, were categorized initially based on early species concepts like morphology, anatomy, and geography. They were often done so in the con-text of contemporary geopolitics and became tied to modern nation states and empires: in identities through their binomial nomenclature and global association, in how they were studied, and in conservation policies. In the past few decades, scientists have turned to genetics to (re)categorize ani-mals, often challenging the earlier species concepts and resulting taxa. The current *Andrias* quandary represents scientific advancement outpacing and conflicting with political developments. Although the concept of the Anthropocene remains debated, the case for the Asian salamanders is clear. Their conservation crises are anthropogenic, and the more recent damages to these creatures and their habitats are only a fraction of the reasons behind this predicament that is much more fundamental in its making, entrenched, and extensive.

Colonial Science and the Creation of the Japanese and Chinese Salamanders

Samuel Turvey, a lead researcher who identified *A. sligoi*, writes, "[i]naccurate taxonomic assessment of threatened populations can hinder conservation prioritization and management" (Turvey *et al.* 2019: 1). Categorization of species taxa is crucial since it determines if and how creatures will be studied and protected. In the past, when the identity of a species became a source of colonial science competition or overlapped with entities such as newly emerging nation states, research and conservation often became politicized. The giant salamanders were discovered to science in the 19th century under such circumstances, when researchers were using territorial borders to claim animal endemism, albeit in conjunction with other units (Cooper 2007).[5] Great Powers were competing to dominate the globe, in part through asserting superior knowledge of fauna and flora in their imperialist target locales. Therein emerged the modern scientific study of the Asian giant salamander, in a manner that for decades privileged the Japanese specimens over the Chinese.

The first habitat associated with a creature's discovery to science often attracts significant attention from global researchers, particularly when the species is a megafauna endemic to limited locales. This was the case for the Asian giant salamanders, with the first European scholars encountering a live amphibian on the Japanese archipelago (Kobara 2000: 28; Siebold 1842: xv; 1969: 166; Ueno 1960: 315–316).[6] When the first live specimen was sent to Leiden, the founding director of the Rijksmuseum van Natuurlijke Historie (RMNH) Coenraad Jacob Temminck (1778–1858) named the creature *Triton japonicus* after the holotype's national origin (Matsui 2001: 76). A few years later, in 1838 when publishing in *Fauna Japonica*, his colleague Hermann Schlegel (1804–1884) used the name *Salamandra maxima*. Both cases linked the creature to the archipelago.

European zoos and museums thereafter competed to procure giant salamanders from Japan, and multiple studies on the creature emerged in the next three decades using those specimens, live and dead. This interest, dubbed a "zoological sensation," is often linked to George Cuvier's (1769–1832) fossil identification a decade prior of an extinct European cousin. Until this finding, the fossil was famous as *Homo diluvii testes*, the diluvialist Johann Jacob Scheuchzer's (1672–1733) evidence of human remains from the Great Flood (Kempe 2003: 129*ff*; Melvin 1969: 202–213). We see the association between these two creatures in contemporary minds, including that of Johann Jakob Tschudi (1818–1889). After observing the Japanese specimen daily at Leiden, Tschudi recommended *Megalobatrachus sieboldi* as its name and *Andrias scheuchzeri* as its extinct European cousin's name (Tschudi 1837).

Leading zoos acquired the creature from Japan by mid-century, often using their imperialist connections: Ménagerie du Jardin du Plantes in 1859, London Zoo in 1860, Berlin in 1862, Frankfurt in 1863, and Vienna in 1865. The Berlin specimen had a particularly symbolic origin, as a trophy of the Eulenburg Expedition that resulted in the unequal Prusso-Japanese treaty. Aquariums also added the salamander to their collection. In 1864, Hamburg Aquarium received one as a gift from the German merchant and diplomat Gustav Overbeck (1830–1894). Overbeck had captured one during his trip to Japan while stationed in Hong Kong (Lubach 1852; Wolf-Eberhard 2012: 6–10). Studies based on these specimens suggested numerous alternate species names, as seen in Tschudi's case (Hoeven et al. 1862: 1–7; Matsui 2001: 77). In 1900, Tokyo Imperial University (Tōdai) zoologist Ishikawa Chiyomatsu (1860–1935) counted seven different names, which he boasted as evidence of the creatures' popularity among European scientists (Ishikawa 1901).

The same could not be said about the continental giant salamanders, whose distinctness was dismissed in the turn-of-the-century colonial science competition. The Chinese specimen first known in Europe led French zoologist Émile Blanchard (1819–1900) to suggest that the creatures in Japan and China were different species (Blanchard 1871c, 1871b, 1871a: 616). Until this time, texts like *Fauna Japonica* had mentioned similar creatures found on the continent but had made little effort to differentiate them from those on the archipelago, echoing Chinese and Japanese *materia medica* texts (Hoeven 1866; Schlegel 1842; Siebold 1842: 135). Blanchard based his 1871 analysis on the skin specimen that Père Abbé Armand David (1826–1900) had sent to the Muséum de Paris following his second voyage around China, when David more famously encountered the giant panda (David 1871: 95; 1872: 119; 1949; Ienaga 2011: chapter 1; 2017: chapter 4; Songster 2018: chapter 1). Comparing David's acquisition, likely in Qinghai, and the two live specimens from Japan the Muséum held at this time, Blanchard declared that "the [morphological] comparison leaves no room for uncertainty" that the two were different species (Blanchard 1871c: 213; 1871b: 79).[7] He named the Chinese creature *Sieboldia davidiana*.

Yet, imperialist rivalry over influence in East Asia, including the knowledge of regional fauna, led Britain-based scientists to question this discovery. John Edward Gray (1800–1875) of the Royal Society of London refuted Blanchard based on a skin and bone specimen sent from Shanghai by Robert Swinhoe (1836–1877) to the British Museum (Gray 1873). At this time, the British were lagging in the competition due to their later and coastal arrival to Qing (Fan 2004: 18). The greater global zoological contest between the two powers had led the British to accuse the French of being species "splitters," and allegations of "species-monger[ing]" flew about (Ritvo 1997: 63–64). Concerning the great amphibian, the British won for the time being, and Blanchard's findings were generally dismissed over the next five decades. Well into the 1920s, most giant salamander researchers in western Europe and the United States deemed the issue inconclusive, or that the Asian giant salamander was one (Boulenger 1882: 81; Chapman 1893; David 1888: 463; Despax 1913a; Geerts 1883: 278–279; Nakanishi 1961: 185; Rein 1905: 263–264; Schlegel 1872; Stejneger 1907: 7).

Meanwhile, Japanese Meiji and Taishō giant salamander researchers strengthened the amphibian's primary link to the archipelago, with hopes to join the European academic circle by studying the famed creature (Kobayashi 1906: 16–17; Tago 1904: 43; 1929). Since early in its founding, researchers at Tōdai realized that proximity to habitats provided an edge on zoological research. The establishment of the Misaki Marine Research Station in 1886 was one such attempt to capitalize on western Pacific fauna in a popular field of marine zoology. Scholars such as Mitsukuri Kakichi (1858–1909) became a leading world expert on sea cucumbers and Iijima Isao (1861–1921) on sea sponges.[8] Access to specimens also attracted international scientists to Japan, for example, the American malacologist Edward Sylvester Morse (1838–1925), who became the first foreigner hired to teach zoology at Tōdai. He and his successor Charles O. Whitman (1842–1910), who later headed Woods Hole, encouraged their Japanese students to study endemic animals including the giant salamander (Isono 1988a: 95–96). The creature remained popular in Europe, and zoos continued to vie for specimens.[9] Fusing the continental and the archipelagic kinds allowed Japanese scientists to study specimens within reach and claim universal expertise. International attention would remain on Japan.

Meiji Japanese publications on the creature reveal these intents in their Eurocentric reference points, their emphasis on the researchers' access to amphibian habitats, and their dismissal of the Chinese kind as distinct. Sasaki Chūjirō (1857–1938) and Ishikawa Chiyomatsu were two scholars who became known for their giant salamander studies. In 1887, Sasaki published the first modern study on the species by a Japanese scientist. He accentuated his relatively large-scale observation of 71 specimens that he conducted for his Tōdai's biology department graduation research in Iga, Ise, and Yamato in 1880–1881, as he reported new findings, including how salamanders spawn rather than give live birth (Ishikawa 1899b: 452).[10] Whitman is said to have handed him this topic (Isono 1988b: 108–109; Ueno 1988: 38–39). Published in English, Sasaki's study introduced the species as *Cryptobranchus japonicus*, Van der Hoeven, choosing one out of several extant names that connected the creature directly with the archipelago. He began his discussion by suggesting that *Cryptobranchus japonicus* was of the same genus as *Homo diluvii testes*. Through the bulk of the article, Sasaki referred to *Fauna Japonica*, whose information he confirmed or corrected (Sasaki 1887).

Ishikawa Chiyomatsu's research on the giant salamander showed a similar pattern. Ishikawa's interest in embryology led him to the megafauna because their large eggs were ideal for his research. In his case, the accessible material turned out to be a jackpot.[11] Ishikawa's giant salamander studies cemented his position as the creature's world authority. When the amphibians in Artis Royal Zoo spawned in 1903, the institution contacted Ishikawa for advice on incubating them and rearing larvae, and a Japanese journal boasted of this feat (Anonymous 1904; Kerbert 1904).

The first to graduate from Tōdai's zoology department and also having studied in Freiburg, Ishikawa saw the study of the creature as a Japanese duty since giant salamanders were extinct in Europe (Ishikawa 1899b: 451–453; Ueno 1988: 39). Like Sasaki, Ishikawa emphasized his proximity to the research material as he described his difficulties during his 1898–1899 Mimasaka salamander egg-hunting trips, and that he succeeded only with help from the locals familiar with the terrain and animal behavior (Ishikawa 1899c).[12] This claim suggested that the excursion's success hinged on Ishikawa's unique position unrivaled by any European scholar: a Tōdai scientist fluent in Japanese and with family ties to the locality (Ishikawa 1899a).[13]

Ishikawa published his first full version of his initial research in a 1901 German language journal as part of his effort to secure his position in the European academe. Like Sasaki, he began by relating the archipelago's giant salamanders to Scheuchzer's fossil (Ishikawa 1901). Ishikawa also presented himself as a peer to his German colleagues and distanced himself from the Japanese public. For example, he mentioned that the habitat locals had served him cooked giant salamanders, and that the thought of Scheuchzer's fossil took his appetite away (Ishikawa 1904: 11). Such a narrative boasted the unique place Japanese scholars held in salamander research: close enough to the local community that would help them in their endeavors and invite them to meals, yet part of the civilized European academe and reluctant to eat scientifically significant animals dwindling in population.

Meiji-Taisho academics also linked the amphibian to the modern Imperial Japanese state, as a national symbol to foster domestic patriotism, and international recognition for its scientific and conservation endeavors. As early as 1873, natural historian and bureaucrat Tanaka Yoshio (1838–1916) included a giant salamander in the Japan pavilion exhibit at the Vienna Expo (Murata 2009: 93–94; Rein and von Roretz 1876: 34; Ueno 1973). Sasaki and Ishikawa published their findings in state organs: the former in Tōdai College of Science journal's first issue, and the latter in the Imperial Household Museum journal (Kobara 1985: 32–33).[14] Early-20th-century scientists, as members of Japan Society for Preserving Landscapes and Historic and Natural Monuments, highlighted the creature as a flagship species when they lobbied for a state-led preservation enterprise (Anonymous 1916). They quickly learnt that when convincing non-specialist political and financial supporters, tugging at their nationalist hearts was more effective than explaining the science. Zoologists, botanists, and geologists all argued that protecting nature was essential if Japan were to be considered civilized and become one of the Great Powers (Miyake 1918: 68–69; Miyoshi 1916: 91–92; Watase 1917: 123–125). Once the law was passed in 1919 to establish state-led Natural Monument preservation, scientists on the designation committee, like Watase Shōzaburō (1862–1929), chose several giant salamander habitats for protection. They were done so as "famous animals of Eastern Asia, although not the special productions of Japan," suggesting that those in China and Japan were the

160   Lisa Yoshikawa

same species (Kaburagi 1938: 110; Watase 1926: 37). Through this designa-
tion, the government officially connected the giant salamander to Imperial
Japan, and broadcasted this connection in foreign language publications
(Watase 1926: 37).

By this time, emphasizing the giant salamander's link to Japan was doubly
significant because the limelight was being recast on the continental creatures
through publication about a possible third Asian giant salamander species.
This time, wars and the ensuing domestic disorders stole its thunder. In
1924, London Zoo's Edward George Boulenger (1888–1946) introduced
Megalobatrachus sligoi, a recent procurement from Hong Kong. He claimed
that this species was different from *Sieboldia davidiana* and *Megalobatrachus
maximus*—which he deemed were identical (Boulenger 1924). Arthur de
Carle Sowerby (1885–1954) followed using specimens secured from Shanghai,
and another from Amoy (Sowerby 1925: 74; 1929). A later source claimed
that at an unspecified date, the Smithsonian curator and renowned American
Museum of Natural History (AMNH) herpetologist Leonhard Stejneger
(1851–1943) had reversed his 1925 conclusion that the Japanese and Chinese
kinds were the same (Liu 1950: 72).[15]

Research on the continental specimens accelerated when Chinese scholars
joined the field in the 1930s. Chinese scientists trailed their Japanese peers by
two to three decades in obtaining European or American training. The first
wave of returnees took teaching and research positions in the late 1910s, and
among them was Bing Zhi (1886–1965), a Boxer fellow who later headed the
Science Society of China's (SSC) Biological Institute in Nanjing (Jiang 2016:
160–167; Wang 2002: 311). The Chinese scientists continued to study taxon-
omy into the 1930s, and it was under Bing's tutelage that Mangven L.Y.
Chang (1902-?) and Hsi-fan Hsü (1912–1996) emerged as renowned herpetologists (Wang 2002: 300; Zhao and Adler 1993: 41). Chang went on to study
in Paris, receiving his doctorate in 1935; Hsü studied in Neuchâtel.

In the 1930s, researchers such as Chang began to situate the Chinese giant
salamander as a subspecies of the Japanese. In the two early-decade papers
published in the international version of the SSC's organ, Chang and Hsü
referred to the giant salamander as *Megalobatrachus japonicus* (Temminck)
and considered the continental and archipelago creatures as one kind (Chang
1933: 306–307; Chang and Hsü 1932: 138–141). By the mid-1930s, Chang
renamed the Chinese variety as *Megalobatrachus japonicus davidi* (Blanchard),
partly by revisiting R. Despax's inconclusive 1913 study that compared
continental and archipelago larvae morphologies (Chang 1935b, 1968: 8–11;
Despax 1913b).[16]

Chang attempted to bring the international academe's attention to his new
subspecies and connect it to China. Unlike Japanese scientists who promoted
their studies by linking the archipelago's creature to its extinct European
cousin, Chang did so by outlining the giant salamanders' history in Chinese
written sources. He was not alone in turning to traditional sources to express
patriotism, and declare national sovereignty and superiority: his teacher Bing
Zhi had done the same along with his other SSC colleagues such as geneticist

Chen Zhen (1894–1957) and botanist Hu Xiansu (1894–1968) (Jiang 2016). Japanese herpetologist Kaburagi Tokio (1890–1968) emulated this behavior belatedly in his 1938 article (Kaburagi 1938: 96–99). Chang introduced his new subspecies to the French reading audience by tracing Chinese knowledge of the creature to *Shanhaijing* (Classic of the Mountains and Rivers), which he dated to 600 BCE (Strassberg 2002: 130). He continued down history citing names likely more familiar to his audience, such as Confucius, ending with late Ming scholar Li Shizhen's *Bencao gangmu* (Compendium of Materia Medica).[17]

In the process, Chang lauded advanced early Chinese knowledge of the creature's habitats, behavior, morphology, or related myths. He also emphasized the historically sophisticated Song Chinese awareness of the creatures' rarity, or their attempt by then to dissect the creature to investigate its diet. In explaining these literatures, Chang added political associations that retroactively place the creature squarely in Republican China: Anhui, Hubei, Hunan, Shaanxi, Shanxi, and Sichuan. He likely did not imagine that some of these regions would soon be under Japanese occupation or that the Japanese would destroy the SSC biology lab amid the Nanjing Atrocities (Chang 1935b: 347–348, 1968: 8–11; Jiang 2016: 201).

Chinese declaration that the continental kind was a distinct species from the Japanese had to wait until 1950 (Liu 1950: 69–77). Herpetological research continued in China during the ensuing wars, but publication stalled. Liu Cheng-chao (1900–1976) headed a research center in Chengdu at Western China Union University after his Suzhou University lab had to evacuate inland following the Japanese invasion. Liu traced his academic lineage to the American biologist Alice M. Boring (1883–1955), who taught at Yenching University. He received his doctorate at Cornell in 1934. In a manuscript completed soon after the Asia-Pacific Wars, Liu, for the first time since Blanchard, unequivocally differentiated the Chinese and the Japanese giant salamanders and argued that the reported multiple Chinese kinds were one (Zhao and Adler 1993: 42–44). Like the early Japanese researchers, Liu claimed authority over the species based on proximity to the habitat in the Xikiang province, ironically resulting from the Japanese imperialist invasion. His conclusion was based on morphological comparison to preserved specimens in Chicago Natural History Museum.[18] He also invoked what we today call the evolutionary species concept, based on reproductive isolation due to geography (Liu 1950: 70–74). The 1968 republishing of Mangven L.Y. Chang's 1934 doctoral dissertation as "the most recent comprehensive treatment of the salamanders of China," with the exception of Liu's studies supported the earlier scientists' works (Chang 1968: iii; SSAR 2019). After Liu's death, his student Er-mi Zhao (1930–2016) continued to facilitate herpetological research, much of which became habitat surveys (Zhao and Adler 1993: 467–469).[19]

Japanese herpetologists meanwhile ignored, were oblivious to, or continued to question, Chinese findings that the continental and archipelago's giant salamanders differed (Iwama 1955: 360–362; Oyama 1919, 1924; Satō Ikio

1943: 346; Satō Kiyoaki 1971: 143; Tago 1931: 39–40). Under postwar Japan's Law for the Protection of Cultural Properties, the creature's preservation continued in a category similar to that in prewar years, as one "not particular to Japan but famously associated with Japan" (Shimizu 2016: 186). Only in the 1960s did Bunkachō begin to list the continental kind as a subspecies of that on the archipelago (Bunka hogo iinkai 1967; Bunkachō 1971: 48).

Categorizing Asian giant salamanders came to a head for the Japanese in the 1970s, when geopolitics crossed with entrepreneurship to make the issue into a practical one. In 1973, an Okayama businessman made headlines for importing 1.5 tons, or about eight hundred giant salamanders from China. Mr. Ōmori sought to capitalize on the China-boom that engulfed Japan following the bilateral normalization, and the culinary demand for giant salamanders in Japan that traced back to the interwar gourmand culture (Kinoshita 1973: chapter 6; Kitaōji 1935).[20] The scheme was to sell the creatures from China to luxury restaurants, to circumvent the prohibition to touch, never mind eat, the Special Natural Monuments of Japan. The plan proceeded until Bunkachō intervened, fearing that the morphological similarities between the continental and archipelago's giant salamanders would lead to consumption of the latter under false pretense. Mr. Ōmori was left with hundreds of unsold amphibians that were expensive to keep alive, and contemplated releasing them into the wild, except he feared that they might endanger the Japanese kinds.[21] Writing of this fiasco, herpetologist Ikoma Yoshihiro (1892–1979) reflected on Blanchard's and Boulenger's studies, and began to consider possibilities of competition between the two kinds, or hybridization that would "prevent preservation of pure [Japanese] blood" (Ikoma 1973: 465).

The 1970s Japanese voices differentiate between the Japanese and Chinese giant salamanders, although how and based on which species concept, or simply myopic nationalism, is unclear. Some Japanese aquariums and zoos continued to house giant salamanders from Japan and China together, with or without knowing individual organism's sex.[22] The Japanese voices also had little regard for the continental kind, except for Mr. Ōmori's reluctance to exterminate the unsold imports. This sentiment and the lack of accurate taxonomic assessment proved catastrophic for the very Japanese kind he wanted to protect, as newly affordable genetic testing at the turn of the century would reveal. The same technology would also reveal the dire state of the continental kind, whose welfare Japanese entrepreneurs like Mr. Ōmori were threatening, inadvertently or nonchalantly.

Conservation in the Genome Age

Recent DNA technology shifted the scientific debate on Asian giant salamanders and their conservation in three ways. First, once the phylogenetic species concept became dominant, scientists came to agree that at the very least two species of the amphibian existed: *Andrias davidianus* and *japonicus*. Second, low genetic variability and hybridization were identified as threats to conservation, in addition to the known issue of general population decline. The concept of biodiversity, if not the term, as a conservation issue had become commonplace by the 1970s, and its understanding became molecular in the genome age. Third, this technological revolution outpaced changes in politics, and institutions and mechanisms created through them. Nationalistic and imperialistic frameworks continued to limit research and conservation, and prejudices to obstruct paths to solutions.

Early genetic studies on the giant salamander differed between China and Japan in their research materials and hence their results, due largely to political circumstances. Kyōto University herpetologist Matsui Masafumi was the first to conduct population genetic studies of Asian giant salamanders.[23] In 1992, Matsui and his team tested DNA samples from 22 *A. japonicus* individuals, all from one river in Mie prefecture. At this time, DNA research required fresh tissue, which was difficult to obtain from an endangered Special Natural Monument. Material only became available when an illegal dumping of waste oil killed individuals; hence the locality of the samples was limited. The results suggested low genetic variability within the subjects, raising conservation concerns (Matsui and Hayashi 1992). This finding set the tone for a subsequent similar study on *A. japonicus*, which had to wait until 2008 for the continuing reason of material availability.

Cytogenetic studies date back to the 1980s for *A. davidianus*, partly because the research material was accessible: an earlier US-based study used specimen purchased in a wet market in Hong Kong (Sessions, Leon, and Kezer 1982: 342). Population genetics research for *A. davidianus* began in 2000, led by Robert W. Murphy and his Canada-based team, with the aid of Chengdu Institute of Biology's Er-mi Zhao. In stark contrast to the situation in Japan, these scientists collected, purchased, or were given nineteen *A. davidianus* live specimens from six localities representing all three of the major river systems: Yellow, Yangtze, and Pearl. They euthanized most and immediately dissected the creatures. With a small number, like Matsui's study, but more geographically diverse samples, Murphy confirmed the general lack of genetic variability in *Andrias*. Yet, his team also found within it substantial variation based on mitochondrial DNA (mDNA) sequences, and significant population sub-structuring. At this point, he did not suggest that these differences constituted separate species. More disturbing from the perspective of conservation, Murphy's research showed a lack of geographical correlation in the sub-structuring. This result suggested that anthropogenic movement of the creatures within the continent had led to genetic intermixing of formerly distinct local populations (Murphy *et al.* 2000).

Murphy's 2000 findings, along with a 2004 study by an East China Normal University team led by Wang Xiao-ming, would frame the research that followed in the next two decades. Wang's study was significant as the first broad-range look at China's giant salamander population, which also made available parts of earlier Chinese language studies for the English reading audience. His team looked at thirteen sites over five provinces and Chongqing municipality, and conducted population surveys and/or undertook questionnaires among giant salamander reserve management personnel and habitat villagers.[24] The former found wild creatures in only one locality in Hunan. The latter revealed continuing illegal hunting for the amphibians, sometimes using pesticides, to supply the luxury food trade. The authors documented a perceived historical decline since the 1950s in the number and size of animals being captured, and their increasing black-market value. They also cited anthropogenic habitat destruction since the 1950s as a culprit to the declining salamander population (Wang *et al.* 2004). A study published a few years later echoed the same two major threats as the most critical to Chinese amphibians in general, as did specific studies targeting *A. davidianus* in other locations such as Qinghai and Guizhou (Pan *et al.* 2016; Pierson *et al.* 2015; Xie *et al.* 2007).

Subsequent studies continued to place the beginning of the giant salamander conservation crises to the Maoist era, a convenient cloak that would obscure what is often perceived as a national embarrassment under imperial exploitation, behind the known domestic destructions of "revolutionary China" (Shapiro 2009). The research method and subjects predetermined this result to a certain extent. Murphy's latest genetic population study would frame the content. Giant salamander farming became a focal point in these discussions for two reasons. First, since the 1960s, captive breeding had been the most common government-led conservation method in China for any species labeled as endangered: Chinese alligators, pandas, south China tigers, and more (Coggins 2003: 81; Harris 2008: 16–17 and 82–86; Ienaga 2011: 103; Thorbjarnarson and Wang 2010: 4–5 and 196–223). The same was true for giant salamanders starting in the 1960s and 1970s in Sichuan, Hunan, Shaanxi, Guizhou, Henan, and Anhui provinces, although reproduction in captivity had only recently seen success. The Chinese government would release the amphibians into the wild, and their survival rate seemed promising based on experimentation (Zhang *et al.* 2016). Lack of coordination between the specimens' lineage locality and release sites, however, exacerbated the anthropogenic genetic mixing among the creatures (Wang *et al.* 2004: 200–201).

Some of these "returns" were often confiscated specimens from commercial salamander farms, the second and equally troubling development for the creature's conservation (Wang *et al.* 2004: 201). The PRC government had been advocating commercial farming of giant salamanders as a protein source for humans since the 1950s (Anonymous 1959). More recently the industry was lauded for helping to alleviate agricultural poverty in Shaanxi province.[25] Interviews conducted during a 2013 giant salamander local ecological knowledge (LEK) survey in Guizhou revealed some local hunting for the amphibian within Natural Reserves and knowledge of existing farms (Pan 2015: 5–6 and 12–16).[26] The region's non-Han informants, however, made interviewing, and hence LEK studies, difficult and likely skewed (Pan *et al.* 2016).

The conservation community began to focus on the commercial salamander industry in 2010, with the first extensive study published six years later.

Zoological Society of London's (ZSL) Andrew A. Cunningham and his team undertook this study based in Shaanxi. The industry is based on mostly culinary and some TCM demand for the creatures, whose local trade value had increased by more than tenfold between the late 1970s and the late 1980s. Farm numbers surged in the late 1990s, likely when the wild population became scarce (Cunningham *et al.* 2016: 267–269). The contemporaneous increase in the publication of giant salamander farming technique manuals supports this finding.[27] The market price of the creatures was USD 200 /kilogram in 2009 according to one source, and CNY 1,500–2,000 per individual in 2011 according to another (Cunningham *et al.* 2016: 271; Zhang *et al.* 2016: 13). The 2011 official records showed 124 licensed farms in Shaanxi housing a total of 2.6 million individuals. Salamander farming had become one of the main industries in Shaanxi by this time. Cunningham suspects that illegal farms outnumbered licensed ones (Cunningham *et al.* 2016: 267 and 270).

Although farming might have alleviated overexploitation of the wild population, it created additional problems of infectious disease and genetic mixing across habitat localities. The former was first reported in 2011, when a ranavirus infection led to mass farmed giant salamander deaths in Shaanxi province. The study suggested that an addition of newly collected animals to the farm environment led to the outbreak. Although this case was likely caused by a species-specific virus, ranavirus is seen as a potential threat to global amphibian biodiversity and incidents of cross-species infection have been reported (Geng *et al.* 2011). A later study has suggested a disease spreading when infected pig frogs were fed to farmed giant salamanders (Cunningham *et al.* 2016: 272). That two separate administrative offices, State Forestry Administration (frogs and toads) and Ministry of Agriculture (salamanders, newts, and caecilians), are responsible for various amphibians in China is problematic in this context (Xie *et al.* 2007: 275). Cunningham's 2016 study has found disease in nearly all of the Shaanxi farms surveyed. Farm wastewater is often discharged back to the wild without treatment, creating possible catastrophes (Cunningham *et al.* 2016: 271).

The exacerbation of cross-locality genetic contamination, suggested initially by Murphy's 2000 study, is traced back in this context to farms purchasing salamanders from outside localities for breeding purposes. Many farms had been unable to breed the second filial generation (F2) animals, and hence needed a steady supply of wild or first filial generation (F1) individuals to continue producing. An industry emerged to supply higher-value breeding individuals rather than lower-value creatures intended for restaurants. Some have dubbed this a pyramid scheme. This reality casts doubt on whether farming truly led to a decrease in poaching. When some of the resulting farm stocks are released into the wild as a conservation measure, as described above, the problem spread. There is no testing for diseases or lineage locality before release. Thus, scientists reason that PRC's conservation method is in reality doing more harm than good (Cunningham *et al.* 2016: 270 and 272). More recent news on the Chinese giant salamanders' conservation crises, discussed in the introduction, broke in this context.

Japan's study of giant salamander conservation in the past two decades took a different turn because of the aforementioned limited research material availability, the historically nationalistic conservation framework, and the international nature of the crises. While some scientists continue to propose solutions to habitat loss and similar general threats, much of the scholarship focuses on the decline in the number of Japanese giant salamanders due to their hybridization with the Chinese kind. Resulting suggestions to conserve the Japanese kind by eradicating both the hybrids and the *A. davidianus* from the archipelago, however, have failed to address the larger issue of protecting global biodiversity.[28]

The story began in 2005, when a DNA study found that giant salamanders matching those from China inhabited Kyoto's Kamo River.[29] Matsui, the lead researcher, thereby confirmed longstanding rumors that had suggested this possibility based on morphology. The result also suggested the limits of morphological identification, since it showed little pattern correlating species and morphology. None of the creatures from other habitats tested positive for Chinese-ness at this time.[30] Nor did Matsui have evidence for hybridization in Kamo River or knew whether cross breeding was possible, although he expressed concerns. He additionally indicated the limits of this study that relied on mDNA that only reveals maternal lineage. No detailed comparative genetic study between the Chinese and Japanese kinds even existed at this time. The original localities of the Chinese kinds found in Kamo River were also inconclusive, showing haplotypes from both the Pearl and Yangtze Rivers. Matsui speculated that they might have been among those imported to Japan after the normalization in 1972. Another concern was the disposal of the now captured and confirmed Chinese kinds in Kamo River, given that they were an exotic and internationally protected species (Matsui 2005: 1–7).[31]

Matsui and his teams followed up this result in two ways. One set of studies explored the genetic distinctions between *A. davidianus* and *A. japonicus*. A 2007 publication, in Japanese, compared nuclear DNA (nDNA) of giant salamanders from Mie prefecture rivers with *A. davidianus* specimen and *A. japonicus* from four other rivers. The results indicated significant genetic differences between the Chinese and the Japanese kinds, and among the Japanese kinds, between those from Gifu and Aichi prefectures to those from Mie (Matsui and Tominaga 2007). Matsui's 2008 publication, this time in English, produced similar results that differentiated the Chinese and Japanese kinds, as well as western and eastern clades in Japan. While claiming that "genetic distance should not be the absolute standard for determining taxonomical relationship," Matsui and his team acknowledged the "widely accepted view" that *A. davidianus* and *A. japonicus* were two different species (Matsui and Tominaga 2008: 323).

In the second and larger set of studies, Matsui and his team examined the possibility of hybridization in Japanese rivers. Most of these studies were initially published and presented orally in Japanese only, suggesting that Matsui and his team saw the issue of hybridization to be of limited local

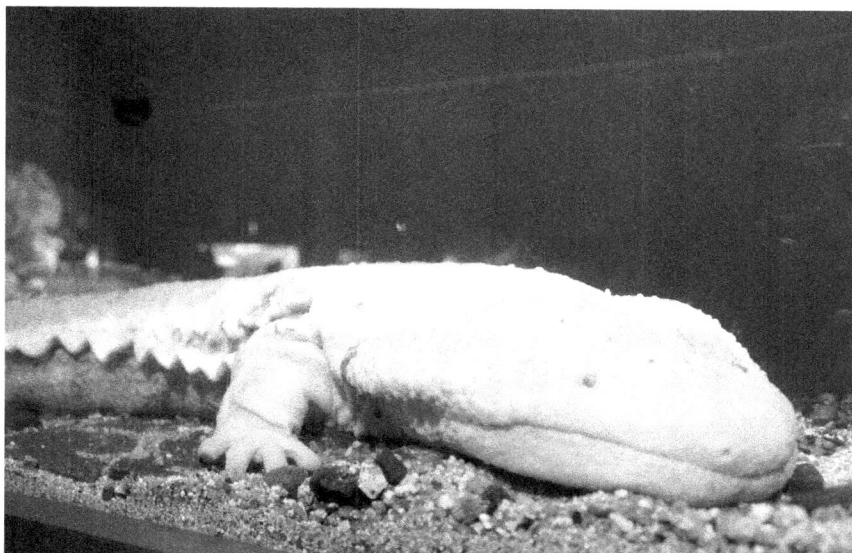

Figure 7.3 Chinese-Japanese hybrid salamander, Kyoto Aquarium, October 2015 (photograph by the author).

business. Early results of the first three-year study suggesting hybridization emerged in 2008, revealing that the majority of Kamo River specimens tested using mDNA and allozymes were hybrids (Figure 7.3). They also suggested that the F1 likely had backcrossed (BC) already (Matsui 2010). A similar investigation the next year confirmed this result, with now three-fourths of the giant salamanders tested as either Chinese or hybrids, suggesting that "they are threatening the survival of pure Japan-born" (Matsui 2011a). The final 2010 result, published the following year, used microsatellite gene analyses in addition to mDNA and allozymes, and came to the direst conclusion yet: in Kamo River "[g]enetic pollution of the Japanese species by alien Chinese species is at its maximum and the Japanese species is now immediately before extinction" (Matsui 2011b).[32] Of the 87 individuals tested, only 1 was deemed Japanese and none Chinese; 86 were deemed hybrids, with 25 F1s and 61 F2s or BCs (Matsui 2011b).

The findings led to at least three tangible actions. First, the hybrids were removed from Kamo River and placed in a swimming pool at Nihon Hanzaki Kenkyūjo (The Hanzaki Research Institute of Japan). The Institute, founded in 2005 by the ex-head of Himeji Aquarium Tochimoto Takeyoshi (1941–2019), is one of its kind in Japan dedicated to the study of giant salamanders. The facility is housed in an abandoned primary school in the mountains of Asago, Hyōgo prefecture (Nihon Hanzaki Kenkyūjo 2019). Second, Matsui and his team publicized the results to domestic and international media, as well as in academic conferences and other venues. Third, a countermeasure discussion committee was established with the cooperation of the Bunkachō, and the Kyoto city and prefecture. Their one immediate decision was to begin removal of all Chinese and hybrid creatures from the wild (Matsui 2012).

Gairaishu chūgoku san ōsanshōuo taisaku kentōkai (Alien Chinese Giant Salamander Countermeasure Discussion Committee, Committee hereon) first met in May 2011 and announced their six-year plan to investigate the giant salamander population in Kyoto's six drainage systems. Matsui headed the Committee, with Tochimoto as the deputy chair (Kyōto-shi 2011). The Kyoto City Official Website published select Committee meeting summaries (2011) and minutes (2012, 2015). Matsui's second three-year study coincided with the first three years of the Committee's existence, and the two groups shared findings (Matsui 2014a, 2017: 16–17).[33]

In the first year, the Committee conducted an "emergency investigation of the Special National Monument giant salamander," the result of which it published on the same website. For the first time, hybrids were found in the Katsura River drainage system, in additional to the already known Kamo system, albeit at a lower rate. In total, 65% of the tested individuals were hybrids or Chinese, and 20% Japanese, with the rest unidentifiable. The former was segregated again to Hyōgo and also to the Kyoto Aquarium; the Aquarium also harbored the *A. japonicus* to prevent future hybridization (Kyōto-shi 2012c).[34] In the same year, Matsui's study focused on Kamo River only, and found that 92% of the individuals were hybrids. He worried about exhausting space to segregate the hybrids, and an alternate measure to deal with the offspring of two protected species (Matsui 2013).

The Committee in 2012 began to discuss similar issues. The year's investigation of the two drainage systems found even higher hybrid and Chinese combined percentage than the previous year, now at 80%. For the first time, the report suggested concern that *A. japonicus* was being outcompeted in their struggle for existence (Kyōto-shi 2012a). The meeting minutes summary published on the Kyoto city website revealed two problems it faced, and the nationalistic framework under which the members understood conservation. First was the removal of hybrids and *A. davidianus* from the Rivers, for which committee members echoed Matsui's concern about availability of housing. They concluded that both could be euthanized under Japanese law, a necessity "to preserve biodiversity." They affirmed that "euthanasia as a means to solve an anthropogenic hybridization problem was inevitable as an anthropogenic final solution" (Kyōto-shi 2012a). The committee report did not differentiate between the hybrids and the Chinese salamanders when suggesting removal of non-Japanese amphibians; that the disposal of the endangered Chinese kind, at this time still known as one species, also threatened biodiversity was not mentioned. It instead discussed what the massive removal of the apex predator would do to the Kamo River ecosystem, and how to accomplish this removal in the first place due to the likelihood of higher undetected hybrid population (Kyōto-shi 2012a, 2012b).

The report also showed concern about public relations. Rumors about the hybrids had been spreading via news media since 2007, even before scientific confirmation.[35] How to convince the public that slaughter was necessary was key, and the committee considered using the euphemism, "increasing the [number of specimen] sample" because the term "euthanize" may "develop a life of its own." They were likely aware of the earlier Taiwan hybrid macaque case that led to massive public outcry. As in the macaque case, the cost of keeping hybrids and *A. davidianus* under captivity was high with at an estimated 100,000 yen a month to feed, with an anticipated lifespan of over a century.[36] The difference with the macaque case was that the Chinese kind was critically endangered. Whether the committee feared backlash against threatening biodiversity for suggesting to euthanize *A. davidianus*, or more generally for taking any life including those of the hybrids, is unclear.

The second issue was repopulation of the rivers with *A. japonicus* once the hybrids and the Chinese kinds were removed. Politics was evident in the committee's conundrum: it struggled to define Japanese-ness and Chinese-ness among giant salamanders. Genetic results suggested that some F1 backcrossed with the Japanese kind, and the offspring were "extremely close to being Japanese." The committee saw a need to define what should be removed, since it was unsure whether the handful that they had deemed purely Japanese could repopulate the rivers. They tentatively decided to retain those that showed genetic characteristics that suggest backcrossing with the *A. japonicus*. In the Committee's eyes, exigency necessitated foregoing the one-drop-rule (Kyōto-shi 2012a).

Matsui, who continued with his team's three-year research and was now investigating four Kyoto drainage systems, concurred with the importance of delineating the hybrids and *A. japonicus*. For one, Bunkachō under the Ministry of Education was responsible for the Japanese-identified individuals and the Environmental Ministry for the alien ones including the hybrids and the Chinese. The latter jurisdiction was based on Japan's 2004 Invasive Alien Species Act (Matsui 2014a). Such political siloing at best obstructed holistic problem-solving of this regional biodiversity crisis. By the end of the study in 2013, given that *A. japonicus* had not been found in the Kamo River system since the 2011–2012 investigation, Matsui feared their extirpation (Matsui 2014b).

The Kyoto Committee continued to investigate and meet annually, but after 2014 stopped publishing their reports online. The official reasoning was to "protect information about the habitat and spawning locations" and to prevent capturing of *A. davidianus* and hybrids and their release into other rivers (Kyōto-shi 2013, 2014, 2016, 2018).[37] Online advertisements had been found selling hybrids, which was not illegal. This real fear also enabled concealing discussions around *A. davidianus* and the hybrids' fates in Japan. The anomaly was in 2015, when the minutes summary published the total results since 2011, and concluded that hybridization was spreading in the Katsura river system. The summary also reported the number and whereabouts of the captives as of February 2015: of the 278 captured, most were hybrids housed in Hanzaki Research Institute of Japan and Kyoto Aquarium. Fourteen *A. davidianus* were also kept at those locations. The hybrids were also being handed to "research institutions … for scholarly uses" (Kyōto-shi 2015).

By this time, the hybridization problem had been found elsewhere in Japan. In 2013, investigation revealed that nearly 60% of giant salamanders in Mie prefecture's Taki River system were hybrids (Shimizu 2016).[38] Four years later, F2 hybrids were found in Okayama prefecture's Yoshii River, and more in 2019, and in late 2020, hybrids were found in the Asahi River.[39] In 2017, Matsui contemplated how to "protect the pure-Japanese born only…there would be criticism against exterminating an internationally protected animal." He was referring to *A. davidianus*, which he labeled "the root of all evils." He also realized that CITES listed the *Andrias* genus, which meant that the hybrids were also internationally protected (CITES 2017: 44). Evidence "clearly indicates that pure Japanese species is completely inferior to pure Chinese ones or hybrids of the two species," he lamented, due to outcompeting for resources and/or heterosis (Matsui 2017: 13). He recommended that the Japanese government should designate the Chinese and hybrids as Special Invasive Species, which would allow removal legally and ensure funding to do so. Only then would the population of *A. japonicus*, which Matsui continued to insist was of one kind based on close genetic distance, recover (Matsui 2017: 18–19).

Conclusion

All researchers and conservationists agree that the *Andrias* remain endangered today due to anthropogenic causes. Most are well intentioned in attempting to identify causes as a means to find solutions, but stop short of looking beyond immediate reasons. Samuel Turvey and his team's 2019 research, the first mentioned in this chapter, began in this way but came to new discoveries as the investigation progressed. They initially identified the 1970s as when exploitation of the Chinese giant salamanders began, including human-mediated translocation. Turvey hence turned to older museum specimens to assess geographically distinct clades among the continental creatures. What emerged was his identification of at least three different species and the dearth of museum samples for the Chinese amphibian (Turvey *et al.* 2019). The 2018 studies that identified the possibility of multiple continental giant salamander clades counted five to eight; Turvey found three (Turvey *et al.* 2018; Yan *et al.* 2018). He links this shortcoming to material availability, but then falls short of investigating the causality of this problem, which was the creatures' early research history that saw the *Andrias* as one and prioritized the Japanese specimens for geopolitical reasons. A quick search of six European and US institutions known to have historically researched the genus reveals that almost 70% of the collections are *A. japonicus* compared to only 27% *A. davidianus*.[40] A *Vertnet* database search for *Andrias* specimens housed in participating institutions, mostly in North America, showed a slightly less skewed ratio of about two *japonicus* to one *davidianus*.[41] Nonetheless the correlation preliminarily stands, and hence the significance of the imperialists' role in this conservation crises.

Second is the more specific role the Japanese researchers and officials have played in this biodiversity crisis. Under Imperial Japan, scientists were at minimum instrumental in prioritizing their native giant salamander population for international research purposes, from personal and professional to nationalistic and imperialistic reasons. Japanese policy makers, following the latter motivation, not only capitalized on this behavior, but also condoned hindering of research on the continental creatures. Allowing the destruction of the Nanjing laboratory, for example, and continuing with the greater Japanese imperialism and colonialism crippled Chinese research for decades. European and US colonial science competition also left the Chinese researchers and amphibians on the sidelines. To point, findings by Blanchard in 1871, Boulanger in 1924, and Chang in the 1930s, muffled by their contemporaries for these reasons, were later vindicated by genetic studies. The excellent research that the SSC and other scientists, and later their students, conducted despite these obstructions is noteworthy.

More recently, Japanese researchers and officials continued to be bound by the geopolitical framework in which they work. Cooperation between continental and archipelago's *Andrias* researchers has only recently started, led on the Japanese side by the younger generation of herpetologists, such as Okada Sumio and Taguchi Yūki.[42] All of the Japanese researchers and officials I have consulted in the last several years are well intentioned and dedicated to conserving biodiversity. Many are frustrated by the limited funding, cumbersome bureaucratic red tape, and domestic and international political frameworks that they see as shackles to their goal. Some struggle with ethics surrounding the meaning of life—including those of the hybrids—that humans created and are now threatening global biodiversity.[43] Researchers outside the giant salamander habitat countries, for example those based in London, have gathered international teams to produce fruitful results. Such successes may serve as a form of restitution for past injustices, but are difficult for Japanese researchers and institutions to replicate.

Current policymakers in both China and Japan are crucial in determining the greater framework in which their domestic *Andrias* crises must be addressed, and both fall far short of being exemplary. Condemnations of PRC's general conservation policies abound and need little repeating: from the immediate problems in regulating habitat destruction and enforcing conservation laws, to the greater issue of creating holistic policies to conserve wild populations and incorporating non-Han actors and voices into ongoing efforts. On the archipelago, the central government obstructs science by drawing politically driven jurisdictions over animals, and underfunding and understaffing conservation efforts.

In the end, however, such a framework based on 19th-century geopolitics must be questioned when discussing global biodiversity, species taxa, and endemism. Species taxa are constantly changing due to speciation, extinction, or new discoveries, and naming and categorizing animals based on the geopolitical unit that represents one point in time and holds political baggage obstructs true conservation of biodiversity. In a similar vein, endemism defined using the same geopolitical unit is problematic. Speciation is a process that occurs over time and is often only identified in retrospect. Extinction is sometimes immediately visible, but makes the species taxa issue moot for the particular conservation purposes since it no longer exists. New species discoveries, which we have been seeing through the newly available molecular technology, are immediate and remain relevant. They force us to confront the question of our historical conceptualizations of species taxa and endemism, and how they have impacted scientific research and conservation efforts. The recent discovery of the cryptic species among the continental giant salamanders, or the yet unresolved multiple clades that were found on the archipelago, fall into this category that speaks to the greater difficulties of biodiversity conservation today. Science allows us to learn of Professor Lew's genetic identity. We must continue to wrestle with what the answer means, and how we act on it, because from where the Asian salamanders swim, anthropogenic harm has been real. On the larger concept of Anthropocene, the jury remains out; the *Andrias* crisis, like the proposed epoch that seemingly and unjustly indicts all of humanity, primarily implicates inhabitants of industrialized nation states and empires.

Addendum: In May 2022, a team of scientists identified *Andrias jiangxiensis*, a new species of giant salamander found in the Jiulingshan National Nature Reserve in Jiangxi province, PRC. The team headed by Jing Chai of the Kunming Institute of Zoology deemed, based on an 18-month field monitoring and experimentation, that the species had "genetic distinctiveness and do not detect genetic admixture with other [Chinese] species" (Chai *et al.* 2022: 469). Although this new finding brings some hope to the conservation of Chinese giant salamanders, the team still concluded that this new species faced "a great extinction threat" and suggested that they be categorized in the same IUCN (Critically Endangered) and PRC (Category II) conservation status as the other Chinese salamanders (Chai *et al.* 2022: 477). The team also confirmed the argument made in this chapter: there has not yet been a "genetically pure" population of live *A. davidianus* or *A. sligoi* found, and the team relied on museum specimens for comparative study among Chinese giant salamanders (Chai *et al.* 2022: 478). Nineteenth- and twentieth-century imperialist science politics and competition continue to haunt us.

Notes

1 Professor Lew's sex was identified as female in a recent ZSL video.
2 The likely lifespan information comes from *A. japonicus* researchers I have interviewed.
3 Survey locations include Anhui, Hunan, Chongqing city, Sichuan, Shaanxi, Henan, Qinghai, Guizhou.
4 A 2007 investigation confirmed this finding. "Ōsanshōuo majiwaru kiki," *Asahi*, February 17, 2007.
5 For more on endemism, see Anderson 1994; Crother and Murray 2011; Doutt 1961; Frank and McCoy 1990; Noguera-Urbano 2016; Peterson and Watson 1998. The term "indigenous" is also ambiguous and problematic.

6 Siebold 1969 is a reprint of second edition, first published in 1897.
7 « La comparaison ne laisse guère subsister d'incertitude » in the original.
8 I discuss this is my book in progress, *The Empire's Menagerie: Mapping Animals in Imperial Japan*.
9 See for example an *Asahi* article: Anonymous and no title, *Asahi*, February 10, 1882, Osaka morning edition.
10 Today's Wakayama and Nara prefectures.
11 For more on the relationship between accessible material and scientific development see, for example, Jiang 2016 and Onaga 2010.
12 For *Tōyō Gakugei Zasshi*, see Takada 1996.
13 Today's Okayama.
14 Tōkyō teishitsu hakubutsu-kan chōsa hōkoku.
15 I have been unable to verify this claim based on an uncited and unpublished manuscript.
16 Chang 1968 was originally published in 1936. See also Chang 1935a and 1935c.
17 The names he identifies with giant salamanders include 鯢魚 (C: niyu, J: geigyo); 鰆魚 (C: teyu, J: teigyo); 人魚 (C: renyu, J: ningyo).
18 Xikang is currently part of Tibet and Sichuan.
19 Examples of such articles include Song and Fang 1982; Shen 1983; Pan, Liu, and Qian 1985; Su 1986; Zhang 1987; Song 1987.
20 Kinoshita 1973 originally published in 1925.
21 "Nageki no chūgoku ōsanshōuo," in *Asahi*, September 28, 1973, Tottori edition.
22 *A. davidianus* and *A. japonicus* were co-housed for example in Ueno zoo and Inokashira park. "Ōsanshōuo jōhatsu: Tōnan? Dassō? Fukamaru nazo," *Asahi*, November 26, 1977; "Ōsanshōuo mata jōhatsu: Nusumareta? Kondo ha Inokashira kōen kara," *Asahi*, November 28, 1977, evening edition; "Tabete minsai sōdō," *Mainichi*, January 11, 1973, evening edition.
23 Life dates of active researchers have been omitted.
24 The provinces include Anhui, Hunan, Sichuan, Shaanxi, and Henan.
25 Jiang Feng, "Dani yangzhi zhu Shaanxi nongmin tuopin zhi fu," *Renmin ribao*, March 19, 2012.
26 These reserves were the Fanjingshan and Leigongshan natural reserves, considered to be important for giant salamander protection. This study suggested, based on low sighting, that the creatures are likely functionally extirpated in Mayanghe Natural Reserve. None were initially established for the purpose of protecting *A. davidianus*.
27 See for example Jin and Wang 1997.
28 See, for example, Taguchi and Natsuhara 2009.
29 Kamo River (賀茂川), not to be confused with Kamo River (鴨川).
30 These habitats included Katsura and Hatake Rivers in Kyōto, Minō and Akuta Rivers in Ōsaka, Ichi River drainage system in Hyōgo, Taki and Kawakami Rivers in Mie, and Kiso River in Gifu.
31 A 2007 investigation confirmed this finding. "Ōsanshōuo majiwaru kiki," *Asahi*, February 17, 2007.
32 From the original English abstract (only) language.
33 For the development of microsatellite markers, see Yoshikawa *et al.* 2011; Yoshikawa *et al.* 2012.
34 The Kamo (鴨) system here include: Kamo (賀茂), Anba, and Takano Rivers. The Katsura system include: Kami-Katsura and Kiyotaki Rivers, and downstream beyond Kugabashi.
35 "Ōsanshōuo majiwaru kiki," *Asahi*, February 17, 2007; "A! Ōsanshōuo, kawa ni modosu no matta," *Asahi*, July 3, 2009, Osaka edition; "Ōsanshōuo koyūshu 'gekihen'," *Asahi*, October 25, 2010, Osaka edition.
36 "Sanshōuo ikiba nashi," *Asahi*, October 27, 2012, Osaka evening edition.
37 The committee met in 2017 but I am unable to locate the acknowledgment upload.

38   "Yakkaimono o kyōzai ni," *Mainichi*, October 27, 2015, Mie edition.
39   "Kuni shitei no tokubetsu tennen kinenbutsu ga chūgoku shu ni: Ōsanshōuo no kōzasshu hakken Okayama," *Sankei nyūsu*, January 27, 2018; "Ōsanshōuo kōzasshu hakken zairaishu hogo ni chikara Okayama," *Mainichi*, January 25, 2018; "Kagamino ni ōsanshōuo kōzasshu Chūgoku chihō hatsukakunin zairaishu no zetsumetsumo," *San'yō shinbun digital*, January 24, 2018; "Okayama ken kita de aratani kōzasshu no ōsanshōuo Kagaminochō no hokaku chōsa de nihiki kakunin," *San'yō shibun digital*, March 9, 2019; "Ōsanshōuo kōzasshu o kakunin, Maniwa de hatsu, zairaishu hogo e hokaku hōshin," *San'yō shinbun digital*, March 2, 2021.
40   The institutions include New York's AMNH (n = 14, j = 10, d = 3, a = 1), Washington DC's National Museum of Natural History (n = 28, j = 17, d = 10, a = 1), Chicago's Field Museum (n = 18, j = 10, d = 6, a = 2), Leiden's Naturalis Biodiversity Center (formerly RMNH; n = 21, j = 20, d = 1), Paris Muséum national d'histoire naturelle (n = 15, j = 11, d = 4), and London's Natural History Museum (n = 13, j = 8, d = 5). n = total recorded; j = *A. japonicus*; d = *A. davidianus*; a = unspecified *Andrias*. The databases were accessed online on November 3, 2019.
41   VertNet is an NSF-funded online database that stores specimen collection and other biodiversity data. It absorbed HerpNet in 2015; HerpNet gathered herpetological specimen collection data from mainly North American universities and museums.
42   Okada has co-authored several articles with an international team of herpetologists including those from China; these include Browne *et al.* 2012b; Turvey 2018. Taguchi, in a 2016 conversation with the author, has recalled his trip to visit Chinese Giant Salamander farms.
43   Experts that I have met for this project include the following. I thank them for sharing their time, insights, and thoughts with me: Fukutomi Masaya, Edo Kaneaki, Tanaka Atsushi, Itō Akihiro, Kuwabara Kazushi, Ōta Noboru, Tochimoto Takeyoshi, Taguchi Yūki, and Takahashi Mizuki. I also thank the following institutions and their staff: Nihon Hanzaki Kenkyūjo (Hyōgo), Asa Zoo (Hiroshima), Nihon Ōsanshōuo Sentā (Mie), City of Maniwa (Okayama), Mizuho Hanzake Kinenkan (Shimane), Nabarishi Kyōdo Shiryōkan (Mie), City of Kyoto, and Japan Agency for Cultural Affairs (Tokyo). All errors and shortcomings are mine.

Bibliography

All books published in Japan are published in Tokyo unless otherwise stated. Chinese and Japanese author names are left in their traditional order when the publications are in those languages. Unless otherwise noted, all websites were last accessed on June 25, 2021.

Newspapers Cited

Asahi shinbun
Asahi shinbun, Osaka edition
Asahi shinbun, Tottori edition
Mainichi shinbun
Mainichi shinbun, Mie edition
Renmin ribao
Sankei nyūsu
San'yō shinbun digital

Other Sources

American Museum of Natural History Vertebrate Zoology Database; accessed November 3, 2019. https://emu-prod.amnh.org/db/emuwebamnh/index.php

Anderson, Sydney. 1994. "Area and Endemism." *The Quarterly Review of Biology* 69, no. 4 (December): 451–471.

Anonymous. 1904. "Geigyo Oranda koku nite sanran su." *Tōyō gakugei zasshi* 273 (June): 291.

Anonymous. 1916. "Kaichō bettei Ōiso kōaien shiiku shinshū Tottori Tōya-san no sanshōuo shinchō yonshaku amari." *Shiseki meishō tennen kinenbutsu* 1:9 (January): cover page.

Anonymous. 1959. "Guizhou sheng siyang wawayu." *Zhongguo shuichan*: 23.

BBC News. 2019. "London Zoo unveils smuggled salamander found in Coventry." Dated April 3, 2019. https://www.bbc.com/news/uk-england-47794220

Blanchard, Émile. 1871a. *Les récentes explorations de la Chine*. Publication place unknown.

Blanchard, Émile. 1871b. "Note sur une nouvelle Salamandre gigantesque (Sieboldia Davidiana Blanch.) de la Chine occidentale." *Comptes Rendus* 73: 79–80.

Blanchard, Émile. 1871c. "On a new gigantic salamander (Sieboldia davidiana) from Western China." *The Annals and Magazine of Natural History* 4th series, 8: 212–214.

Boulenger, E. G. 1924. "On a new Giant Salamander, living in the Society's Gardens." *Proceedings of the Zoological Society of London* 94, no. 1: 173–174.

Boulenger, George Albert. 1882. *Catalogue of the Batrachia Gradientia S. Caudata and Batrachia Apoda in the Collection of the British Museum*. London: British Museum.

Browne, Robert K., Hong Li, Zhenghuan Wang, Paul M. Hime, Amy McMillan, Minyao Wu, Raul Diaz, Zhang Hongxing, Dale McGinnity, and Jeffrey T. Briggler. 2012a. "The giant salamanders (Cryptobranchidae): Part A. paleontology, phylogeny, genetics, and morphology." *Amphibian and Reptile Conservation* 5:4 (September): 17–29.

Browne, Robert K., Hong Li, Zhenghuan Wang, Sumio Okada, Paul Hime, Amy McMillan, Minyao Wu, Raul Diaz, Dale McGinnity, and Jeffrey T. Briggler. 2012b. "The giant salamanders (Cryptobranchidae): Part B. Biogeography, ecology and reproduction." *Amphibian and Reptile Conservation* 5:4 (September): 30–50.

Bunkachō ed. 1971. *Tennen kinenbutsu jiten*. Daiichi hōki shuppan.

Bunkazai hogo iinkai ed. 1967. *Tokubetsu shiseki meishō tennen kinenbutsu zuroku*. Daiichi hōki shuppan kabushiki gaisha.

Cai, Shouqiu ed. 2000. *Huanjing ziyuan faxue jiaocheng*. Wuhan: Wuhan daxue chubanshe.

Chai, Jing, Chen-Qi Lu, Mu-Rong Yi, Nian-Hua Dai, Xiao-Dong Weng, Ming-Xiao Di, Yong Peng, Yong Tang, Qing-Hua Shan, Kai Wang, Huan-Zhang Liu, Hai-Peng Zhao, Jie-Qiong Jin, Ru-Jun Cao, Ping Lu, Lai-Chun Luo, Robert W. Murphy, Ya-Ping Zhang, and Jing Che. 2022. "Discovery of a wild, genetically pure Chinese giant salamander creates new conservation opportunities." *Zoological Research* 43, no. 3: 469–480.

Chang, L. Y. 1933. "On the Salamanders of Chekiang." *Contributions from the Biological Laboratory of the Science Society of China. Zoological series* 9, no. 8: 305–328.

Chang, L. Y. 1935a. "Sur les larves de quatre espèces de salamandres de Chine." *Bulletin du Muséum national d'histoire naturelle* series 2, no. 7: 172–175.

Chang, L. Y. 1935b. "Sur la salamander géante de la Chine." *Bulletin de la société zoologique de France* 60, no. 1 (April): 347–353.

Chang, M. L. Y. 1935c. "Note préliminaire sur la classification des salamandres d'Asie orientale." *Bulletin de la société zoologique de France* 12 (November): 424–427.

Chang, L. Y. 1968. *Amphibiens Urodèles de la Chine: With a new list of Chinese salamanders by Arden H. Brame, Jr.* Ann Arbor: Society for the Study of Amphibians and Reptiles.

Chang, L. Y. and Hsü, H. F. 1932. "Study of some Amphibians from Szechuan." *Contributions from the Biological Laboratory of the Science Society of China. Zoological series* 8, no. 5: 137–178.

Chapman, Henry C. 1893. "Observations on the Japanese salamander, Cryptobranchus maximus (Schlegel)." *Proceedings of the Academy of Natural Sciences of Philadelphia* 45: 227–233.

Chicago Field Museum Zoological Collections Database; accessed November 3, 2019. https://collections-zoology.fieldmuseum.org

CITES. 2017. "Appendix I, II, & III (04/04/2017)." Dated October 4, 2017. https://cites.org/sites/default/files/eng/app/2017/E-Appendices-2017-10-04.pdf

Coggins, Chris. 2003. *The Tiger and the Pangolin: Nature, Culture, and Conservation in China*. Honolulu: The University of Hawai'i Press.

Cooper, Alix. 2007. *Inventing the Indigenous: Local Knowledge and Natural History in Early Modern Europe*. Cambridge: Cambridge University Press.

Crother, Brian I. and Christopher M. Murray. 2011. "Ontology of areas of endemism." *Journal of Biogeography* 38: 1009–1015.

Cunningham, Andrew A., Samuel T. Turvey, Feng Zhou, Helen M. R. Meredith, Wei Guan, Xinglian Liu, Changming Sun, Zhongqian Wang, and Miyao Wu. 2016. "Development of the Chinese giant salamander *Andrias davidianus* farming industry in Shaanxi Province, China: Conservation threats and opportunities." *Oryx* 50, no. 2: 265–273.

David, Armand. 1949. *Abbé David's Diary*. Cambridge: Harvard University Press.

David, Armand. 1871. "Rapport adressé a MM. Les professeurs-administrateurs du muséum d'histoire naturelle." *Nouvelles archives du Muséum d'histoire naturelle* 7: 75–100.

David, Armand. 1872. "Journal d'un voyage dans le centre de la Chine et dans le Thibet oriental." *Nouvelles archives du Muséum d'histoire naturelle* 8: 117–119.

David, Armand. 1888. "La faune chinoise." In *Congrès scientifique international des catholiques*, 451–467. Paris: Bureaux des *Annales de philosophie chrètienne*.

Despax, R. 1913a. " Sur le genre Megalobatrachus en Chine." *Bulletin de la sociéte zoologique de France* 38, no. 1 (April): 134–136.

Despax, R. 1913b. "Sur une Larve de *Megalobatrachus* Tschud de provenance chinoise." *Bulletin du Muséum national d'histoire naturelle* 19: 183–184.

Doutt, Richard L. 1961. "The dimensions of endemism." *Annals of the Entomological Society of America* 54: 46–53.

Fan, Fa-ti. 2004. *British Naturalists in Qing China: Science, Empire, and Cultural Encounter*. Cambridge and London: Harvard University Press.

Frank, J. H. and E. D. McCoy. 1990. "Introduction to attack and defense: behavioral ecology of predators and their prey: Endemics and epidemics of shibboleths and other things causing chaos." *Insect Behavioral Ecology* 73, no 1. 1 (March): 1–9.

Fujimoto, Yoshihiro and Kamishima Yoshihisa. 2004. "Tokubetsu tennen kinenbutsu ōsanshōuo seisokuchi nai ni okeru seisoku kankyō chōsa: kasen kōzōbutsu no seisoku ni oyobosu eikyō ni tsuite." *Chūgoku gakuen kiyō* 2 (June): 89–95.

Geerts, A. J. C. 1883. "Notice sur la grande salamandre du Japon Cryptobranchus japonicus V.D. Hoeven." In *Nouvelles archives du Muséum d'histoire naturelle, deuxième série*, edited by G. Masson 2:5, 273–290. Paris: Librairie de l'académie de médecine.

Geng, Y., K. Y. Wang, Z. Y. Zhou, C. W. Li, J. Wang, M. He, Z. Q. Yin, and W. M. Lai. 2011. "First report of a ranavirus associated with morbidity in farmed Chinese giant salamanders (*Andrias davidianus*)." *Journal of Comparative Pathology* 145: 95–102.

Gray, J. E. 1873. "On a Salamander (Sieboldia) from Shanghai." *The Annals and Magazine of Natural History* Series 4, 12: 188.

Groening, Gert and Joachim Wolschke-Bulmahn. 1992. "Some notes on the mania for native plants in Germany." *Landscape Journal* 11, no. 2: 116–126.

Harris, Richard B. 2008. *Wildlife Conservation in China: Preserving the Habitat of China's Wild West*. Armonk and London: M. E. Sharpe.

Hoeven, van der J. 1866. "Notes on the Genus *Menobranchus* and its Natural Affinities." *The Annals and Magazine of Natural History* 3rd series, volume 18: 363–375.

Hoeven, van der J., F. J. J. Schmidt, and Q. J. Goddard. 1862. *Aanteekeningen over de anatomie van den Cryptobranchus japonicus*. Haarlem: De Erven Loosjes, Natuurkundige Verhandelingen van de Hollandsche Maatschappij der Wetenschappen te Haarlem.

Ienaga, Masayuki. 2011. *Panda gaikō*. Kabushikigaisha media fakutorī.

Ienaga, Masayuki. 2017. *Kokuhō no seijishi: Chūgoku no kokyū to panda*. Tōkyō daigaku shuppankai.

Ikoma, Yoshirō. 1973. *Nihon hanzaki shūran*. Tsuyama: Tsuyama kagaku kyōiku hakubutsukan.

International Union for Conservation of Nature (IUCN). 2019. "IUCN: A brief history." Accessed September, 2019. https://www.iucn.org/about/iucn-a-brief-history

Ishikawa, Chiyomatsu. 1899a. "Geigyo no hanashi." *Tōyō gakugei zasshi* 217 (October): 401–407.

Ishikawa, Chiyomatsu. 1899b. "Geigyo no hanashi." *Tōyō gakugei zasshi* 218 (November): 451–454.

Ishikawa, Chiyomatsu. 1899c. "Geigyo no hanashi." *Tōyō gakugei zasshi* 219 (December): 512–518.

Ishikawa, Chiyomatsu. 1901. "Über den Riesen-Salamander Japan's." *Mitteilungen den Deutschen Gesellschaft für Natur- und Völkerkunde Ostasiens* 9, no. 1: 79–94.

Ishikawa, Chiyomatsu. 1904. *Beiträge zur Kenntniss des Riesensalamanders*. Imperial Household Museum.

Isono, Naohide. 1988a. "Chāruzu O Hoittoman." In *Kindai nihon seibutsugakusha shōden*, edited by Kihara Hitoshi, Shinotō Yoshito, and Isono Naohide, 94–99. Hirakawa shuppansha.

Isono, Naohide. 1988b. "Sasaki Chūjirō." In *Kindai nihon seibutsugakusha shōden*, edited by Kihara Hitoshi, Shinotō Yoshito, and Isono Naohide, 107–110. Hirakawa shuppansha.

Iwama, Haruo. 1955. "Ōsanshōuo no seichō to tōa tairiku kara no *Sieboldia davidiana* ni tsuite." *Nihon seibutsu chiri gakkai kaihō* 16–19: 360–362.

Jiang, Lijing. 2016. "Retouching the Past with Living Things: Indigenous Species, Tradition, and Biological Research in Republican China, 1918-1937." *Historical Studies in the Natural Sciences* 46, no. 2: 154–206.

Jin, Licheng and Wang Jianguo eds. 1997. *Dani shengwuxue yu yangzhi shiyong jishu.* Taipei: Shuichan chubanshe.

Kaburagi, Tokio. 1938. "Ryōseirui no tennen kinenbutsu." In *Tennen kinenbutsu chōsa hōkoku, dōbutsu no bu, dai 2 gō,* edited by Monbushō, 92–103. Monbushō.

Kempe, Michael. 2003. *Wissenschaft, Theologie, Aufklärung: Johann Jakob Scheuchzer (1672-1733) und die Sinfluttheorie.* Epfendorf: Bibliotheca academica Verlag.

Kerbert, C. 1904. "Zur Fortpflanzung von Megalobatrachus maximus Schlegel." *Zoologischer Anzeiger* 27, no. 10 (23 February): 305–320.

Kinoshita, Kenjirō. 1973. *Bimi kyūshin, vol. 1.* Itsuki shobō.

Kitaōji, Rosanjin. 1935. "Kuidōraku yonjūnen o kataru 3: Chinmi nidai sanshōuo to gamagaeru no aji." *Hoshigaoka* 57 (June): 12–15.

Kobara, Jirō. 1985. *Osanshōuo.* Dōbutsusha.

Kobara, Jirō. 2000. "Shī-boruto no ōsanshōuo." *Ryōseirui-shi* 4: 28–30.

Kobayashi, Makie. 1906. "Dōbutsuen." *Tōkyōshi kyōikukai zasshi* 18 (February): 13–17.

Kyōto-shi. 2011. "'Dai ikkai gairaishu Chūgoku san ōsanshōuo taisaku kentōkai' no gaiyō." Meeting date May 11, 2011. https://www.city.kyoto.lg.jp/bunshi/page/0000104005.html

Kyōto-shi. 2012a. "'Dai nikai gairaishu Chūgoku san ōsanshōuo taisaku kentōkai' gijiroku yōshi." Meeting date February 16, 2012. https://www.city.kyoto.lg.jp/bunshi/page/0000120784.html

Kyōto-shi. 2012b. "'Dai sankai gairaishu Chūgoku san ōsanshōuo taisaku kentōkai' gijiroku yōshi." Meeting date August 29, 2012. https://www.city.kyoto.lg.jp/bunshi/page/0000184799.html

Kyōto-shi. 2012c. "Heisei 23 nendo tokubetsu tennen kinenbutsu ōsanshōuo kinkyū chōsa/chōsa jisseki." http://www.city.kyoto.lg.jp/bunshi/page/0000120788.html

Kyōto-shi. 2013. "Dai yonkai gairaishu Chūgoku san ōsanshōuo taisaku kentōkai." Meeting date March 28, 2013. https://www.city.kyoto.lg.jp/templates/shingikai_kekka/bunshi/0000211002.html

Kyōto-shi. 2014. "Dai gokai gairaishu Chūgoku san ōsanshōuo taisaku kentōkai." Meeting date March 18, 2014. https://www.city.kyoto.lg.jp/templates/shingikai_kekka/bunshi/0000211008.html

Kyōto-shi. 2015. "Dai rokkai gairaishu Chūgoku san ōsanshōuo taisaku kentōkai." Meeting date February 3, 2015. https://www.city.kyoto.lg.jp/bunshi/page/0000182095.html

Kyōto-shi. 2016. "Dai nanakai gairaishu Chūgoku san ōsanshōuo taisaku kentōkai." Meeting date February 25, 2016. https://www.city.kyoto.lg.jp/templates/shingikai_kekka/bunshi/0000211010.html

Kyōto-shi. 2018. "Dai kyūkai gairaishu Chūgoku san ōsanshōuo taisaku kentōkai." Meeting date March 7, 2018. https://www.city.kyoto.lg.jp/templates/shingikai_annai/bunshi/0000232615.html

Liu, Ch'eng-chao. 1950. *Amphibians of Western China.* Chicago: Chicago Natural History Museum.

Lubach, D. 1852. "Een Geologisch Raadsel de Fossile Mensch Van Scheuchzer." In *Albuum der natuur: een werk ter verspreiding van natuurkennis onder beschaafde lezers van allerlei stand,* 22–32. Haarlem: BIJ A.C. Kruseman.

Matsui, Masafumi. 2001. "Ōsanshōuo no zokumei ni tsuite." *Hachū ryōseirui gakkai-hō* 2001, no. 2 (September): 75–78.

Matsui, Masafumi. 2005. *DNA kaiseki ni yoru gairaishu Chūgoku ōsanshōuo jigyō hōkokusho.* Kasen zaidan. Dated 2005. http://public-report.kasen.or.jp/171215025.pdf

Matsui, Masafumi. 2010. "Gairaishu chūgoku ōsanshōuo no seitai risuku hyōka: 2008 nendo jisseki hōkokusho." Last modified April 21, 2016. https://kaken.nii.ac.jp/ja/report/KAKENHI-PROJECT-20510215/205102152008jisseki/

Matsui, Masafumi. 2011a. "Gairaishu chūgoku ōsanshōuo no seitai risuku hyōka: 2009 nendo jisseki hōkokusho." Last modified April 21, 2016. https://kaken.nii.ac.jp/ja/report/KAKENHI-PROJECT-20510215/205102152009jisseki/

Matsui, Masafumi. 2011b. "Kagaku kenkyūhō hojokin kenkyū seika hōkokusho: Gairai Chūgoku ōsanshōuo no seitai risuku hyōka, 2008-2010." Dated May 18, 2011. https://kaken.nii.ac.jp/ja/file/KAKENHI-PROJECT-20510215/20510215seika.pdf

Matsui, Masafumi. 2012. "Gairaishu chūgoku ōsanshōuo no seitai risuku hyōka: 2010 nendo jisseki hōkokusho." Last modified April 21, 2016. https://kaken.nii.ac.jp/ja/report/KAKENHI-PROJECT-20510215/20510215seika/

Matsui, Masafumi. 2013. "Gairaishu ni yoru ōsanshōuo no idenshi osen no jittai haaku: 2011 nendo jisshi jōkyō hōkokusho." Published July 10, 2013. https://kaken.nii.ac.jp/ja/report/KAKENHI-PROJECT-23510294/235102942011hokoku/

Matsui, Masafumi. 2014a. "Gairaishu ni yoru ōsanshōuo no idenshi osen no jittai haaku: 2012 nendo jisshi jōkyō hōkokusho." Published July 24, 2014. https://kaken.nii.ac.jp/ja/report/KAKENHI-PROJECT-23510294/235102942012hokoku/

Matsui, Masafumi. 2014b. "Gairaishu ni yoru ōsanshōuo no idenshi osen no jittai haaku: Kagaku kenkyūhi josei jigyō kenkyū kekka hōkokusho 2011-2013." Dated May 6, 2014. https://kaken.nii.ac.jp/ja/file/KAKENHI-PROJECT-23510294/23510294seika.pdf

Matsui, Masafumi. 2017. "Chūgoku ōsanshōuo ga zairai no ōsanshōuo ni ataeru eikyō." *Chiiki shizenshi to hozen* 39, no. 1: 13–19.

Matsui, Masafumi and Terutake Hayashi. 1992. "Genetic Uniformity in the Japanese Giant Salamander, *Andrias japonicus.*" *Copeia* 1992, no. 1 (Feb. 3): 232–235.

Matsui, Masafumi and Atsushi Tominaga. 2007. "Ōsanshōuo chiiki kotaigun hozen no tame no AFLP ni yoru iden tayōsei chōsa." *Ōyō seitai kōgaku* 10, no. 2: 175–184.

Matsui, Masafumi, Atsushi Tominaga, Wan-zhao Liu, and Tomoko Tanaka-Ueno. 2008. "Reduced genetic variation in the Japanese Giant Salamander, *Andrias japonicus* (Amphibia: Caudata)." *Molecular Phylogenetics and Evolution* 49: 318–326.

Melvin, John. 1969. "Some Notes on Dr. Scheuchzer and on Homo diluvii testis." In *Toward a History of Geology*, edited by Cecil J. Schneer, 202–213. Cambridge, MA and London: The MIT Press.

Miyake, Tsunekata. 1918. "Konchūgaku jō yori mitaru wagakuni." *Shiseki meishō tennen kinenbutsu* 2, no. 9 (September): 68–69.

Miyoshi, Manabu. 1916. "Tennen kinenbutsu no hozon ni tsuite." *Shiseki meishō tennen kinenbutsu* 1, no. 12 (July): 91–92.

Murata, Mariko. 2009. "Myūjiamu no juyō: Kindai nihon ni okeru hakubutsukan no shatei." *Kyōto seika daigaku kiyo* 35: 84–122.

Murphy, Robert W., Jinzhong Fu, Darlene E. Upton, Thales de Lema, and Er-mi Zhao. 2000. "Genetic variability among endangered Chinese giant salamanders, *Andrias davidianus.*" *Molecular Ecology* 9: 1539–1547.

Muséum National D'Histoire Naturelle Database; accessed November 3, 2019. https://science.mnhn.fr/institution/mnhn/search

Nakanishi, Kei. 1961. "Meiji ishin ni yoru kikō kaikaku." In *Nagasaki igaku no hyakunen*, edited by Nagasaki daigaku igakubu, 181–195. Nagasaki: Nagasaki daigaku igakubu.

Natural History Museum (London) Data Portal; accessed November 3, 2019. https://data.nhm.ac.uk/?_gl=1*3uq5l1*_ga*MTMxMTk2ODcxMS4xNjA4ODI zNDU5*_ga_PYMKGK73C4*MTYyNDY1ODg0Ni4xLjEuMTYyNDY1OTE wNy4wMw==*

Naturalis Biodiversity Center BioPortal; accessed November 3, 2019. https://bioportal. naturalis.nl/?language=en&back

Nihon Hanzaki Kenkyūjo. 2019. "Tokutei hieiri katsudō hōjin (NPO hōjin) Nihon Hanzaki Kenkyūjo setsuritsu." Accessed November 5, 2019. https://www.hanzaki. net/nihonhanzakikenkyuujyo-ni-tsuite/npo法人設立/

Noguera-Urbano, Elkin A. 2016. "Areas of endemism: travelling through space and the unexplored dimension." *Systematics and Biodiversity* 14, no. 2: 131–139.

Onaga, Lisa. 2010. "Toyama Kametaro and Vernon Kellogg: Silkworm inheritance experiments in Japan, Siam, and the United States, 1900–1912." *Journal of the History of Biology* 43: 215–264.

Oyama, Junji. 1919. "Shina san hanzaki no shinshu." *Dōbutsugaku zasshi* 374 (December): 400–402.

Oyama, Junji. 1924. "Shina san no ikko no hanzaki ni tsuite." *Dōbutsugaku zasshi* 36, no. 426 (June): 256.

Pan, Jionghua, Liu Chenghan, and Qian Xiongguang. 1985. "Guangdong sheng dalu liangqilei de diaocha ji quxi yanjiu." *Liangqi paxing dongwu xuebao* 4, no. 3 (September): 200–208.

Pan, Yuan. 2015. "Comparing the status of Chinese giant salamanders between three nature reserves in Guizhou province using local ecological knowledge." Master's thesis, Imperial College London.

Pan, Yuan, Gang Wei, Andrew A. Cunningham, Shize Li Shu Chen, E. J. Milner-Gulland, and Samuel T. Turvey. 2016. "Using Local ecological knowledge to assess the status of the Critically Endangered Chinese giant salamander *Andrias davidianus* in Guizhou Province, China." *Oryx* 50, no. 2: 257–264.

Pauly, Philip J. 1996. "The Beauty and Menace of the Japanese Cherry Trees." *Isis* 87, no. 1 (March): 51–73.

Peterson, A. Townsend and David M. Watson. 1998. "Problems with areal definitions of endemism: the effects of spatial scaling." *Diversity and Distribution* 4: 189–194.

Pierson, Todd W., Yan Fang, Wang Yunyu, and Theodore Papenfuss. 2015. "A survey for the Chinese giant salamander (*Andrias davidianus*; Blanchard, 1871) in the Qinghai Province. *Amphibian and Reptile Conservation* 8, no. 1: 1–6.

Rein, Johannes Justas. 1905. *Japan nach Reisen und Studien, Vol. 1.* Leipzig: Verlag von Wilhelm Engelmann.

Rein, Johannes Justas and A. von Roretz. 1876. "Beitrag zur Kenntniss des Riesensalamanders (Cryptobranchus japonicus)." *Der Zoologische Garten* 2 (February): 32–37.

Ritvo, Harriet. 1997. *The Platypus and the Mermaid and Other Figments of the Classifying Imagination.* Cambridge: Harvard University Press.

Sasaki, Chūjirō. 1887. "Some Notes on the giant salamander of Japan (Cryptobranchus Japonicus, Van der Hoeven)." *The Journal of the College of Science, Imperial University, Japan* 1: 269–274.

Satō, Ikio. 1943. *Nihonsan yūbirui sōsetsu*. Nihon shuppansha.

Satō, Kiyoaki. 1971. "Ōsanshōuo." *Okayama seishin gakuen kiyō*, 143.

Schlegel, H. 1842. "De Groote Salamander van Japan." In *De Diergarde de het Museum van het Genootschap Natura Artis Magistra te Amsterdam*, 9–16. Amsterdam: M. Westerman & Zoon.

Schlegel, H. 1872. "De Salamanders." In *De dierentuin van het Koninklijk Zoologisch Genootschap Natura Artis Magistra te Amsterdam*, 61–63. Amsterdam: Gebr. Van Es.

Sessions, Stanley K., Pedro E. Leon, and James Kezer. 1982. "Cytogenetics of the Chinese giant salamander, Andrias davidianus (Blanchard): The evolutionary significance of cryptobranchoid karyotypes." *Chromosoma* 86: 341–357.

Setoguchi, Akihisa. 2003. "Inyūshu mondai to iu sōten: Taiwan zaru konzetsu no seijigaku." *Gendai shisō* 31, no. 13 (November): 122–134.

Shapiro, Judith. 2009. *Mao's War Against Nature: Politics and the Environment in Revolutionary China*. Cambridge: Cambridge University Press.

Shen, Youhui. 1983. "Hunan sheng liangqi dongwu diaocha ji quxi fenxi." *Liangqi paxing dongwu xuebao* 2, no. 1 (March): 49–58.

Shimizu, Zenkichi. 2016. "Yodogawa suikei Kizu gawa (Mieken/Naraken) ni okeru ōsanshōuo no hogo." *Hachū ryōseirui gakkaihō* 2016, no. 2: 186–192.

Siebold, Philipp Franz Von. 1842. *Fauna Japonica*. Düsseldorf: Apud a Arnz et Socios.

Siebold, Philipp Franz Von. 1969. *Nippon: Archiv zur Beschreibung von Japan, vol. 1*. Osnabrück: Biblio Verlag.

Smithsonian National Museum of Natural History Museum Collection Record; accessed November 3, 2019. https://collections.nmnh.si.edu/search/

Society for the Study of Amphibians and Reptiles (SSAR). 2019. "About SSAR." Accessed October, 2019. https://ssarherps.org/about-ssar/

Song, Mingao. 1987. "Shaanxi liangqi paxing dongwu quxi fenxi." *Liangqi paxing dongwu xuebao* 6, no. 4 (November): 63–73.

Song, Ming-tao and Fang Rong-Shen. 1982. "Shaanxi Shangluo diqu liangqi paxing dongwu de diaocha." *Liangqi paxing dongwu xuebao* 1, no. 1: 88–89.

Songster, Elena E. 2018. *Panda Nation: The Construction and Conservation of China's Modern Icon*. Oxford: Oxford University Press.

Sowerby, Arthur de Carle. 1925. "The Giant Salamander of China." In *A Naturalist's Note-Book in China*, edited by Arthur de Carle Sowerby. Shanghai: North-China Daily News & Herald, LTD.

Sowerby, A. 1929. "Giant Salamander at Amoy." *The China Journal of Science & Arts* 11, no. 2: 104–105.

Stejneger, Leonhard. 1907. *Herpetology of Japan and Adjacent Territory*. Washington, DC: Washington Government Printing Office.

Strassberg, Richard E. 2002. *A Chinese Bestiary: Strange Creatures from the Guideways through Mountains and Seas*. Berkeley, LA, and London: The University of California Press.

Su, Zhongxi. 1986. "Guangxi Yulin diqu Liangqi dongwu chubu daiocha." *Liangqi paxing dongwu xuebao* 5, no. 3 (August): 235–237.

Tago, Katsuya. 1904. "Sanshōuo no hanashi." *Dōbutsugaku zasshi* 16, no. 184 (February): 41–46.

Tago, Katsuya. 1929. "Notes on the Habits and Life History of Megalobatrachus japonicus." In *Xe Congrès International de Zoologie Tenu à Budapest du 4 au 10 Septembre 1927*, edited by E. Csiki, 328–338. Budapest: Imprimerie Stephaneum S. A.

Tago, Katsuya. 1931. *Imori to sanshōuo*. Kyoto: Kyōto unsōdō.

Taguchi, Yūki and Natsuhara Yoshihiro. 2009. "Ōsanshōuo ga gyakujō kanō na seki no jōken." *Hozen seitaigaku kenkyū* 14: 165–172.

Takada, Seiji. 1996. "Butsurigaku to shūhen: Kagaku zasshi no senzen to sengo." *Nihon Butsurigaku Kaishi* 51, no. 3 (March): 189–193.

Thorbjarnarson, John B. and Xiaoming Wang. 2010. *The Chinese Alligator: Ecology, Behavior, Conservation, and Culture*. Baltimore: The Johns Hopkins University Press.

Tschudi, I. J. 1837. "Über den Homo diluvii testis, Andrias Scheuchzeri." *Neues Jahrbuch für Mineralogie, Geognosie, Geologie und Petrefaktenkunde, Jahrgang 1837*: 545–547.

Turvey, Samuel, Melissa M. Marr, Ian Barnes, Selina Brace, Benjamin Tapley, Robert W. Murphy, Ermi Zhao, and Andrew A. Cunningham. 2019. "Historical museum collections clarify the evolutionary history of cryptic species radiation in the world's largest amphibians." *Ecology and Environment* 2019; 00: 1–15.

Turvey, Samuel T., Shu Chen, Benjamin Tapley, Gang Wei, Feng Xie, Fang Yan, Jian Yang, Zhiqiang Liang, Haifeng Tian, Minyao Wu, Sumio Okada, Jie Wang, Jingcai Lü, Feng Zhou, Sarah K. Papworth, Jay Redbond, Thomas Brown, Jing Che, and Andrew Cummingham. 2018. "Imminent extinction in the wild of the world's largest amphibian." *Current Biology* 28 (May 21): R592–R594.

Ueno, Masuzō. 1960. "Shīboruto no Edo sanpu ryokō no dōbutsu-gaku teki igi." *Jinbun* 6: 309–325.

Ueno, Masuzō. 1973. *Nihon hakubutsugaku-shi*. Heibonsha.

Ueno, Masuzō. 1988. "Meiji-Taishō-ki no Dōbutsugaku." In *Kindai nihon seibutsugakusha shōden*, edited by Kihara Hitoshi, Shinotō Yoshito, and Isono Naohide, 36–46. Hirakawa Shuppansha.

VertNet. Accessed November 3, 2019. http://www.vertnet.org/

Wang, Xiao-ming, Ke-jia Zhang, Zheng-huan Wang, You-zhong Ding, Wei Wu, and Song Huang. 2004. "The decline of the Chinese giant salamander *Andrias davidianus* and implication for its conservation." *Oryx* 38, no. 2: 197–202.

Wang, Zuoyue. 2002. "Saving China through science: The Science Society of China, scientific nationalism, and civil society in Republican China." *Osiris* 17: 291–322.

Watase, Shōtarō. 1917. "Shizenbutsu no hozon ni tsuite." *Shiseki meishō tennen kinenbutsu* 1:16 (February): 123–125.

Watase, Shozaburo. 1926. "Zoological Natural Monuments." In *Preservation of Natural Monuments in Japan*, edited by The Department of Home Affairs, 37–40. Department of Home Affairs.

Wolf-Eberhard, Von Engelmann. 2012. "Zur Geschichte der Haltung von Riesensalamandern in Europa." *Sekretär* 12, no. 1/2: 3–23.

Xie, Feng, Michael Wai Neng Lau, Simon N. Stuart, Janice S. Chanson, Neil A. Cox, and Debra L. Fischman. 2007. "Conservation needs of amphibians in China: A review." *Science in China Series C: Life Sciences* 50, no. 2 (April): 265–276.

Yan, Fang, Jingcai Lü, Baolin Zhang, Zhiyong Yuan, Haipeng Zhao, Song Huang, Gang Wei, Xue Mi, Dahu Zou, Wei Xu, Shu Chen, Jie Wang, Feng Xie, Minyao Wu, Hanbin Xiao, Zhiqiang Liang, Jieqiong Jin, Shifang Wu, CunShuan Xu, Benjamin Tapley, Samuel T. Turvey, Theodore J Papenfuss, Andrew A Cunningham, Robert W. Murphy, Yaping Zhang, and Jing Che. 2018. "The Chinese giant salamander exemplifies the hidden extinction of cryptic species." *Current Biology* 28 (May 21): 590–594.

Yoshikawa, Natsuhiko, Masafumi Matsui, Azusa Hayano, and Miho Inoue-Murayama. 2012. "Development of microsatellite markers for the Japanese giant salamander (*Andrias japonicus*) through next-generation sequencing, and cross-amplification in its congener." *Conservation Genetics Resources* 4: 971–974.

Yoshikawa, Natsuhiko, Kaneko Shingo, Kuwabara Kazushi, Okumura Naoko, Matsui Masafumi, Isagi Yuji. 2011. "Development of Microsatellite Markers for the Two Giant Salamander Species (*Andrias japonicus* and *A. davidianus*)." *Current Herpetology* 30, no. 2 (December): 177–180.

Zhang, Lu, Wei Jiang, Qi-Jun Wang, Hu Zhao, Hong-Xing Zhang, Ruth M. Marcec, Scott T. Willard, and Andrew J. Kouba. 2016. "Reintroducing and post-release survival of a living fossil: The Chinese giant salamander." *PLoS One* 11, no. 6 (3 June): 1–15.

Zhang, Yuxia. 1987. "Guangxi liangqilei de jaiocha ji quxi yanjiu." *Liangqi paxing dongwu xuebao* 6, no. 1 (February): 52–58.

Zhao, Er-mi and Kraig Adler. 1993. *Herpetology of China*. Ithaca: Society for the Study of Amphibians and Reptiles.

Zoological Society of London (ZSL). 2019. "Professor Lew: An Unexpected Journey." Published April 2, 2019. https://www.zsl.org/blogs/zsl-london-zoo/professor-lew-an-unexpected-journey

Zoological Society of London (ZSL). 2021. "Meet our Chinese Giant Salamander, Professor Lew." Published April 15, 2021. https://www.zsl.org/videos/behind-the-scenes/meet-our-chinese-giant-salamander-professor-lew

8 The Destiny of China's Honeybees in the *Anthropocene*

Keokam Kraisoraphong

Introduction

By way of its sweetest naturally occurring product, honeybees have played a significant part in China's civilization for centuries. Interactions with honeybees began since the Chinese people hunted for wild colonies to collect honey. Further through civilization dependent on agriculture China's advances in animal husbandry saw its interactions with honeybees evolve into the more human-domineering activity of beekeeping. Even though it began late relative to those in Egypt and the Levant (Kritsky, 2017: 257–258), ancient Chinese apiculture did develop over many thousand years. Today, China is a major honey producer with over eight million managed bee colonies (Teichroew et al., 2016: 1). China's path to becoming the world's leading producer and exporter of honey thus follows a fascinating journey throughout its civilization.

Whether the Anthropocene began with the process of agricultural expansion some 7,000 or 8,000 years ago, or whether it began with the industrial revolution of the 18th century, China's ecological footprint has long been visible in this global process of human ecosystem domination. More so over the past 30 years, when China accelerated its speed of economic development by rapidly increasing its control and technological mastery of nature. Viewed on the one hand as an achievement, China's inventiveness to push back the limits of nature and replace ecosystems function by human technology has not gone unchallenged. Paralleling remarkable achievements in agriculture and rural economy, China's honeybees are under threat and at risk of decline in population and diversity. Ironically, being the leading producer and exporter of honey China is finding its farmers forced to hand-pollinate their own crops due to disappearances of wild honeybees (Goulson, 2012), while Chinese consumers now prefer to purchase imported honey – from fear of Chinese fake honey on the market (Ministry of Agriculture P.R. China, 2018; Root, 2019).

Following increased rates of economic activity and their ensuing environmental challenges, the Chinese government has now called for higher quality growth emphasizing green reforms toward environmentally sustainable

DOI: 10.4324/9781003212089-8

development. Nonetheless, it has also committed to fully modernize agriculture and the rural economy by 2050 (Huang and Rozelle, 2018). Under the Anthropocene narrative where "more human managerial action, rationalized domination, and active technological stewardship" (Fremaux, 2019: 170) is the rule of the game, China's accomplishment of a fully modern agriculture and rural economy could have significant implications for the existence of other species – particularly the honeybees known for their deliverance of irreplaceable benefits to human. Commonly known for their services in honey production and essential crop pollination, honeybees are key to maintaining a healthy ecosystem. On a global scale, where China stands as the world's largest honey producer and largest provider of pollination ecosystem services – how it chooses to act are extremely valuable lessons for those facing similar situations in other parts of the world (Teichroew et al., 2016: 1–8).

This chapter begins part one, following this introduction, to examine interactions of the Chinese people with honeybees – back to the time when Chinese references to bees and honey appeared, noting how China had developed toward beekeeping rather than hunting and how honeybee products have had their nutritional and medicinal importance throughout Chinese civilization. It then looks, in part two, at developments of the human–honeybee interactions under the new geological circumstances of the Anthropocene and how that has laid the path for China to become a world leading honey producer and exporter. This part also observes China's honeybee situation – the implications of its position as the largest supplier of the world's major honey consumer markets – and examines as to why some Asian countries have become the alternative route for China's global honey trade. In part three the chapter discusses the consequences of China's human–honeybee interactions from the perspectives that human, as the name-bearer, is the cause as well as the subject of the Anthropocene. It looks at the honeybee conservation framework proposed by concerned scholars to reconcile conservation and agriculture and discusses human agency's role and ambiguous structure as intentional power and unintentional force in the Anthropocene (Horn and Bergthaller, 2020: 68). The final part of the chapter concludes by discussing what could likely transpire and what it would take to shift China to a new paradigm toward becoming a redeeming agent in its interactions with its honeybees – rather than for Anthropocene to be just another process by which China becomes more like the West in its process of modernization (Lo, 2019: 14).

China's History with Honeybees

The preserved remains of the honeybee in amber estimated to be 80 million years old is evidence that "honeybees are one of the oldest forms of animal life still in existence from the Neolithic Age" (Head, 2012). For China, the oldest reference of honeybees dates back to 1000 BCE when pictographs of the Chinese word for bees, *feng*, were carved on animal bones, and on tortoise ventral shells (Kritsky, 2017: 257–258; Lau, 2012). The Chinese word for honey, *mi*, was later recorded around 300 BCE, under the dietary recommendation in the *Book of Manner* known as *Li Ji* (Lau, 2012).

Like most Asians, the Chinese's interactions with honeybees are extremely ancient. For many Asian society hunting for wild honeybee colonies to collect their honey developed into complex traditions – many of which continue to this day. But for China – even though hunting for wild honey had been described back in the Chin dynasty (265–290 CE) by a record of wild honeybees, *A. dorsata* combs being taken from a rock wall (Kritsky, 2017: 257–258) – the Chinese's agricultural and hunting practices of honeybees had advanced toward beekeeping rather than bee hunting. Such practice was determined by two factors which greatly influence apicultural traditions in Asia.[1] First, the lack of easily hunted wild honey but on the other hand the relative ease of the domestication of cavity-nesting honey bee species, *Apis cerana*, encouraged the Chinese to develop practices in beekeeping rather than bee hunting (Oldroyd and Wongsiri, 2006: 209–210).[2] Throughout several thousand years of civilization dependent on agriculture along the Yellow River and the Yangtze River Chinese farmers have learned and mastered their honey bee husbandry practices as they came to understand bee behavior through their experience in collecting bee products from the wild (Lau, 2012: 78–81).

Second, technical advances in beekeeping tended to spread quickly through the influences of Chinese and Mogul civilization, and the religions of Islam, Buddhism, and Hinduism. (Oldroyd and Wongsiri, 2006: 209–210). Over the millennium following the first detail description of the 3rd-century CE

Figure 8.1 Close-up of Honeybees at Work

(Public Domain Image CC1.0 from Rawpixel image 5925090).

ancient Chinese beekeeping, Chinese knowledge of beekeeping expanded. Dating back to 158–166 CE, a biography of a Confucian scholar who kept bees provides one of China's first record of a professional beekeeper (Kritsky, 2017: 257–258; Lau, 2012). Honey harvesting became a common business practice since the end of the Tang dynasty in the 9th century CE (Lau, 2012). The Mogul government in 1273 produced the book *Fundamentals of Agriculture and Sericulture* with a chapter on bees describing the beekeeper's tasks over the course of a year (Kritsky, 2017: 257–258). The honey harvest in the Ming Dynasty (1368 CE–1644 CE) which occurred during the sixth month of the Chinese calendar then became a nationally recorded event (Lau, 2012). Written records by about a thousand Chinese writers have described the nature of honeybee hives and China's long history of beekeeping (Crane, 1999), providing a detailed collection of written records on colonies management to obtain bee products (Lau, 2012).

Prior to the introduction of movable frame hives, honey for its rarity was regarded widely more as medicine than food (Oldroyd and Wongsiri, 2006: 222). While references to the recognition of honey's nutritional and medicinal values by the Chinese date back to no less than 3,500 to 4,000 years (Kritsky, 2017: 257–258), China has practiced apitherapy using bee products in human treatments for 3,000 to 5,000 years (El-Wahab and Eita, 2015: 19–27). Benefits of bee products were also recorded by many ancient Chinese writings over a thousand years ago to narrate their function in medicine, pharmacology, dietetic therapy, the preservation of one's health and for anti-decrepitude (Zhu and Wongsiri, 2008: 303–312). Studies on apitherapy health care list some of these written records:

Table 8.1 Ancient Chinese records narrating the functions of bee products medicine, pharmacology, and dietetic therapy

Period	Records on functions of bee products
3rd century BC	China's ancient prescription book: *Fifty-two Prescriptions* – two prescriptions involving bees, one of which uses honey to treat disease.
Zhou dynasty (around 300 BCE)	*Li Ji (Book of Manner)*: recorded that 2,300 years ago sweet honey used to show filial piety and respect for the elderly. Bees and the larvae of cicada, which have a high nutritional value, were the food of monarchs and nobilities. Honey was recorded as a dietary recommendation.
2,000 years ago	*Sheng Nong's Herbal*: records 365 kinds of medicinal materials, of which honey, beeswax, and bees are considered higher-grade drugs capable of treating diseases and maintaining health.
25~88 AD	*Prescriptions for Diseases*: records that honey was the main ingredient in honey pills and an instant herbal mixture for treating cough associated with asthma.

(Continued)

Table 8.1 (Continued)

Period	Records on functions of bee products
Han dynasty (206 BC–220 AD)	*Treatise on Cold-induced Febrile Diseases* (by Zhang Zhongjing): documents a kind of honey suppository to treat frail patients' constipation. *Medical Treasures of the Golden Chamber* (by Zhang Zhongjing) a prescription book on treatment for miscellaneous diseases: contains 262 prescriptions, about 20 of which are pills and 80% of them are honey pills. From that time on for nearly 1,800 years, honey pills have always kept this proportion. It records that honey soup with liquorice powder was used to treat stomachache caused by bellyworms and regulates Qi with beeswax for treating diarrhea.
Jin dynasty (266~420 AD)	*Prescription for Emergent Reference* (by Ge Hong)
452~536 AD	*Focuses of Shen Nong's Herbal* (by Tao Hongjing)
499~583 AD	*Collection of Proved Recipes* (by Yao Sengyuan)
541~643 AD	*Treatise on Property of Drugs* (by Zhen Qian)
581~682 AD	*Supplement to Invaluable Prescriptions and Supplements to Invaluable Prescriptions for Ready References* (by Sun Simiao)
621~713 AD	*Treatise on Dietetic Therapy* (by Meng Xi)
841 AD	*Chuan Xin Fang* (by Liu Yuxi)
Song dynasty (992 AD)	*Taiping Royal Prescriptions* (compiled by the Official Institute of the Imperial Physicians)
1180~1251 AD	*Treatise on Food* (by Li Gao)
1406 AD	*Prescriptions for Universal Relief* (the largest prescription book in China)
1518~1593 AD	*Compendium of Materia Medica* (by Li Shizhen)
1584 AD	*Chi Shui Xuan Zhu* (by Sun Yikui & others)
About 1,200 years ago. Recompiled every dynasty and in the 11th century AD was recompiled and amended by the author's 14th offspring	*Si Bu Yi Dian* (by Tibetan expert: Yu Tuo Yuan Dan Gongbu – amendments by Tibetan physicians): records drugs made of honey, yellow wax, hydromel, wild bees, and bumblebees; it contains 100 prescriptions for treating diseases by eating honey or only using honey as component.
	Dang Ha Ya Long, the Dai nationality ancient medical book on medicine and pharmacology records treating diseases by using folk prescriptions: honey comb of *Apis dorsata* and *Apis florea*, branches of beehives. *Nie Su Nuo Qi*, the Yi nationality book on medicines and pharmacology narrates the medical functions of honey, honeycomb, and bees, as well as antidotes to poisoning caused by poisonous honey. Zi Feng liquor, produced by the institute on national medicine and pharmacology in Simao, Yunan Province, is used to treat rheumatic diseases arthropathy.

Source: Compiled, organized, and tabulated by author from: Zhu, Fang and Wongsiri, Siriwat. 2008. "A Brief Introduction to Apitherapy Health Care." *Journal of Thai Traditional & Alternative Medicine*, Vol. 6 No. 3 September–December 2008: 303–312.

Honeybee venom therapy, using bee venom into specific points of human body as complementary and alternative therapy, has been practiced in China for 3,000 years (Zhang et al., 2018). The medical books of Yijing (Morality book), which was found in a site near Chang Sha, Hunan province during an archaeological expedition in 1963, has recorded honeybee venom treatment in the laws of longevity (Koo's Acupuncture and Herb Clinic, 2018). Written records on the ancient Chinese methods of using natural products from honeybees such as honey, royal jelly, and bee venom have allowed apitherapy techniques to develop in treating "depression, neurological disease, muscle cramps, skin tumors, and nerve pain" – making honeybees a significant instrument in Chinese medicine (El-Wahab and Eita, 2015: 19–27).

Among the natural products from honeybees that have long been a significant part of Chinese medicine, honey is the major trade commodity that has brought China to the world's top exporter. Honey has been described as "a miraculous product" for reasons that it is the fruit of a unique interaction between the plant and animal kingdoms' (García, 2018; De-Melo et al., 2018; Da Silva et al., 2016):

> Plants and bees co-evolved over 100 million years to create this complex and healthy product, which contains around 200 different substances. Honey is mainly composed of sugars, but also contains many other substances, such as proteins and enzymes, amino acids, organic acids, vitamins, minerals, phenolic, and volatile compounds.

Some important properties of honey – viscosity, crystallization, color, flavor, taste, specific gravity, solubility, and conservation – are maintained by honey's second largest component – water (Escuredo et al., 2013; García, 2018).

Honey as an essential ingredient in Chinese medicine has always had authenticity as its key value – determined by the purity of the transformation of nectar into honey. But the core components of capitalism – in this case, excessive exploitation of nature, the rise of industrialism, the self-destructive over-confidence in human-technical power, and the anthropocentric mindset in denial of ecological limits (Fremaux, 2019: 1–2) – have driven humans to imitate – even attempt to replace – natural processes of ecosystem services. To keep up with consumers' demand for natural and healthy foods the honey industry, short of authentic honey, has been driven by the ever-growing honey markets to maintain an uninterrupted supply with adulterated honey.

Honeybees and China's Web of Honey Trade under the New Geological Circumstances of the Anthropocene

China's diverse geo-climatic spread across a vast area – hot and humid in its southeast while cooler and drier toward its northwest – allows honeybees to enjoy the biological diversity of over 10,000 varieties of flowering plants. For this reason, China's honey flow is known to move toward a northwest direction according to the blooming of its nectar plants (Tang-Dong, 2006: 4–5). The Qinghai region of north-west China is thus where small beekeepers of the *Apis cerana* are known to be the source for Chinese honey (Gentlemen Marketing Agency (GMA), 2019).

The *Apis cerana* is the honeybee species native to China. It is also the common species native to Asia with its habitat range expanding from Japan's eastern region to the western boundaries of Asia in Afghanistan (Egelie et al., 2015). In their wide habitat range they have been able to withstand and survive in many different ecosystems (Li, 2019). Being a cavity-nesting honeybee species the *Apis cerana* also provided the ease for domestication and encouraged the Chinese to master the technology of beekeeping and to develop their beekeeping skills into a commercial apiculture industry. Beekeeping techniques similar to modern beekeeping skills have been used in China since the Ming Era (1368 CE–1644 CE) such that "they knew how to regularly clean up pests like spiders, wasps and ants and to care for the weak post-swarmed colonies" (Lau, 2012).

China's long history of commercial apiculture has made it into one of the world's largest beekeeping countries with its leading number of bee colonies and amount of honey production (Li et al., 2012). Managed apiaries in China today account for approximately 10% of global supply – a number no less than three to four times higher than that of the United States (Teichroew et al., 2016: 2). Honeybees used by beekeepers in China are now those of two distinct species: the *Apis cerana*, which is the indigenous semi-domesticated honeybee and the *Apis mellifera*, the non-native Western honeybee first introduced to China since 1896 (Li et al., 2012).

Along growing worldwide concerns of declines in pollinators and colony losses in managed honeybees, there have been reports of increased losses in the *Apis mellifera*, which is the sole managed species on which the United States and Europe rely from a narrow genetic stock (Teichroew et al., 2016: 2). Colonies of managed honeybee – *Apis mellifera* – are reported to have decreased by 25% over 20 years in Europe and 59% over 58 years in North America (Miller-Struttmann, 2016; Nagy, 2019).

Coinciding with reports suggesting the number of honeybee colonies has been declining in the United States and Europe over the past decade, honey consumption in both continents have been on the rise. The US total consumption over the years since 2010 has increased overall by 35%. Americans consume approximately 450 million tons of honey each year, an amount three times the 150 million tons of honey it produces annually (Wills, 2015). With 35% consumed in homes, restaurants, and institutions, and the industry buying the remaining 65% for use in the bulk of different processed foods (Schneider, 2011), the United States has the highest deficit in the international trade of natural honey. In 2017 the United States imported 67% of the

Figure 8.2 Modern Bee Farming

(Public Domain Image CC1.0 from Rawpixel image 6023186).

honey it consumed and by 2018, import increased to approximately 75% with only 25% produced locally (Root, 2019). Because the US growing demand for honey cannot be matched by its dwindling production due, in large part, to the collapse of domestic bee colonies, honey prices have been rising – consequently leading to honey laundering in the industry, whereby products "labeled as pure honey in fact may be honey blend or honey syrup – honey adulterated with cane sugar or corn syrup – or product that contains antibiotic residue" (Wills, 2015).

At annual production of 230,000 tons, the EU is the world's second largest honey producer, next to China. The EU in 2018 had 17.5 million hives, run by approximately 650,000 beekeepers. Despite being one of the world's leading honey producer, the EU still needs to cover local demand with imports as overall it is just about 60% self-sufficient (Root, 2018). At the highest dollar value of purchase, the EU's honey import in 2018 reached USD 1.2 billion, of which China is the main source of the increase in imports (García, 2018), accounting for up to 50% of Europe's imported honey (Tamma, 2018). Europe's apiculture has also partly been affected by massive losses of the honeybee *Apis mellifera* population such that up to 50% of all bee colonies collapsed in some region – the main trigger being the infestation with Varroa mites, widespread blood-sucking parasites, and the transmission of deformed wing virus by the mites (University of Vienna, 2016). Such plagues that have killed commercial honeybees at high rates in Europe also coincide with honey laundering (Wills, 2015) that has raised concerns of local honey producers and consumers who claim that Europe's local markets are being flooded with substandard imported honey (Tamma, 2018).

Chinese honey has been a culprit in the claims from both the United States and Europe for laundered honey imports flooding their markets. The world's largest honey producer and exporter, China has an annual production volume of over 400,000 metric tons (Ma, 2020). At approximately 40% share of the world market, China is in international terms "by far the largest honey producing nation in the world" (William Reed Business Media, 2017). According to the Food and Agriculture Organization, China's honey production, driven by rise in exports, increased by 88% between 2000 and 2014 – earning it USD 276.6 million in 2016 (Tamma, 2018). China, then during 2018, earned the highest dollar value worth from natural honey exports at USD 249.3 million, or 11.2% of total world exports (Root, 2019).

Looking back, China experienced a sharp drop in its honey export between 2000 and 2002 when Chinese honey was heavily taxed or banned by its two most important markets – the United States and Europe – for impurities and pollution of heavy metal and antibiotics and for adulteration (Wu et al., 2015: 4377–4394). To these markets, Chinese-origin honey are known for containing the banned antibiotic chloramphenicol, which Chinese farmers use to treat unhealthy bee colonies and control disease in bees (William Reed Business Media, 2017).[3] An even more serious health threat also discovered in Chinese-origin honey was lead; because heavy metals are accumulative, they are absorbed by organs and are retained.[4] Dilution also became an issue, when Chinese-origin honey were found to have been "concocted without the help of bees" but instead were "made from artificial sweeteners and then extensively filtered to remove any proof of contaminants or adulteration or indications of precisely where the honey originated" (Schneider, 2011). Honey adulteration by this refined method of ultrafiltration (UF) has allowed Chinese honey exporters to mask their contaminated product, by removing or concealing indicators of added sweeteners or contaminants and all floral fingerprints which normally indicate the honey's geographical origin (García, 2018). In fact, UF rather than contaminated honey is believed to be the real threat to the purity of honey internationally (Durham, 2004).

The United States began in 2001 with 221% antidumping duties imposed on Chinese honey for being sold in the United States at less than fair-market value (Wills, 2015). But with steadily increasing per capita consumption – indicating that imported honey, for the most part, is acceptable to US customers (Root, 2019) – Chinese-origin-products have found their way around the punitive import tariffs into the United States market. By a growing multimillion-dollar laundering system, a network of Asian countries has been used to "wash" Chinese-origin products with new packaging and false documents about its origin – generating new routes for transshipping to the United States (Leeder, 2011). In their investigations on illegal importation of honey from China in 2008, the US authorities "launched a series of indictment and arrests of 23 German, Chinese, Taiwanese and American corporate officials and their nine international companies" – who were then charged with "conspiracy to smuggle more than US$ 70 million worth of Chinese honey into the U.S. by falsely declaring that the honey originated from countries other than China" which "allowed them to avoid paying stiff anti-dumping charges imposed on China" (García, 2018, Schneider, 2011). The lack of a US standard for honey and the need to update the official controls for honey adulteration (García, 2018; Schneider, 2011; Wills, 2015) have left the United States plagued by this honey laundering scheme for more than a decade (Schneider, 2011) – for it has opened "the doors to a risky shift in U.S. consumer honey market in terms of honey quality and protection" (García, 2018: 3). Although a changing pattern in the US honey imports has been observed, those in the honey trade claim that up to 2011, millions of pounds of Chinese honey transshipped from India and Vietnam were still coming into the United States (Schneider, 2011) – while in 2017 it has been reported that 53% of total US honey imports equivalent to 107,104 tons were provided by India, Vietnam, Ukraine, Thailand, and Taiwan (García, 2018: 3). Continued laundering of honey transshipped from China has been detected by how some countries have increased their total export volumes of low-priced honey to the United States without parallel growth in their number of hives nor improved hive productivity – but have, on the other hand, increased their imports of honey from China (García, 2018; Schneider, 2011; Wills, 2015). Adulteration also obviously explains such increase in export volume given the environmental degradation and water systems in India and China are unlikely to allow such increase in honey production (Garcia and Phipps, 2018).

On the other continent, Chinese honey was banned by the EU on health grounds in 2002 for chloramphenicol contamination (Durham, 2004). Then between 2002 and 2004, a ban was imposed on grounds of a lack of origin and a risk that it contained lead – only to eventually be lifted "due to increased demand, which Europe could not satisfy elsewhere" (Tamma, 2018). This has, however, allowed adulteration to flourish – since provided with the availability of low-cost Chinese honey, where the "heterogeneity of prices according to their geographic origin" could be an economic advantage, several European countries took it as an incentive to import honey from China only to then re-export it as locally produced (García, 2018: 3). In such cases, importation of relatively inexpensive honey and its plausible re-exportation had increased the chances of adulteration to mask their geographical origin (Garcia and Phipps, 2018: 3). This apparently explains the dramatic increase in honey exports by a number of European countries (Tamma, 2018).

China, on the other hand, has its own honey standard[5] for its mode of honey production, which is not in accordance with the Codex Alimentarius.[6]

Chinese beekeepers employ the "Quick Honey" model which harvests unripe honey with high water content. While this model brings higher yields and diminished costs, it is a production system which requires the in-factory processing of honey. Honey products from such processes have proved to lack the cited positive properties of honey – reconfirming the fact that "the production of honey by bees is indeed a long and laborious process that man can imitate but never emulate" (García, 2018).

As the situation between the two major honey importing markets goes, the EU's decision to ban honey over lead and other contaminates has increased the odds that more of those honey suspected to be of Chinese origin were bound for US borders (Leeder, 2011) – causing the United States to claim that Chinese honey banned in Europe had been flooding US grocery shelves (Schneider, 2011). On the other hand, the EU has been reported to have absorbed much of what used to be sold in the United States (Root, 2019). Either way, the situation with Chinese honey imports has some observable consequences in both the United States and the EU: first, oversupply has caused a downward pressure on pure honey prices. Second, significant decreases in export volumes of several traditional honey producing countries over the past ten years reflect the disincentives for these countries to produce and export honey. Third, new exporters are emerging to re-export relatively inexpensive imports as locally produced honey – straight or in blends (García, 2018; Leeder, 2018). Sadly, honest beekeepers have had to bear the brunt of this entire process as their pure honey have been losing out, causing many of them to cease their operations (García, 2018: 5).

On China's domestic front, per capita consumption of natural honey has been growing in recent years making its local honey market one of the fastest growing since 2014 – with up to 19.6% rate of growth (Root, 2019; William Reed Business Media, 2017). Chinese consumers use honey as both food and medicine for which it is known for antibacterial and healing sore throat cough or insomnia. Although honey has become easier to find than sugar in Chinese supermarkets, Chinese consumers are concerned about counterfeit domestic honey. With honey demand far exceeding honey supply Chinese consumers have turned to consume imported honey (Gentlemen Marketing Agency (GMA), 2019). China customs statistics indicate that two of the main sources of honey imports to China are New Zealand and Australia – particularly New Zealand, known for its premium product: manuka honey (DCCC, 2017).

A study on consumer attitudes and behavior toward honey in China points out that since honey is the by-product of the honeybee's plant pollination, consumption of honey partially compensates for the honeybee's pollination services. This means that honey consumption not only benefits the honey industry but also provides support to the domestic agricultural sector since the honeybee also plays an important role in the pollination of fruit and vegetable plants. Stable pollination services provided by honeybee colonies in this sense depend to a certain extent on the economic viability of the honey market. The values generated by honey demand dynamics thus extends far beyond honey itself because without honey production, the pollination cost would be much higher (Zhang, 2018).

China's current position in the global honey trade seen in the two major markets – the United States and the EU – has caused some honey experts to be concerned that new technologies will make detection of adulterations even more difficult (Schneider, 2011). On the one hand, it is an ethical question of human responsibility and a political question of collective action among humans – while on the other hand, there is a question of "the ethical obligations of humans toward the non-human" which arises from our 'entanglement with other vulnerable species in the midst of an unstable nature' (Horn and Bergthaller, 2020: 11, 73) – evident in the case of human interactions with honeybees.

Looking to the Future: What Will It Take for China to Become a Redeeming Agent?

China is one of the world's two only countries with a population of more than 1 billion, accounting for roughly one-fifth of global population. Home to an exceeding 1.4 billion people, it is known to be the most populous country in the world (World Population Review, 2021). In terms of resources, it has 8% of the world's total arable land and about 5% of the world's total water availability (ChinaPower, 2021; Huang, 2017). Despite increase in demand from population growth, China is now able to reach nearly 96% self-sufficiency in food. This has been achieved by an annual agricultural growth of about 4.6% for 35 years, which has meant significant changes in the agricultural structure – a rapid agricultural and rural transformation from a grain-based economy to cash crop – due to demand changes as income increased (Huang, 2017). Like elsewhere in the world, China's urban economy has accelerated its consumption of resources to feed its cities and consequently recreated its environment through tremendous rural disruptions.

As much as agricultural and rural transformation is an economic phenomenon it is also an ecological process that is reshaping the context for human action in this globalized era. When defined as "the health status of ecosystems and the ability of ecosystem service supplies for humans," China's ecological and environmental security is evidently under threat from massive transformations in land-use changes, loss of biodiversity, the re-routing of rivers, and climate change impacts (Zhang and Xu, 2017). Throughout this entire process China's ecological footprint has been immense and so has been its impacts on its coexistence with other species. Honeybees domesticated and wild have been one of the most significant species bearing the weight of this process. Interactions between human and honeybees in China clearly portray the status of humans as the cause as well as the subject of the Anthropocene. In other words, human agency in the Anthropocene is "the paradoxical combination of immense power and lack of control, of goal-oriented action and unintended side-effects" (Horn and Bergthaller, 2020).

China is known to harbor an extremely high diversity of wild bees at the species as well as genetic/colony level. At the same time, its increase in the number of managed pollinators since 1961 has been as high as over 160% – the highest number than in any other regions (Teichroew et al., 2016: 2). The Asian honeybee, *Apis cerana* – the main species Chinese beekeepers use – can still be found in the wild (Egelie et al., 2015; Li, 2019). But they are in need of conservation efforts (Egelie et al., 2015) because their population size has been rapidly decreasing due to the threats of habitat degradation, overuse of agrochemicals, climate change, and the introduction of non-native bees (Li, 2019; Teichroew et al., 2016: 3–5). Land-use changes from massive infrastructure projects, open-pit mining, deforestation, and the spread of intensive agriculture (Stevenson, 2016) – particularly agricultural land expansion into monoculture fields and urban areas – have significantly changed China's landscape composition (Liu et al., 2014).

China's land-use changes of such scale have resulted in the reduction in floral and nesting resources for wild bees – radically reducing the areas in which wild bee species can thrive and keep their habitat intact. With the disappearance of wild honeybees reported as resulting from habitat destruction in many areas (Teichroew et al., 2016), the decline in honeybees as economically important pollinators in several parts of China has been explained by the shortage in their food availability combined with their exposure to the heavy use of agrochemicals for monocropping (Li, 2019; Li et al., 2012; Stevenson, 2016). Agricultural pesticides such as neonicotinoids contribute to the extinction of bees such that it affects the brains and bodies of bees evidenced by the changes in their behavior and the reduction in their fertility and lifespan (Harper, 2018). China is the world's largest pesticide producer, where the use of pesticides has been subsidized and promoted through government policies leading to over-application of agro-chemicals which has extensively contaminated its air, soil, and groundwater (Zhang et al., 2011). The disappearance of honeybee population in some fruit-producing regions such as the Maoxian region of Sichuan Province in the southwest of China has been attributed to the lack of natural habitat and excessive pesticide use by large-scale cultivation starting since the early 1980s (DW, 2014; Goulson, 2012; Liess, 2015; Partap and Ya, 2012; Thibault, 2014). Under conditions of pollinator decline, farmers who are direct beneficiaries from honeybee pollination are also at risk of being greatly impacted in their daily agricultural practices. Farmers have been forced to use "human pollinators" to hand-pollinate their apples – working to "replace what human nearsightedness has wiped out" (Williams, 2016).

Honeybees have also been affected by climate change, another facet of the new circumstances of the Anthropocene in which China's ecological footprint from its industrial development has been significant. Due to their narrow distribution ranges and hence limited climate adaptation, Chinese wild honeybee populations would be highly vulnerable to extreme climate change impact on their "geographical distribution, phenological traits, foraging behavior, and physiology, as well as their interactions with diseases" (Hegland et al., 2009). Rising temperature from climate change could affect the fertility of bees – but more importantly it can lead to fluctuations in flowering periods of plants as well as the seasons which alter the conditions for honeybees to find food to raise their worker bees – thus creating additional stress in beehives (Gebhardt, 2014).

China's managed apiaries are mainly of two honeybee species: the semi-domesticated Asian honeybee, *Apis cerana*, and the non-native Western honeybee, *Apis mellifera*. Both species share many common forging habitats over a huge geographical area and are now present throughout China (Li et al., 2012). In recent years, beekeepers in China have been transitioning to *Apis mellifera* because of its high honey yields. But though *Apis mellifera* colonies may be more profitable for honey production, they can survive only under intense care and protection offered by the beekeepers and are less suitable pollinators of native plants (Egelie et al., 2015). The reasons being: *Apis cerana* could pollinate a wide range of plant species, while *Apis melifera* pollinates concentrating only on limited kinds of flowers (Chen, 2017). Therefore, the mass decline in the Asian honeybee, *Apis cerana* and the replacement of their management by *Apis mellifera* management in many areas could affect the native flora in addition to the bee population (Egelie et al., 2015). This could result in fundamental changes in the ecological balance because many native plants may not be pollinated (Chen, 2017). In fact, since the introduction of the Western honeybee *Apis mellifera* through the first European bee farm established in China in the late 19th century,[7] the distribution area of the Asian honeybees, *Apis cerana*, has been reduced by over 75%, and their populations have been reduced by over 80% from competition with the Western honeybee, *Apis mellifera* (Teichroew et al., 2016).

The Chinese government's commitment to fully modernize agriculture and the rural economy within the next three decades brings to question future ecological sustainability for honeybees as pollinators. Under common practice that beekeepers always try to make up for the losses in their honeybee population, the decline in other pollinators species would result in honeybees being more and more important for pollination because other pollinators are gone (Tamma, 2018). Reports indicating China's bee diversity is under threat and possibly already in decline have led concerned scholars to propose ways to conserve China's bees (Stevenson, 2016; Teichroew et al., 2016). Contextualizing the conservation solutions, the proposed hybrid land management framework identifies two approaches to reconcile conservation and agriculture: land sparing and land sharing. Land sparing intends to minimize farmland area by increasing yield on existing farmland – thus sparing land for habitat conservation or restoration – particularly for those species and habitats whose persistence is incompatible with farming. Land sharing, on the other hand, involves integrating biodiversity into the farmed landscape without reducing yields (Teichroew et al., 2016:7).

China's Conversion of Cropland to Forest Program (CCFP), also known as the Grain for Green (GfG) program or Slope Land Conversion Program, SLCP), is one of China's most important reforestation, ecological restoration, and rural development programs that could provide the means to implement hybrid land management for bee conservation (Teichroew et al., 2016:8). Farmers are the prime managers of the reforestation processes and are compensated for their labor and loss of agricultural land in this "program with the largest investment, greatest involvement and broadest degree of public participation in history." Directly involving 124 million people or 32 million households in 1,897 counties of 25 provinces (Delang and Yuan, 2015), the program constitutes the world's largest payment for ecosystem services scheme with a funding volume of almost USD 30 billion (Teichroew et al., 2016: 7). It has enabled China since 1999 to restore forests landscapes across more than 28 million hectares of farmland and land classified as barren or degraded (Dayne, 2017). But heavy reliance on monoculture tree reforestation of the program has not benefited pollinator conservation. Perhaps with a different mindset and appropriate policy change such program could support the selection of preferred floral plants and the use of a wider diversity of flowering tree species to provide extensive pollinator pasture and habitat across the country (Teichroew et al., 2016: 7–8). At the same time, another very pressing issue that must not be overlooked – but rather requires high government priority in policy change – is eliminating the intensified use of agrochemical, particularly the overuse of pesticides. Such policy change involves the urgent need to phase out subsidies for pesticides (Stevenson, 2016).

Another side of China's human–honeybee interactions related to managed honeybees brings to question how the economic system, forms of consumption, lifestyles, and technologies have caused and continue to drive the ecological changes of the Anthropocene. The entire process within the global web of honey trade – driven by opportunistic exploitations of the ever-growing consumers' demands – is another illustration of how humans have become the dominant species that have dramatically changed the living conditions of the honeybee species (Horn and Bergthaller, 2020: 67). Although certain conditions of managed bees unique to China have, to a certain extent, set it apart from the situation in the United States and Europe (Teichroew et al., 7), the trend that Chinese consumers are turning away from local honey may be construed as a cause for China to be concerned. Honest beekeepers in the United States and Europe – faced with a dramatic decline in honeybee population from colony collapse disorder (CCD) and fierce competition from low-cost Chinese honey – have lost out with their pure honey, "causing many to abandon their operations" (García, 2018: 5). Similar threats and pressures on pollinator communities in other parts of the world could possibly eventually take effect in China, making beekeeping unattractive when the economics of beekeeping are gone and there is no incentive for beekeepers to reconstitute the colonies after the losses (García, 2018; Tamma, 2018).

Discussions

China's environmental concerns, whether "an altruistic awareness about well-being for all" or "simply the anxiety of future impossibility to exploit nature for the use of the Chinese people" (Lo, 2019: 22), have apparently driven such programs as Grain for Green which now serves as a source of lessons for ecological restorations. Researchers under global efforts, which have "turned to restoration as a way to mitigate climate change," have looked to China for lessons on how to achieve the goal of restoring 150 million hectares of deforested and degraded land by 2020, and 350 million hectares by 2030 (Dayne, 2017). While many have praised this program as an impressive achievement (Dayne, 2017; Delang and Yuan, 2015), others have pointed out how heavily relying on monoculture tree reforestation have masked continued biodiversity conservation needs. Such reforestation efforts have likely reduced floral and nesting resources for wild bees and, as a result, disturbed their populations and caused the reduction of pollination ecosystem services as a whole (Teichroew et al., 2016: 3). China's ecological restoration of this manner reflects its vision centered on human's increasing control and technological mastery of nature.

China's standing as the world's largest honey producer and provider of pollination ecosystem services with almost unparalleled diversity of both managed and wild honeybee species makes it a valuable source for lessons to be learned and serve as global examples in various ways. Furthermore, its current situation of "unparalleled pollinator diversity and massive-scale threats" combined on the one hand, and the "large available funding volumes for environmental management" on the other, makes pollinator conservation an issue of urgency for China but fortunately also a financial possibility (Teichroew et al., 2016: 3). However, the success of a pollinator conservation program will likely require another set of approaches different from those for ecological restorations in the Grain for Green program. For such approaches to be relevant, they must take into account the two complimentary conceptions – that at the same time humans are cultural and social beings they are also a biological species among other forms of life. Such dual conception explains how human agency in the Anthropocene is characterized by the combination of an immense power such as that seen in their inventiveness to push back the limits of nature and replace ecosystems function by human technology – and the frightening loss of control over unintentional occurrences such as that of the tipping points where a quantitative change of the honeybee populations will jolt their ecosystem services into a qualitatively different state (Horn and Bergthaller, 2020: 8 and 153).

The challenge for such tensions pervading the epochal consciousness of the Anthropocene is, however, not to resolve them but rather to account for them as precisely as possible. Epochal consciousness is therefore about trying to "articulate their preconditions" and indicate "a shared destiny which defines our situation" and determine "what sustains our horizon of action" (Horn and Bergthaller, 2020: 4–13).

From the perspective that humans are also cultural and social beings, the climate and environmental crisis in the Anthropocene can also be explained along the line of a crisis of culture since it stretches humans' "capabilities to imagine and think of the impending catastrophe and its possible solution" (Lo, 2019: 7). In this regard, human agency entangled in mutual dependencies among other biological species like the honeybees has become the domineering species benefiting from the honeybees' ecosystem services in pollination and honey production. Under the situation of dramatic decline in China's honeybee population of some areas, farmers have turned to use human pollinators to hand-pollinate their crops. The question of "economic systems, forms of consumption, lifestyles and technologies that have caused and continue to drive the ecological changes of the *Anthropocene*" abounds the impacts of our ecologically and environmentally damaging actions on the honeybees (Horn and Bergthaller, 2020: 8):

> The geological force of humanity is a cumulative effect of innumerable uncoordinated individual actions across the globe. Collective human action, in contrast, is only possible in culturally and politically differentiated groups. However, this differentiation of humanity into groups with conflicting interests also constitute the chief obstacle to the measures which are necessary in order to limit human impact on the Earth system.

China's human interactions with both managed and wild honeybees underpin China's global position as a major provider of pollination ecosystem services. When further considered in connection to China's role as the world's largest honey producer and exporter, the impact of China's actions imposes its ecological footprint not only on its own natural ecosystems but also on those within its web of honey trade. It reflects how in an attempt to exploit their own kind for more, humans with technological masteries of nature have been driven to devise methods to imitate the product originally provided by nature's ecological services. "A large number of entrepreneurial chemists" in the complex web of fraud and deceit have been "providing their own version of "ecosystem service" to generate a little opportunistic profit" (Dodwell, 2017). The ruthless side of competition driving the global honey trade holds some significant implications for the threats to food safety, food security, and ecological sustainability that could likely undermine China's future with its honeybees. In this regard, China would need to rethink its human–honeybee interactions – which would require nothing short of a paradigm shift in order to escape the race to its own demise.

Conclusions

In discussing human–honeybee interactions it is important to recognize that human, considered from the perspective of their power to transform their ecological environment, must be viewed from the status of being the cause as well as the subject of the Anthropocene. Under nature's self-regulating system humans have depended on honeybees – as one of the world's most economically valuable pollinators – to maintain a healthy ecosystem and provide the pollination services to one-third of the crops that feed the world. On the other hand, humans have developed the ability to create their own environment to manage the honeybees for their services in the production of honey – to spin it into the product of global demand – natural and counterfeit. Under such circumstances China's role as human agent in the Anthropocene is manifested by its intentional power in creating the enabling environment for its goal-oriented actions in becoming the world's largest producer and exporter of honey. But paradoxically, control over the unintentional force of side effects that China's extremely ambitious economic development has had on the ecosystem health upon which the honeybees depend is what China is without.

China's human–honeybee interactions understood as the climate and environmental crisis in the Anthropocene are also about the crisis of culture – when it becomes a matter of how humans think and imagine of the likely calamity and what possible solution they can devise. China's ecological restoration efforts through its Grain for Green program are known to constitute the world's largest payment for ecosystem services scheme, with the largest investment and greatest involvement by the broadest degree of public participation in history. It has been praised by those under global efforts turning to restoration as a way to mitigate climate change – as a major driver of China's success and a source of lessons from which the development community and national governments should learn. But from another side to this success – from the pollinator community perspective – the monoculture nature of the program may likely have, as a whole, reduced the pollination ecosystem for wild bees. How China sees human within nature's fragile self-regulating system underpins its interactions with honeybees and their conservation.

Being the largest provider of pollination ecosystem services globally – and the world's largest beekeeping country – China's global importance as a resource of genetic diversity in bee pollinators could be a tremendous source of knowledge for those in other parts of the world facing similar threats and pressures on pollinator communities. But such wealth of knowledge could only come about were there a shift from China's development visions centered around human technological mastery of nature to a new paradigm compatible with the epochal consciousness of the Anthropocene – a redefinition of its place as an integral part within planet's complex system of self-regulating process in relations to the honeybees. Human interactions with honeybees indicate to us, thus far, that their destiny is partly of our own making. More significantly, the fate of the honeybees will in part determine ours. With an epochal consciousness it is time that we redefine our relationship to them.

Notes

1 Diamond 1997 states that "in general, the agricultural and hunting practices of humans are greatly influenced by two important factors: the availability of wild species that are suitable for domestication and the geopolitical impediments to the spread of domesticated animals and farming technologies."
2 According to Crane 1999 there are four honeybee species native to China: *Apis florea*, *Apis dorsata* (the giant honeybee), Apis mellifera (introduced to China from the West), and Apis cerana (a specie native to the region).
3 According to Schneider (2011), in 2001 "Chinese beekeepers saw a bacterial epidemic of foulbrood disease race through their hives at wildfire speed, killing tens of millions of bees. They fought the disease with several Indian-made animal antibiotics, including chloramphenicol. Medical researchers found that children given chloramphenicol as an antibiotic are susceptible to DNA damage and carcinogenicity. Soon after, the FDA banned its presence in food."
4 Schneider (2011) attributes lead contamination in some honey from China to the mom-and-pop vendors who use small, unlined, lead-soldered drums to collect and store the honey before it is collected by the brokers for processing. According to Schneider, there are tens of thousands of tiny operators spread from the Yangtze River and coastal Guangdong and Changbai to deep inland Qinghai province.
5 National Standards of People's Republic of China GB 16740-2014, 2015: defines honey as "a natural sweet substance produced through fully brewing when the nectar, secretion and sweet deposits from plants are gathered, mixed with the secretion of their own, modified and stored in the honeycomb by honeybees."
6 The Codex Alimentarius (1981) prohibits the intrusion or extraction of any substance from honey, as opposed to the activities of the bees themselves. Codex Alimentarius, which is the internationally accepted standard for foods, defines honey as "the natural sweet substance produced by honeybees from the nectar of plants or from secretions."
7 The Western honeybee was first introduced to China as early as 1896 (Teichroew et al. 2016).

Bibliography

Chen, Stephen. 2017. "Why China's Asian honeybees are losing out to their Western counterparts." *South China Morning Post*, April 9, 2017. Accessed August 12, 2019. https://www.scmp.com/news/china/society/article/2083725/why-chinas-asian-honeybees-are-losing-out-their-western
ChinaPower. 2021. "How does water security affect China's development?" Accessed June 29, 2021. https://chinapower.csis.org/china-water-security/
Crane, Eva. 1999. *The World History of Beekeeping and Honey Hunting*. New York: Routledge.
Da Silva, P., Gauche, C., Gonzaga, L., Oliviera Costa, A., and Fett, R. 2016. "Honey: Chemical composition, stability and authenticity." *Food Chemistry*, 196: 309–323.
Dayne, Suzanna. 2017. "'Grain for Green': How China is swapping farmland for forest." *Forest News*, November 28, 2017. Accessed June 28, 2021. https://forestsnews.cifor.org/52964/grain-for-green-how-china-is-swapping-farmland-for-forest?fnl=
DCCC. 2017. "Honey export to China surge: Chinese importers focus." Posted March 8, 2017. Accessed June 26, 2021. https://www.dccchina.org/news/honey-export-to-china-surge-chinese-honey-importers-focus/

Delang, Claudio O. and Yuan, Zhen. 2015. *China's Grain for Green Program: A Review of the Largest Ecological Restoration and Rural Development Program in the World*. London: Springer.

De-Melo, A.A.M., de Almeida-Muradian, L.B., Sancho, M.T., and Pascual-Mate, A. 2018. "Composition and properties of *Apis mellifera* honey: A review." *Journal of Apicultural Research*, 57: 5–37.

Dodwell, David. 2017. "The honeybee is facing existential threat – and it could be very bad for humans." *South China Morning Post*, July 9, 2017.

Durham, Michael. 2004. "A bitter taste of honey." *The Guardian*, July 21, 2004. Accessed June 28, 2021. https://www.theguardian.com/profile/michaeldurham

DW (Deutsche Welle). 2014. "Pollinating by hand: Doing bees' work." *Environment*, July 31, 2014. Accessed June 27, 2021. https://www.dw.com/en/pollinating-by-hand-doing-bees-work/a-17822242

Egelie, Ashley A., Mortensen, Ashley N., Gillett-Kaufman, Jennifer L., and Ellis, James D. 2015. "*Apis Cerana*." Entomology and Nematology Department, University of Florida, January 2015. Accessed August 25, 2019. https://www.entnemdept.ufl.edu/creatures/misc/bees/Apis_cerana.htm

El-Wahab, Safaa Diab Abd and Eita, Lamiaa Hassnein. 2015. "The effectiveness of live bee sting acupuncture on depression." *IOSR Journal of Nursing and Health Science*, 4(4) Ver. III (July-August 2015): 19–27.

Escuredo, O., Míguez, M., Fernández-González, M., and Seijo, M. 2013. "Nutritional value and antioxidant activity of honeys produced in a European Atlantic area." *Food Chemistry*, 138, 851–856.

Fremaux, Anne. 2019. *After the Anthropocene: Green Republicanism in a Post-Capitalist World*. Cham: Palgrave Macmillan.

García, Norberto L. 2018. "The current situation on the international honey market." *Bee World*, 3 July 2018. Accessed August 12, 2019. https://doi.org/10.1080/0005772X.2018.1483814

Garcia, N. and Phipps, R. 2018. "Honey market report." *American Bee Journal*, 158: 23–30.

Gebhardt, Ulrike. 2014. "Bees and humans need each other," *Global Ideas*, January 28, 2014. Accessed June 27, 2021. https://www.dw.com/en/bees-and-humans-need-each-other/a-17391639

Gentlemen Marketing Agency (GMA). 2019. "Honey market shares in China." Brand in China, Food & Beverage, March 20, 2019. Accessed September 14, 2019. https://ecommercechinaagency.com/honey-market-shares-in-china/

Goulson, Dave. 2012. "Decline of bees forces China's apple farmers to pollinate by hand." *China Dialogue*, October 2, 2012. Accessed June 23, 2021. https://chinadialogue.net/en/food/5193-decline-of-bees-forces-china-s-apple-farmers-to-pollinate-by-hand/

Harper (Krynica), Jo. 2018. "Hi honey, I'm not from home." *DW Business*, October 7, 2018. Accessed August 13, 2019. https://www.dw.com/en/hi-honey-im-not-from-home/a-45403408

Harrington, Cameron and Lecavalier, Emma. 2014. "The Environment and emancipation in critical security studies: The case of the Canadian Arctic." *Critical Studies on Security*, 2(1): 105–119.

Head, Vivian. 2012. *The Beekeeping Handbook: A Practical Apiary Guide for the Yard, Garden, and Rooftop*. East Petersburg: Fox Chapel Publishing.

Hegland, S.J., Nielsen, A., Lázaro, A., Bjerknes, A.L., and Totland, Ø. 2009. "How does climate warming affect plant-pollinator interactions?" *Ecology Letters* 2009, February 12(2): 184–195. Accessed June 29, 2021. https://pubmed.ncbi.nlm.nih.gov/19049509/

Horn, Eva and Bergthaller, Hannes. 2020. *The Anthropocene: Key Issues for the Humanities*. London: Routledge.

Huang, Jikun. 2017. "Agricultural Transformation in China: Pathway, Driving Forces and Consequences." Paper presented at the World Food Policy Conference: January 16–17, 2017, Anantara Hotel, Bangkok, Thailand.

Huang, Jikun and Rozelle, Scott. 2018. "China's agricultural reform grows bolder." *East Asia Forum*, September 5, 2018. Accessed August 24, 2019. https://www.eastasiaforum.org/2018/09/05/chinas-agricultural-reform-grows-bolder/

Koo's, Acupuncture and Herb Clinic P.C. 2018. "Honeybee sting acupuncture." Traditional Chinese Medicine (TCM). Accessed September 3, 2019. https://www.koosacupuncture.com/honey-bee-sting-acupuncture

Kritsky, Gene. 2017. "Beekeeping from antiquity through the middle ages." *The Annual Review of Entomology*, 2017, 62: 249–264.

Lau, Constantine W. 2012. "Ancient Chinese apiculture." *Bee World* 89(4): 78–81. Accessed August 13, 2019. http://www.ibrabee.org.uk/component/k2/item/2686

Leeder, Jessica. 2011. "Honey laundering: The sour side of nature's golden sweetener." *The Globe and Mail*, April 29, 2011. Accessed: June 29, 2021. https://www.theglobeandmail.com/technology/science/honey-laundering-the-sour-side-of-natures-golden-sweetener/article562759/

Li, Christopher. 2019. "Chinese noneybees face endangerment." *Planet Bee Foundation*, February 20, 2019. Accessed August 12, 2019. https://www.planetbee.org/planet-bee-blog//the-chinese-honey-bee

Li, J., Qin, H., Wu, J., Sadd, B.M., Wang, X., et al. 2012. "The prevalence of parasites and pathogens in Asian honeybees *Apis cerana* in China." *PLoS ONE*, 7(11): e47955. Accessed August 12, 2019. http://europepmc.org/article/PMC/3492380#free-full-text

Liess, Stuart. 2015. "After Bee Die-off, Chinese apple farmers resort to hand pollination." *The Epoch Times*, April 30, 2015. Accessed August 12, 2019. https://www.theepochtimes.com/after-bee-die-off-chinese-apple-farmers-resort-to-hand-pollination_1321746.html

Liu, J., Kuang, W., Zhang, Z., et al., 2014. "Spatiotemporal characteristics, patterns, and causes of land-use changes in China since the late 1980s." *Journal of Geographical Systems* 24, 195–210. Accessed August 12, 2019. https://link.springer.com/article/10.1007/s11442-014-1082-6

Lo, Kwai-Cheung. 2019. "Introduction: Impoverishing Anthropocene with Chinese characteristics." In *Chinese Shock of the Anthropocene: Image, Music and Text in the Age of Climate Change*, edited by Lo, Kwai-Cheung and Yeung, Jessica, 1–17. Singapore: Palgrave Macmillan.

Ma, Yihan. 2020. "Annual production volume of honey in China from 2009 to 2019." *Statista*. Accessed June 25, 2021. https://www.statista.com/statistics/946665/china-honey-production-volume/

Miller-Struttmann, Nicole. 2016. "The complex causes of worldwide bee declines." *PhysOrg*, January 12, 2016. Accessed June 24, 2021. https://phys.org/news/2016-01-complex-worldwide-bee-declines.html

Ministry of Agriculture P.R. China. 2018. "Chinese market of honey and imported honey." 2018 World Honey & Bee Products Show. Accessed August 25, 2019. http://www.chinaagtradefair.com/chinesehonemarket.html

Nagy, Istvan. 2019. "European beekeeping in crisis." *EU Observer*, November 18, 2019. Accessed June 24, 2021. https://euobserver.com/opinion/146608

Oldroyd, Benjamin P. and Wongsiri, Siriwat. 2006. *Asian Honey Bees: Biology, Conservation, and Human Interactions*. Cambridge: Harvard University Press.

Partap, Uma and Ya, Tang. 2012. "The human pollinators of fruit crops in Maoxian County, Sichuan, China: A case study of the failure of pollination services and farmers' adaptation strategies." *Mountain Research and Development*, 32(2): 176–186.

Root, A.I. 2019. "2018 Annual honey report." *Bee Culture*, July 2, 2019. Accessed September 14, 2019. https://www.beeculture.com/2018-annual-honey-report/

Schneider, Andrew. 2011. "Asian honey, banned in Europe, is flooding U.S. grocery shelves." *Food Safety News*, August 15, 2011. Accessed June 25, 2021. https://www.foodsafetynews.com/2011/08/honey-laundering/

Stevenson, Andrew. 2016. "Report: China's native bees at risk." *GoKunming*, September 26, 2016. Accessed August 13, 2019. https://www.gokunming.com/en/blog/item/3818/report-chinas-native-bees-at-risk

Tamma, Paolo. 2018. "Honeygate: How Europe is being flooded with fake honey." *EURACTIVE.com*, July 19, 2018. Accessed August 7, 2019. https://www.euractiv.com/section/agriculture-food/news/honey-gate-how-europe-is-being-flooded-with-fake-honey/

Tang-Dong, Jin. 2006. "Honey flows upwards across China." *Bees for Development Journal*, 79: 4–5. Accessed August 14, 2019. http://www.beesfordevelopment.org/journal/issue-79/

Teichroew, Jonathan L., Xu, Jianchu, Ahrends, Antje, Huang, Zachary Y., Tan, Ken, and Xie, Zhenghua. 2016. "Is China's unparalleled and understudied bee diversity at risk?" *Biological Conservation*, 26 May 2016. Accessed August 7, 2019. https://www.sciencedirect.com/science/article/abs/pii/S0006320716302038

Thibault, Harold. 2014. 'When humans are forced to replace the bees they killed.' *Le Monde: WorldCrunch*, May 5, 2014. Accessed June 27, 2021. https://worldcrunch.com/tech-science/when-humans-are-forced-to-replace-the-bees-they-killed

University of Vienna. "New findings about the deformed wing virus, a major factor in honeybee colony mortality." *ScienceDaily*, November 11, 2016. Accessed August 12, 2019. http://www.sciencedaily.com/releases/2016/11/161111120731.htm

William Reed Business Media. 2017. "China continues to dominate world honey production." Beverage daily.com, March 16, 2017. Accessed August 13, 2019. https://www.beveragedaily.com/Article/2004/07/02/China-continues-to-dominate-world-honey-production#

Wills, Rick. 2015. "Catch the buzz: Illegal honey, again." *Bee Culture*, March 8, 2015. Accessed June 25, 2021. https://www.beeculture.com/catch-the-buzz-illegal-honey-again/

World Population Review. 2021. "China population 2021." Accessed June 29, 2021. https://worldpopulationreview.com/countries/china-population

Wu, S., Fooks, J.R., Messer, K.D., and Delaney, D. 2015. "Consumer demand for local honey." *Applied Economics*, 47(41), 4377–4394.

Zhang, Hongqi and Erqi, Xu. 2017. "An evaluation of the ecological and environmental security on China's terrestrial ecosystems." *Scientific Reports*, 11 April 2017. Accessed September 6, 2019. https://pubmed.ncbi.nlm.nih.gov/28400605/

Zhang, Minzhu. 2018. "Consumer Attitudes and Behavior Towards Honey in China." International Master of Science in Rural Development Thesis, Academic Year 2017–2018. Ghent University, Belgium. Accessed September 6, 2019. https://libstore.ugent.be/fulltxt/RUG01/002/482/264/RUG01002482264_2018_0001_AC.pdf

Zhang, Shuai, Liu, Yi, Ye, Yang, Wang, Xue-Rui, Lin, Li-Ting, Xiao, Ling-Yong, Zhou, Ping, Shi, Guang-Xia, and Liu, Cun-Zhi. 2018. "Bee venom therapy: Potential mechanisms and therapeutic applications." *Toxicon*, 148, June 15, 2018: 64–73.

Zhu, Fang and Wongsiri, Siriwat. 2008. "A brief introduction to apitherapy health care." *Journal of Thai Traditional & Alternative Medicine*, 6(3) September–December 2008: 303–312. Accessed September 6, 2019. http://arabbeeresearchers.net/wp-content/uploads/2017/06/A_Brief_Introduction_to_Apitherapy_Health_Care.pdf

9 Human–Animal Interface and Sociocultural Ecology of Zoonotic Disease Outbreaks in Anthropocene China

Sungwon Yoon

Introduction

The emergence and re-emergence of pandemics occurred in recent decades underscore the role that human–wildlife interface played in precipitating the spread of infectious diseases. It is reported that 75% of all infectious diseases are potentially zoonotic and of wildlife origin (Salyer et al., 2017). Animal diseases that cross over to human population (and vice versa) are considered as zoonoses. Outbreaks of Ebola virus infection, Middle East respiratory syndrome (MERS), and severe acute respiratory syndrome (SARS) virus infections are some of the examples that clearly illustrate the adverse consequences of the incursion of humans into wilderness. Impoverished areas, particularly in developing countries, often act as "hotspots" for these zoonoses. Although the role of human–wildlife interface in predicting emergent zoonosis (i.e., the transmission of an infectious disease from animals to humans) is well documented, little attention has been paid to the sociocultural dimension of zoonosis.

In other words, sociocultural variations in attitude toward wildlife species as well as perceived risks when interacting with animals have not been well understood. Wildlife consumption is a widespread practice in Asian region – ranging from Indonesia to Indochina to China. There is demand for not only the processed products such as leather, hide, or bones but also for the parts or meat of the animals. The basis for this form of consumption stems from misguided belief in the medicinal properties of these animals, cultural proclivities, and tendencies to egoistic consumption. Drawing on zoonotic disease infections and the responses in China, this chapter aims to examine the cultural and behavioral milieu in which human consumption of wildlife species as appetite and aspiration takes place in market capitalism and how such context contributes to the pathogenic crises in anthropogenic China and Asia.

This chapter argues that human's anthropogenic activities threaten the capacity and resilience of ecosystems. More specifically, human–animal interactions through the consumption habits of some Chinese people or the way animals are handled have a significant ramification for the ecosystems, environment, and, most importantly, the population health. The SARS virus from palm civets and avian influenza from infected chickens are cases in

DOI: 10.4324/9781003212089-9

point. The chapter describes the complex relations that involve construction of animals (wildlife or domesticated livestock) as a substitute in Chinese culture, socioeconomic drivers in contemporary everyday life, and anthropogenic implications. The chapter then discusses the global discourse of One World One Health, a collaborative effort at local, national, and global levels to guarantee an optimal healthy status for humans, animals and environment, and presents how the discourse has been translated into the local level in China and wider Asia in light of emerging zoonotic disease outbreaks. The chapter maintains that in order to optimally address these challenges, it is not enough for government to simply draw up laws or strengthen public health capacity to contain the zoonotic diseases outbreaks. Instead, effective multisectoral partnerships should be developed within China and with other national governments to help address the underlying sources that contribute to zoonoses. One important aspect is to fundamentally alter the alleged cultural beliefs and attitudes toward wildlife consumption and handling of wildlife or livestock through scientific education at the grassroots level.

Traditional Chinese Cuisine and Wildlife: Appetite for the Exotic

Cantonese cuisine is probably the most famous and globalized of all the different culinary schools of China. Across the Chinatowns of the world, rows of roast duck, braised chicken and char siew being adorned the smudged-up windows of the restaurants testify the resilience and popularity the food of migrants from Guangdong provinces since the 19th century. The overseas Chinese, however, were only able to conjure their staple dishes from the ingredients made available to them in their adopted homeland, popularizing the most common dishes that have made Cantonese cuisine as popular as it is today.

In the heartland of Pearl River Delta in Southern Chinese province of Guangdong today, Cantonese cuisine is considered one of the most popular and refined cuisine in the People's Republic today. Within Chinese nation, it is joked that there is nothing under the sky that Cantonese would not eat, often referring to the adventurous and undiscriminating palate of the Cantonese people. This is true, as Southern Chinese cuisine every so often included a range of animals that may shock most people. From snake broth to salamander soup, from spicy rabbit heads to fried crickets, there is an entire subset of exotic foods that locals relish and consume (Mcleish, 2018). The Cantonese have a dish known as 龙虎凤 [Literally Dragon, Tiger and Phoenix which are symbols for the snake, cats, and chickens]. Using the meat of these animals, the Cantonese would cook this dish in late autumn to nourish themselves. Listed ingredients often include shredded meat of not only chicken, snakes, and cats, but also civet cats and dried fish stomach (Baidu, n.d.). One, for instance, can find supper of fried field mice or snacks of cured mice bacon (though this is not confined to China – Vietnam, Myanmar, Laos, Mozambique are countries where mice is considered a delicacy) (Zhang et al., 2012).

Surveys of wildlife markets in Guangzhou, China, invariably describe the diverse range of wildlife species such as masked palm civets, ferret badgers, barking deer, wild boar, bamboo rats, endangered leopard cats, and various species of hedgehogs, foxes, squirrels, gerbils, and snakes (Li, 2020). In particular, the civet cats figure prominently in Chinese cuisines as the main ingredient in the exotic wildlife dish, posing significant challenges for China's ecology and environment (Zhang et al., 2012). To be fair, the eating of the "exotic" is not just a Cantonese sin alone – in each and almost every one of the Chinese provinces, there are delicacies suited to local palettes that would often provoke disgust among non-Chinese. Many of these foods and dishes are considered by traditionalists to be particularly healthy or nourishing (O'Reilly, 2018). The consumptions of shark fins, bird nest, and frogs, for instance, are other examples.

The dining experience, however, is not only confined to restaurants. There have been reports of a Chinese businessman traveling from Guangdong to Guangxi who buy tigers and bring them home in order to conjure up feasts for his friends (Reuters, 2015). Today, even as the fear of diseases has prompted government to shut down markets and vendors selling exotic meat, this practice still goes on. There are places in rural areas where people could visit to eat game – and these places usually look more like farms than anything where exotic wildlife are caught and kept in cages. These places offer many of those visitors a dining experience they would never find in mainstream restaurants – and people visit for a variety of reasons: curiosity; a belief in the restorative or enhancing effects of eating this exotic wildlife; a desire to be adventurous or simply because of egoistic consumption. These beliefs override the common sense and the inclination to pay attention to the law in many people's consumption behavior (Robson, 2017).

Beyond the question of taste or the novelty of it, the consumption of exotic wildlife in reality does not bring about the kind of restorative or nutritional value most vendors or consumers of these dishes suggest them to have. Due to the changing demographic patterns, increased wealth and varying consumption patterns, the way animals are fitting into the modern Chinese diet has changed (Flitton, 2016). Not only are people consuming more of the traditional livestock such as chicken, pork, or beef, people often become more adventurous as they seek out new food elsewhere, particularly forbidden ones. There are reports of Chinese traveling to Myanmar or Vietnam border towns to the wildlife markets there to consume "exotic" animals that are forbidden in China. There is every possibility that these cross-border visitors also bring zoonotic diseases back to the country.

The consumption of wildlife, however, might not just be limited to those who willingly seek out to eat these meats. It is entirely possible that many in China are not aware of what they are eating in their meals. Just like many people who eat sausages in the West are often unaware of what goes into the sausages, it is well known in China that processed products and meats served at cheap eateries might not necessary be what the consumer anticipates.

The Animals Strike Back: Emerging Zoonotic Disease Outbreaks

Food Safety Concerns

The first immediate health challenge that many Chinese people face is the fact that they might be consuming animals which they are unaware of. How is this possible? Fraud. Due to the price increases of some livestock such as beef or mutton, unscrupulous vendors have begun to use methods to fraudulently pass off the meat of animals for beef or mutton. There have been numerous reports of vendors using artificial coloring to color pork to make them taste like beef, and so as beef. There are also reports emerging of how vendors are using meat of animals normally not eaten such as rats (see Figure 9.1), foxes, or ducks and disguising them as mutton. Beyond the question of fraud, such practices also raise public health concerns regarding the chemicals used to change the taste and color, as well as the question of consuming meats that might not be safe for human consumption. All in all, such consumption brings to the fore food safety concerns with the consumption of animals in China.

Figure 9.1 Field Mice Dish in Vietnam

(Photo by Cookie Nguyen; CC 4.0 International https://commons.wikimedia.org/wiki/File: Cooked_field_mouse_in_Van_Giang.jpg).

Emerging Zoonotic Disease Outbreaks

Over the course of the last two decades, China has seen a series of infectious disease outbreaks related to animals and birds. Never witnessed before in history, at the center of the outbreaks is animals that have lived in harmony with humans in the centuries before. Some categories of these animals are livestock that humans have reared and consumed for centuries while others are wild and exotic animals that have coexisted with human beings. There are concrete cases: the SARS outbreak due to the handling and consumption of civet cats in Guangdong (see Figure 9.2); the avian influenza outbreak; and the swine flu outbreak (Bloomberg, 2014; Doucleff, 2017). Mostly recently, we have witnessed the global pandemic of the COVID-19, where the origins have not been authoritatively established. We shall examine the cases in-depth in the following sections. These infectious disease outbreaks and pandemics now truly define our era. For the first time in recent history, we see the consequences of human–animal interactions becoming extremely lethal in a short span of time. We are also seeing increasingly frequent bouts of global disease outbreaks, caused when humans handle or consume different animals or wildlife (Court, 2018). The consequences are becoming more deadly (both qualitatively and quantitatively). If anything, it can be argued that this marks the beginning of a new era in human–animal interaction in Asia, and if we are not careful, the animals that we eat would end up having the last laugh.

Figure 9.2 Example of a Civet Cat in Cage for Kopi Luwak Coffee production in Indonesia

The coffee seeds passed out from the civet cats are collected and processed to be made into premium coffee for Indonesia and beyond.

(Photo by SUTR; CC2.0 Generic http://flickr.com/photo/29297680@N00/5609840328).

The Chinese Government's Response

The response of the Chinese government in these events is important. Infectious disease outbreaks and pandemics are not something that can be defeated easily, nor do they heed to the government's call, no matter how authoritarian the government is. The 2003 global SARS outbreak is probably the first major test case of how the government handles the disease outbreaks – and gaining from experience the government is able to move to handle the avian influenza and swine flu cases.

In response to emerging zoonotic disease outbreaks, the Chinese government has actively tried to close down markets that sell exotic meats and wildlife and has banned the consumption of some dishes like the "Dragon, Tiger, Phoenix" in Guangzhou in the aftermath of the SARS outbreak. When cases of avian influenza or swine flu are reported, the suspected stocks which might be contagious are culled. The medical community in China has been mobilized to study these stocks, and national attempts are being moved to develop the antivirus. Yet, the problem never goes away. In the years after SARS, we are seeing the patterns of behavior becoming more prevalent. The problem is that when the threat receded, the vendors come back in full force. Unless something is done to provide the vendors a different livelihood, it becomes almost impossible to prevent the vendors from hawking their ware.

The SARS Outbreak as Negative Consequence of Human–Animal Interface

SARS provides a pertinent case for understanding the negative consequences of human–animal interactions in China because, for the first time, the outbreak brought together state and non-state actors to address what Fidler (Fidler, 2004) describes as the first zoonotic pandemic of the 21st century (Bloom, 2003). SARS is a newly identified human infection caused by a type of coronavirus unlike any other known animal virus in its family. An analysis of epidemiological information from the various outbreak sites shows that the overall case fatality ratio approached 11%, with much higher rates among elderly people (Anderson et al., 2005).

Transmission mainly occurred person-to-person through exposure to infected respiratory droplets expelled during coughing or sneezing or following contact with body fluids during certain medical interventions (WHO, 2003a). Contamination of the environment, arising from fecal shedding of the virus, is thought to have played a small role in disease transmission. Management of SARS relied on standard epidemiological interventions: identification of those fitting the case definition, isolation, infection control, contact tracing, active surveillance of contacts, and evidence-based recommendations for international travellers. Though demanding and socially disruptive, particularly when a large number of people were placed in quarantine, these standard interventions, backed by high-level political commitment, were said to be sufficiently effective to contain the global outbreak less than four months after the initial alert (Leung and Ooi, 2003).

The earliest cases of SARS are now thought to have emerged in November 2002 in China when the first patient was hospitalized and treated in Foshan No. 1 Hospital. The patient went on to infect eleven other people, including nine medical workers (Pan, 2003). On December 15, two SARS patients were hospitalized in the city of Heyuan and five medical workers were then infected. Driven by a rumor that Heyuan was being attacked by an unknown and highly contagious virus, this prompted panic buying of vinegar and anti-viral medicine in late December (Chen, Wang and Duan, 2003). The atypical pneumonia cases were reported to the Guangdong Health Bureau on 2 January 2003. The next day the Heyuan Centre for Disease Control and Prevention (CDC) became the first official Chinese voice discussing the out-break by publishing a notice in the *Heyuan News*. It claimed that there was no epidemic in Heyuan, and that the symptoms, which were caused by unu-sually cold weather, could be cured with immediate medical treatment. On January 16, a patient from Zhongshan, whose symptoms were similar to the Heyuan atypical pneumonia cases, was transferred to the Guangzhou People's Liberation Army Hospital. However, a CDC official in Zhongshan denied the existence of any cases of atypical pneumonia in the city and called the news "mere rumour." On January 21, the Guangdong Health Bureau produced, but did not release to the public, a full report on the outbreak situation.

The initial phase of the Guangdong outbreak, characterized by small, independent clusters and sporadic cases, was subsequently followed by a sharp rise in cases during the first week of February 2003. This was thought to result from amplification during the care of initial patients in hospitals. Cases gradually declined thereafter. Altogether, some 1,512 clinically con-firmed cases occurred in the Guangdong outbreak, with healthcare workers in urban hospitals accounting for up to 27% of cases. This pattern was repeated as the disease began to spread outside Guangdong Province to other areas in China, and then internationally. On February 10, global partners identified reports of a "strange contagious disease" with respiratory symp-toms affecting health workers in Guangdong hospitals and causing wide-spread panic. An urgent alert was sent to network members. The next day, the WHO received reports from the Chinese Ministry of Health confirming the outbreak of an acute respiratory syndrome with 300 cases and five deaths in Guangdong Province, with the government claiming that the situation was coming under control. Concern intensified on February 19, when authorities in Hong Kong reported an outbreak of avian influenza in members of fami-lies who had recently travelled to southern China. The WHO alerted its col-laborating laboratories and activated its influenza pandemic plans (WHO, 2003e).

After the disease had spread from the southern Chinese province of Guangdong, Hanoi, Hong Kong, Singapore, and Toronto became "hot zones" (WHO, 2003b) of SARS, characterized by a rapid increase in the number of cases, especially among healthcare workers and their close contacts. In these locations, SARS first took root in hospital settings where staff, unaware that a new disease had surfaced, exposed themselves to the infectious agent without barrier protection. All of these initial outbreaks were subsequently characterized by chains of secondary transmission within and outside, beyond the healthcare environment. Secondary attack rates of greater than 50% were observed among healthcare workers caring for patients with SARS in both Hong Kong and Hanoi (US CDC, 2003). Within two weeks, similar outbreaks occurred in various hospitals in Singapore and Toronto (WHO, 2003b). This eventually culminated in a global alert and the institution of worldwide surveillance measures, as declared by the WHO on 12 March 2003 (WHO, 2003a).

On 12 March 2003, the WHO issued a global alert after being notified of mounting cases of severe atypical pneumonia among staff in Hanoi and Hong Kong hospitals. After receiving additional reports of cases in Singapore and Toronto over the next three days, the WHO responded by issuing a series of emergency travel recommendations to alert health authorities, physicians, travellers, airlines, and the public to the potential threat which this then unidentified illness could have on healthcare systems all over the globe (WHO, 2003h). These travel recommendations marked a turning point in the early course of the SARS pandemic (WHO, 2003g). Areas with cases detected before the WHO recommendations were issued – namely Vietnam, Hong Kong, Singapore, and Toronto – experienced the largest and most severe outbreaks. After the recommendations were issued, all countries with imported cases, with the exceptions of Taiwan and the interior provinces of China, were able to either prevent further transmission or keep the number of additional cases at a level manageable with existing resources through prompt detection and isolation of cases, strict infection control measures, rigorous contact tracing, and quarantine measures in some circumstances (WHO, 2003d).

During the last week of April, the outbreaks in Hanoi, Hong Kong, Singapore, and Toronto showed signs of peaking. On April 28, Vietnam became the first country to stop local transmission of SARS. However, new probable cases, including cases among hospital staff, additional deaths, and first cases imported to new areas continued to be reported from several countries. The cumulative total number of cases surpassed 5,000 on 28 April 6,000 on May 2, and 7,000 on May 8. By then, cases had been reported from 30 countries on 6 continents. Most new cases were reported from Beijing and increasingly other parts of Mainland China. Of the cumulative global total of 7,761 probable cases and 623 deaths reported on May 17, 5,209 cases and 282 deaths had occurred in Mainland China. Also of concern was a rapidly growing outbreak in Taiwan with a cumulative total, on May 18, of 344 cases, including many hospital staff, and 40 deaths (WHO, 2003b). This was particularly worrying given that Taiwan was not a recognized member state of the WHO and thus had limited access to information or international support.

On 5 July 2003, the WHO reported that the last human chain of transmission of SARS had been broken (WHO, 2003f). On the same day, the WHO declared that outbreaks of SARS had been contained worldwide. While there have been a few cases of infection with the SARS virus in Singapore, Taiwan, and China since the summer of 2003, the WHO did not put forward recommendations concerning travel or other restrictions (WHO, 2003c). Nevertheless, SARS demonstrated dramatically the widespread global effects that can be created by a single, yet deadly, emerging zoonotic disease (US National Intelligence Council, 2003).

Global-local Discourse on Human–Animal Interface: One World One Health

One of the most important topics which have emerged in the discussions over the recent global disease outbreaks is the role of zoonosis in these outbreaks and the mutation and transmission of zoonotic pathogens. Whether it is the civet cats consumed by Chinese southerners in Guangdong Province or the birds carrying the H5N1 virus, the realization that viruses can mutate and be transmitted from animals to humans has raised alarm bells around the world. This growing awareness of the interconnection between animal and human health has led to a new global discourse of "One World, One Health" (OWOH) (Scoones and Forster, 2008; Mackenzie et al., 2013). The concept of OWOH is built on the understanding that disease risks resulting from interactions between animals, humans, and the environment (see Figure 9.3 below) need to

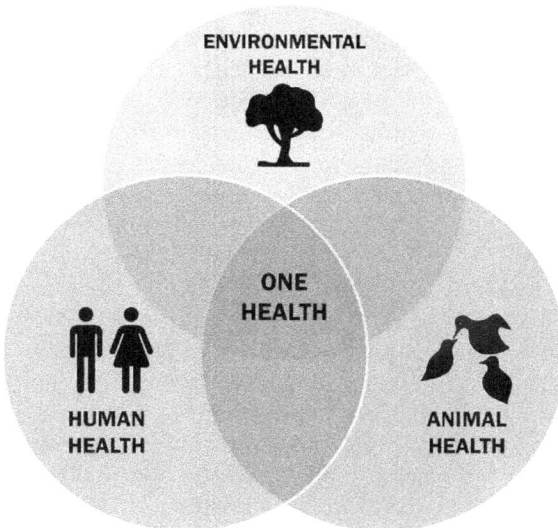

Figure 9.3 Animal-Environment-Human Health Triad based on WHO's One World One Health Concept

(Diagram Created by Thddbfk / CC License ASA 4.0 International; https://commons.wikimedia. org/wiki/File:One-Health-Triad-en.png).

be addressed through greater inter-sectoral collaboration in both science and its application (World Bank, 2021).

The idea of OWOH first emerged in 2004 at a symposium in New York that focused on the potential movements of diseases among human, domestic animal, and wildlife populations. The event was organized by an alliance of several international organizations, including the Food and Agriculture Organization of the United Nations (FAO), the World Organisation for Animal Health (OIE), the WHO, the World Bank, and the United Nations System Influenza Coordination (UNSIC), that sought to develop and advocate for the implementation of the concept. The conference produced the "Manhattan Principles," which lists 12 recommendations for establishing a more holistic approach to preventing epidemic/epizootic diseases and for maintaining the integrity of ecosystems for the benefit of humans, their domesticated animals and foundational biodiversity.

The concept of OWOH recognizes that there is a need to improve measures for the prevention and control of emerging and zoonotic infectious diseases by applying multidisciplinary, cross-sectoral approaches at the human–animal-environment interface. The driving force behind the slogan can be found in the following opening statements made at the conference in Manhattan,

> Recent outbreaks of West Nile Virus, Ebola Haemorrhagic Fever, SARS, Monkeypox, Mad Cow Disease and Avian Influenza remind us that human and animal health are intimately connected. A broader understanding of health and disease demands a unity of approach achievable only through a consilience of human, domestic animal and wildlife health.... To win the disease battles of the 21st Century ... requires interdisciplinary and cross-sectoral approaches to disease prevention, surveillance, monitoring, control and mitigation as well as to environmental conservation more broadly.... We are in an era of One World One Health and we must devise adaptive, forward-looking and multidisciplinary solutions to the challenges that undoubtedly lie ahead.
>
> (Manhattan Principles, 2004)

The aims of the strategic framework of OWOH became more concrete with the announcement of the "Beijing Declaration" in 2006 where a range of international organizations – the FAO, OIE, WHO, World Bank, European Commission – and representatives from more than 100 countries met to mobilize and help coordinate financial support from the donor community for national, regional, and global responses to Highly Pathogenic Avian Influenza (HPAI) and to support efforts at all levels to prepare for a possible human influenza pandemic. Various initiatives arising from the Beijing Declaration have since led to a number of high-level international advocacy meetings. For instance, the *International Ministerial Conference on Avian Influenza* held on October 2008 at Sharm El Sheikh in Egypt carried out a thorough review of the progress of the Beijing Principles

(FAO). The meeting discussed and identified the importance of wildlife epidemiology in HPAI and concerns surrounding the wide-ranging socio-economic impacts of emerging infectious diseases. As the meeting document noted,

> Recognizing the multidimensional nature of HPAI and EID, which involves different health domains and socioeconomics, there is a need for both a wide range of stakeholders and to promote strategic collaboration and partnerships across various disciplines, sectors and department, ministries, institutions and organizations at the country, regional and international levels.
>
> (UNSIC 2008)

The concept of OWOH was endorsed at the *International Ministerial Conference on Animal and Pandemic Influenza* in Vietnam in April 2010. At the meeting, the FAO, OIE, and WHO agreed upon and signed a strategic framework to work closely together to address the human-animal-ecosystem interface, resulting in a tripartite agreement (FAO, OIE, and WHO, 2010). Subsequently, in May 2010, a meeting hosted by the US Centers for Disease Control (US CDC) was held in Atlanta where selected national policymakers, the European Commission, the UN, the World Bank, and other diverse institutions from policy and economic sectors discussed progress in terms of practices related to OWOH. In February 2011, the first *International One Health Congress* was held in Melbourne, Australia (One Health Congress, 2008), to discuss the benefits of working together to promote a One Health approach to human, animal, and environmental health. Subsequently, the promotion of "Whole of Society" multi-sector approaches to pandemic preparedness was affirmed at the International High-Level Technical Meeting organized by UNSIC held in Mexico in November 2011 (Office of the Senior UN System Influenza Coordinator, 2012).

In Asia, the *APEC One Health Forum* was held in Hong Kong in May 2011 to participate in the global One Health dialogue, build networks, and increase cross-sectoral functioning to address the threat of emerging infectious disease (including zoonoses) (Health-APEC). The outcome of the forum was agreement on an *APEC One Health Action Plan*, which was designed to encourage commitment to progressing the implementation of the OWOH approach in APEC member economies. Additionally, ASEAN has partnered in many initiatives to address HPAI and other pandemic issues. The ASEAN +3 Emerging Infectious Disease Programme and the ASEAN Sectoral Group on Livestock played an increasingly stronger role in pandemic influenza preparedness and response by working closely with member countries to support regional coordination mechanisms and to promote partnerships between the human and animal health sectors (Kamdtscott and Yoon, 2010).

In reviewing the OWOH dialogue, it is apparent that OWOH has become a new normative framework that fosters collaboration between many different actors working within the broader area of human and animal health in recent years. However, within the Asian context, there is anecdotal evidence that suggests that the OWOH does not seem to fully take into account the meaning of locality in undertaking the global actions and responding to them. Whether this is due to the lack of effective governance arrangements among different stakeholders, unrealistic global narratives, conflicting resource priorities, or absence of political commitments, the disjuncture between the global discourse and local reality appears to be pervasive.

From the fieldwork interviews in the course of the author's research, it is apparent that despite this successive implementation, the program is under a serious threat of closure due to high costs as well as the labor-intensive nature of operating such a program.

More fundamentally, the absence of clarity in the operational definition of OWOH on the ground remains a challenge. In order to achieve greater effectiveness, it was suggested that leadership be required for the integration of OWOH elements into the operation perhaps in a more conventional "top down" model with a clearly delineated command and control approach. For example, although a great deal of effort has been put into the establishment of multi-sectoral plans between the Ministry of Health and the Ministry of Agriculture, there appears to have been no clear guidance on the way in which collaboration is conducted to translate the plans into operational ones. Working across ministries and other government agencies in terms of coordination without clear protocols and overarching multi-ministerial agreements was commonly observed.

As a consequence, joint programs involving both the health and agricultural sectors in practice tend to operate as separate programs with minimal crossover. For instance, joint courses in education for veterinary and medical professionals to promote multi-sectoral collaboration were rarely conducted. One informant stated that:

> There was an attempt to organise a joint integrated epidemiology training program and there was a big push that this training program should include both medical epidemiologists and veterinary epidemiologists. In the end, it was just not possible for the Government. It was pushed by international agencies here, the same agencies [i.e., FAO and WHO] that are talking about One World, One Health. The official decision was that the Ministry of Health didn't recognise that the qualifications of veterinary epidemiologists were adequate to enrol in this program.
>
> (Interview)

The above statement indicates that Asian governments struggle to conceive of the wider coordination and the roles of other Ministries beyond where they are needed to support the response from the health sector. As things stand, the OWOH concept that addresses the development of truly wide-reaching capabilities in engaging stakeholders from a wide variety of sectors appears to be lacking.

This was a theme that was replicated across interviews. One external observer in China noted, for instance:

> Inter-sectoral collaboration between Ministry of Health and Ministry of Agriculture is progressing but not very well established in China because different sector has a different interest. The Ministry of Agriculture in China – though this may be true to elsewhere – has a very different perspective than the Ministry of Health. In China, at a certain moment, Ministry of Health had their own pandemic plan, and the Ministry of Agriculture had their own pandemic plan. But what they didn't have was actually how they would respond together at a higher phase of pandemic. In the case of Avian Influenza for example, because it was a poultry disease, there was no trigger which could get the State Council involved and organize a systematic multi-sectoral system.
>
> (Interview)

Correspondingly, a lack of understanding of the cultural aspects of the human–animal interface was identified as one of the primary barriers hindering the concept of OWOH to be firmly implemented on the ground. One external observer commented that

> There are so many structural issues between health and agriculture. But it could be cultural and really extend 50-100 years in terms of human civilisation. When I listen to people talking about the reform of the agricultural sector, I'm worried that it's much bigger and it's really a philosophical question that's decades down the line and that relates to the way human society relates to other animals, mostly domestic but also wildlife. So, it's cultural and political and not just technical. Ironically, there is a huge fuss in Western industrialised countries that say that Asia doesn't really have a comprehensive response to zoonosis.
>
> (Interview)

In interviewing government officials and external observers, it is apparent that by and large, issues such as poverty, which enables infectious diseases to take advantage of their environment and to spread, remain somewhat in the shadows despite clear indications of the importance of understanding the sociocultural context of endemic zoonotic diseases in the concept of OWOH (UNSIC, 2008). One informant contended that without addressing the socioeconomic backdrop of pandemic influenza, the concept OWOH would only be rhetoric:

> We are as strong as our weakest link. My frustration on working on avian influenza is that I do not see development partners moving clearly enough in the distancing between animals and humans. My strong view is that if we can make sure that the chickens and the people don't sleep in the same

house in China or Indonesia, it will be an enormous risk reduction. Why can't we invest in a chicken run? On those things, we don't get any attention. We don't get any support from development partners and donors, and that's an area where I still think that they're [development partners] missing it. If there is anything in prevention, that's what we need to be doing.

(Interview)

As observed above, the lack of adequate biosecurity, especially at the village level where backyard farming is common, stands in stark contrast to the OWOH approach of emphasizing the enhancement of farm biosecurity. At the same time, the OWOH approach neglects to embrace equity as an underlying principle in policy development, resource allocation, and implementation. Issues including people's livelihoods, the farming system and the role of poultry in them, and the patterns of poultry trade and movement have not been given much attention. The experience and conception of OWOH represents how the global approach to zoonotic disease outbreaks, largely constructed around a top-down and compassionate model, fails to understand the complexity of responses and cultural dimensions within the local context in China and other Asian countries.

Figure 9.4 Illicit Endangered Wildlife Trade in Möng La, Shan, Myanmar

(Photo by Dan Bennett, CC 2.0 Generic https://commons.wikimedia.org/wiki/File:Myanmar_Illicit_Endangered_Wildlife_Market_04.jpg).

Conclusion: Ways Forward

Sociocultural and Behavioral Responses

From a public health perspective, things are quite clear-cut. Public health professionals simply advocate not eating wildlife, and not coming into contact with the wildlife for good reasons. Exotic game meat and folk recipes should best be avoided for these reasons. Human beings should best stick to domesticated livestock. This, however, is easier said than done as there are cultural and political-economic factors at play here.

In Cantonese culture, animals are conceived to be food. It is not only the Cantonese that are guilty of this. For most part, East Asian cuisine is not known to be vegetarian-based diets. The ability of modern man to harness technology like fire and refrigeration and to develop culinary advancement is an important hallmark of civilization. Even though society is not expected to regress in these developments, mankind as a whole should understand that a further step in evolution is taking care of how we manage the consumption of wildlife as exotic and game meat.

The individuals who are engaged in the hunting or handling of game meat should also gain some knowledge of understanding how diseases may be transmitted from animals to humans. The main problem is that these communities or individuals who tend to consume "exotic meat" are often ignorant of the public health risks the exotic animals or wildlife can pose to human communities. In the most recent case in China's Inner Mongolia, the zero patient who had bubonic plague through the consumption of wild rabbits probably had little idea of how he might possibly have contracted the disease. As of April 2022, the Inner Mongolia announced a level 6 bubonic plague warning, which advised residents to stay away from wild animals, including rates and foxes (Global Times, 2022).

For cases like avian or swine influenza, the sources of transmission are usually from domesticated animals. Avian influenza is caused by a mutated strain of virus that resulted in chicken stock getting infected by overflying migratory birds. In the latter, it is mutated from pigs living in unhygienic conditions. Most of the places in which the livestock animals are kept are unsightly and often unhygienic. The reason is simple: there is no need to invest in the housing for these animals who are about to be slaughtered. If the investment is made, cost will go up. The fact of the matter remains: keeping a better place for the animals will result in better quality livestock. There is a reason why air-flown free-range chickens cost more than the ones kept in cages. Even though the market allows for this differentiation, from the public health perspective, this should be made standard (or as close as possible).

Such a change can come only if stakeholders and the public change their views of the planet and perceive the animals they eat (or not eat) as part of the planetary ecological system we share. It might not be practical to hope that human beings stop the consumption of animals, but it is fundamentally important that all of us re-evaluate what we are doing here. Education aimed at eradicating skewed beliefs about the nutritional value of wildlife and exotic animals, or the public health aspects of handling wildlife or livestock, needs to be transmitted to those who work in these areas. The younger generation should also be taught the idea that the animals who live on this planet are indeed our cohabitants, and that doing our utmost to keep our environment pristine and whole for them would also mean that we will benefit from this too.

Policy Responses

In an increasingly globalized world we live in, zoonotic disease outbreaks become no longer considered to be national affairs. Animals and viruses do not respect national boundaries; the transmission of virus between animals and humans could result in the rapid spread of diseases across national borders.

From the case of global SARS outbreak, it is evident that national responses often impede, rather than facilitate, efforts that can help mankind overcome global calamity. The SARS outbreak demonstrates that traditional divisions between national and global health policy were inadequate. In fact, the outbreak prompted governments to pursue collective actions to tackle the common problems. The SARS outbreak also illustrates that global collaboration against zoonotic diseases is not only a matter of institutional and technical interventions, but also a reminder of how important it is for governments to implement policies that consider human–animal ecosystem and cultural dimensions.

It is generally found that a cross-sectoral or sector-wide approach has proved less enticing and that much of the effort on the ground has remained fragmented and at times redundant. The dialogue between animal and human health sectors appears to have involved political wrangling, which reflects different interests of the respective sectors. More fundamentally, the exercise of OWOH does not effectively embrace the vision of ameliorating the poverty of backyard farmers, understanding the different stakeholders, and addressing enduring problems with harmonization of independent funding streams at the national level. This means that more discussion and ongoing evaluation are needed for OWOH to become a meaningful and solid framework and a practical basis for the future roles of international agencies in the best interests of resource-poor Asian countries, including China. A more systematic forum and open discussion apparatus within the regional architecture might be needed for better diplomacy and information sharing on health and animal issues.

Lastly, the national and global policy response should not be just limited to the combating of public health emergencies of international concern such as emerging zoonotic disease events and pandemics. It should extend to more mundane fields of livestock management techniques that include, among other things, improving health management and living conditions of livestock and handling of animals. It is important that governments such as those in China step in to extend a hand to markets and farms to help build a better infrastructure. While this may be costly, it would still cost less than combating a pandemic in the long run. Recognizing that animals, whether wild and exotic or domesticated, are our coinhabitants on this planet would be the best way for the future ahead.

Bibliography

Anderson R et al. (2005), "Epidemiology, transmission dynamics and control of SARS: The 2002-2003 epidemic," in McLean A et al. (eds.), *SARS: A Case Study in Emerging Infections*, 61–80. Oxford: Oxford University Press.

Baidu, (n.d.), Write up on "Dragon, Tiger, Phoenix", https://baike.baidu.com/item/ 龙虎凤

Bloom BR (2003), Lessons from SARS. *Science* 300: 701.

Bloomberg, (2014), From Beijing to Hong Kong, Tracking H7N9's spread, Bloomberg, 27/01/14, https://www.bloomberg.com/graphics/infographics/tracking-h7n9-bird-flu-in-china.html

Chen G, Wang C and Duan G (2003), Calm response to settle the crisis – historical records of Guangdong's battle against atypical pneumonia, Southern Daily, 20/02/2003.

Court E, Over 600 Deaths As China Suffers Bird Flu Outbreak?: The "terrifying" epidemic kills over one third of the infected humans, Plant Based News, 17/06/2018, https://www.plantbasednews.org/lifestyle/over-600-deaths-china-bird-flu-outbreak

Doucleff M, Deadly Bird Flu in China Evolves, Spreads to New Regions, 07/09/2017, https://www.npr.org/sections/goatsandsoda/2017/09/07/549069924/ deadly-bird-flu-in-china-evolves-spreads-to-new-regions

FAO, OIE, and WHO, (2010), The FAO-OIE-WHO Collaboration: Sharing responsibilities and coordinating global activities to address health risks at the animal-human-ecosystems interfaces, A Triparitite Concept Note. https://cdn.who.int/ media/docs/default-source/ntds/neglected-tropical-diseases-non-disease-specific/ tripartite_concept_note_hanoi_042011_en.pdf?sfvrsn=8042da0c_1&download= true

Fidler D, (2004), *SARS, Governance and the Globalisation of Disease*, London: Macmillan.

Flitton D, (2016), Seven deadly sins: The rare animals the Chinese middle class love to eat, The Sydney Morning Herald, June 2, 2016, https://www.smh.com.au/world/ seven-deadly-sins-the-rare-animals-the-chinese-middle-class-love-to-eat-20160526-gp4qvw.html

Global Times, (2022), North China's Inner Mongolia reports plague among mice, issues level IV warning, 03 April 2022. https://www.globaltimes.cn/page/ 202204/1257468.shtml

Kamradt-Scott A and Yoon S, "The global health governance of pandemic influenza: A snapshot of Asia's contribution," in *Governing Global Health in Asia*, Routledge.

Lee K, Pang T, and Tan Y (eds.) (2013), *Asia's Role in Governing Global Health*, London: Routledge. pp. 99–113.

Leung PC and Ooi EE, (2003), *SARS War: Combating the Disease*, London: World Scientific Publishing.

Li P, (2020), Reopening the trade after SARS: China's wildlife industry and the fateful policy reversal. *Environmental Policy and Law*, 50(3): 251–267.

Mackenzie J, Jeggo M, Daszak P, and Richt J, (eds.) (2013), *One Health: The Human-Animal-Environment Interfaces in Emerging Infectious Diseases: The Concept and Examples of a One Health Approach*, New York: Springer.

McLeish E, The Chinese Diners eating a rare songbird into extinction, and the conservationists fighting to save the yellow-breasted bunting, South China Morning Post,05/05/2018,https://www.scmp.com/lifestyle/article/2153735/chinese-diners-are-eating-yellow-breasted-bunting-rare-songbird-extinction

Office of the Senior UN System Influenza Coordinator, (2012), 2011 Annual Report, http://un-influenza.org/files/UNSIC%202011%20Annual%20Report%20-%20Final_0.pdf

O'Reilly S, Four animals being eaten into extinction by gourmets in China and around the world, South China Morning Post, 05/07/2018, https://www.scmp.com/lifestyle/article/2153833/four-animals-being-eaten-extinction-gourmets-china-and-around-world

Reuters, Chinese Man gets 13 years of jail time for alleged eating tigers, 1st Jan 2015, https://www.businessinsider.com/r-chinese-man-jailed-for-eating-tigers-2014-12

Robson D, Should the world eat more like the Cantonese? BBC Future, 14/06/2017, https://www.bbc.com/future/article/20170612-should-the-world-eat-more-like-the-cantonese

Salyer SJ et al. (2017). "Prioritizing zoonoses for global health capacity building – themes from one health zoonotic disease workshops in 7 countries", 2014-2016, *Emerging Infectious Diseases*. 23(13): s55–s64.

Scoones I and Forster P, (2008), *The International Response to Highly Pathogenic Avian Influenza: Science, Policy and Politics*, Brighton: STEPS Centre.

The Manhattan Principles as Defined During the Meeting Titled Building Interdisciplinary Bridges to Health in a "Globalized World" held in 2004.

UNSIC, (2008), Contributing to One World, One Health: A Strategic Framework for Reducing Risks of Infectious Diseases at the Animal-Human-Ecosystems Interface. A Joint Publication by FAO, OIE, WHO, UNSIC, UNICEF, and the World Bank. UNSIC: New York. http://un-influenza.org/files/OWOH_14Oct08.pdf

US Centre for Disease Prevention and Control, (2003), Outbreak of severe acute respiratory syndrome – Worldwide, 2003, *MMWR*. 52(11): 226–228. http://www.cdc.gov/mmwr/preview/mmwrhtml/mm5211a5.htm

US National Intelligence Council, (2003), *SARS: Down But Still a Threat, Intelligence Community Assessment ICA 2003-09*, Washington, DC: National Intelligence Council.

World Bank, (2021), Safeguarding animal, human and ecosystem health: One health at the world bank. 3 June 2021. https://www.worldbank.org/en/topic/agriculture/brief/safeguarding-animal-human-and-ecosystem-health-one-health-at-the-world-bank

World Health Organization, (2003a), *Acute Respiratory Syndrome in Hong Kong Special Administrative Region of China and Vietnam*, Geneva: World Health Organization. http://www.who.int/csr/don/2003_03_12/en/

World Health Organization, (2003b), *SARS: Status of the Outbreak and Lessons for the Immediate Future*, Geneva: World Health Organization. http://www.who.int/csr/media/sars_wha.pdf

World Health Organization, (2003c), Update. SARS in Taiwan, China, 17 December 2003. Available at http://www.who.int/csr/don/2003_12_17/en/index.html

World Health Organization, (2003d), *Update 83. One hundred days into the outbreak*, 12/03/03. http://www.who.int/csr/don/2003_06_18/en/

World Health Organization, (2003e), *Update 95 – SARS: Chronology of a Serial Killer*, Geneva: World Health Organization. www.who.int/csr/don/2003_07_04/en/print.html

World Health Organization, (2003f), Update 96. Taiwan, China: SARS Transmission interrupted in last outbreak area, 5 July 2003. http://www.who.int/csr/don/2003_07_05/en/

World Health Organization, (2003g), *WHO update 92. Chronology of travel recommendations, areas with local transmission*, Geneva: World Health Organization. http://www.who.int/csr/don/2003_07_01/en/index.html

World Health Organization, (2003h), *World Health Organization issues emergency travel advisory*, Geneva: World Health Organization. http://www.who.int/mediacentre/news/releases/2003/pr23/en/

World Health Organization, (2004), Update 7. China's latest SARS outbreak has been contained, but biosafety concerns remain, 18 May 2004.

Zhang Q, Jiang G, and Liu X, Eating Habits in South China Driving Endangered Animals to Distinction, China Dialogue, 18/12/2012, https://www.chinadialogue.net/article/show/single/en/5506-Eating-habits-in-south-China-driving-endangered-animals-to-extinction

10 Tiger Parks in Anthropocene China

Culture, Capitalism, and the Limits of Conservation

Victor Teo

Tiger Parks in Anthropocene China: Culture, Capitalism, and the Limits of Conservation

From plains of Manchuria to the warm forests of Zhejiang and Sichuan down South, China has been home to at least five subspecies of the tiger (Siberian/Manchuria, Indochinese, Southern China, Bengal, Caspian). Today, the Caspian tiger (found in Central Asia and Xinjiang China) is regarded as extinct, while the South China tiger is classed as severely endangered. Like the United States, China has a greater number of tigers in captivity than in the wild. To a large extent, it would appear that Chinese conservation efforts have revived the fortunes of at least one sub-species of tiger that of the Amur tiger (Siberian/Manchurian) in the Northeast.

Tigers have long mesmerized the Chinese nation with their beauty and the magnificence. Alongside the mythical dragon and phoenix, the tiger's significance in the Chinese historical imagination and cultural importance probably even exceeds the status of the panda. The tiger is featured prominently in the everyday life of the Chinese, whether it's the proverbs people use, the folktales told (such in Water Margins 水浒传), or as characters in visual performance arts. Chinese artistes have cast the tiger as a mainstay because of the symbolic of vitality, life, success, or achievement. Contemporary painters such as Hu Zhaobin (1897–1942) have gone further to portray tiger alongside with other beasts to convey loyalty, patriotism, and righteousness. The tiger also is known to symbolize power, and virility. Consequently, since time immemorial, folk medicines based on the tiger (or parts thereof) have been promoted as the cure for various ailments. Whether it's for the reinforcement of one's kidneys or revitalization of virility, there is always a tiger-based remedy based to the rescue. Ironically, due to the tiger's symbolic and mythical importance in Chinese culture, as well as expanding human settlements and wealth, tigers are the most hunted down species in China. The sad reality is that nature of human–tiger interactions in China has largely been a negative experience in recent history, and a great illustration of the Anthropocenic nature of expanding human activities. The conservation of tigers in China has therefore become somewhat of an urgent task for conservationists abroad and also for the Chinese government and people. Yet, despite this common agreement,

DOI: 10.4324/9781003212089-10

tiger conservation efforts have become somewhat controversial due to practices of certain zoological parks and facilities. Western conservationists have
criticized the efforts of both the Chinese government and these tiger parks,
arguing that the latter are little more than farms that breed tigers that would
further fuel the illicit tiger trade, and at the same time arguing that tigers born
or bred in these facilities cannot be rewilded.

   This chapter focuses on the case study of the conservation efforts in China
with particular emphasis on the tiger zoological parks. The chapter begins
with a survey of the factors that have led to the decimation of the tiger population in China. It then examines the political economy behind the tiger trade,
before going on to investigate the efforts of both the Chinese government and
private actors in the tiger conservation efforts. The chapter would engage in
the critique of the work of the tiger parks, and close with several recommendations to improve the existing conservation infrastructure of tigers in China.
The chapter argues that there is certainly room for improvements for both the
government and the tiger parks to make, but at the same time advocates that
there is a need for conservationists to understand that their demands need to
be balanced with the exigencies of development and other pragmatic interests
of the state. The government (state) and the private tiger parks are important
actors in the conservation ecosphere, and due credit should be given for their
work even if not all their practices conform with international ideals. Engaging
and helping them improve rather than harshly criticizing them would be more
helpful in the reversing of the fate of tigers in China.

Changing Human–Animal Interaction and Sources of Declining Tiger Population

This section begins with an examination of the historical forces at work that
have facilitated the decimation of the tiger population. As with most of the
Anthropocene stories regarding wildlife, the case of the Tiger population in
China is not very dissimilar. Population expansion, cultural myths, illicit
trade, and misguided state policies have all contributed to the decline of the
Tiger population in China.

   Expanding Settlements and Human–Animal Conflict – As Chinese settlements expanded and evolved, humans shifted from cohabitants to become
competitors for access to the space, and finally as dominators of the tiger
species. This evolution of evolution has to be contextualized against two centuries of rapid modernization. Urbanization has undergone so rapidly to the
extent in many areas of China many residents would find that neighborhood
transformed beyond recognition within their lifetime. Rapid urbanization
that occurred with expanding human settlements has often led to a decrease
in the natural habitat for tigers, setting the tone for human–animal conflict.
There is evidence to suggest that this started as far back seven centuries ago.
During the Ming and Qing dynasties there were around 514 attacks in
Southeast China (Zhao, 2021). A historical study posited that the intensifica-
tion of these attacks (479 recorded here) in Southeast China was due to the

rapid ecological destruction of the tiger's habitat during that period (Ming, 2003). By the 1950s, the South China Tiger population had grown to about 4,000 strong. Between 1952 and 1962 more than 2,000 people were killed by the tigers in Mao's hometown province of Hunan (Zhao, 2021). There is record that one man-eating tiger attacked and killed 32 people alone. Also within this period (1952–1962), oral history/accounts suggest that a substantial number of the South China tigers were eradicated in Hunan province (Mo, 2008). All these accounts stand as evidence that historically as China's population grow and settlements expanded, the extent of human-wildlife conflict increases. Across the Indian sub-continent today, there are many accounts and videos of Indian settlements becoming under siege by leopards. This situation is not unlike what had transpired before in China. Most of these human–animal conflicts again are attributed to the expanding human presence in the traditional habitat of the animals.

In Mao's China, the eradication of tigers was seen historically as a good thing as it prevented attacks on settlements and saved lives. There was very little notion of conservation during this period, and the taming or subjugation of wild animals was seen as an inherent good. In the longer-term perspective, the bigger issue was not just the persistent human–animal conflicts in themselves. Ever expanding settlements came at the cost of diminishing home ranges of the tiger permanently. This in turn created pressure on the tiger population in various ways. Increasingly, the traditional primary jungles that tigers roam have mostly been transformed into a vast expanse of interconnected urban metropolis interspersed with smaller cities and rural settlements. Reduced home range means less prey and more competition for food, and consequentially more territorial fights between the tigers themselves. Additionally, it would also create barriers to mating as tiger population might be cut off from each other by expanding settlements, and greatly increase human–animal conflicts across the board.

The Lack of Wildlife Protection Regulations and Misguided State Policies – For most part of China's history, the laws and regulations governing and protecting wildlife have been nonexistent. In rural areas, hunting and trapping are not regulated, particularly in areas where wild animals such as wolfs or tigers come into conflict with the local population. The lack of a wildlife management bureaucracy, animal protection laws, or knowledge governing human–animal interaction meant that for most part, the tiger population dwindled without much attention. Beyond the question of defending one's turf arising out of human–animal contact or conflict, there is also evidence that there were hunting expeditions organized for sport or consumption. At the University of Hong Kong library, there are old photo plates of colonial officers who organized tiger hunting expeditions in Southern China. As tigers embody exceptional vitality and ferocity in Chinese and Western culture, this has delegated the tiger to mythical status in terms of its prowess. Hunters prized the tiger for its ferocity and vigor, and in reality, many of those who have gone hunting are doing so for an ego boost, a trophy, and a public recognition more than a public service.

There are five tiger species intimately associated with China. The first three are the Caspian tiger (*Panthera tigris virgata* 中亚虎), the South China tiger (*Panthera tigris amoyensis* 华南虎) (also known as the Amoy tiger), and the Amur tigers (*Panthera tigris altaica* 东北虎) (also known as the Siberian or Manchurian tigers) that are native to China. The Caspian tiger that was native to Central Asia, Caspian Sea region, and Xinjiang China is now considered largely extinct since the 1970s.

The South China tiger is native to central and Southern China. Its traditional range stretches from the Eastern seaboard provinces of Zhejiang, Fujian, down to Guangdong, Hunan, Jiangxi, and also westward extending to Guizhou and Yunnan provinces. The Amur tiger populates the Siberian forests of Northeast Three provinces and Russian Far East. The South China tiger species was reported to hover around 4,000 during the 1950s, but the numbers dwindled after they were targeted by Mao's "anti-pest" campaigns during the Great Leap Forward. By the 1980s, it is reported that there were 30–40 tigers in the wild, even though none were directly observed (Koehler, 1991). WWF estimates that today there are less than 50 South China tigers left in the wild in China, although other sources suggest that there are less than 12 of these. Chinese state authorities, however, are not so sure since there have been very few sporadic sightings of these tigers directly in the wild. For all intents and purposes, the Southern China Tiger is considered functionally extinct in the wild. Their counterparts up North fare better. The National Forestry and Grassland Administration Feline Research believes that only about 40 Amur (Siberian) tigers live in China in the wild.

Two other important species are found in China – the Indochinese tiger (*Panthera tigris corbetti* 印度支那虎) and the Bengal tiger (*Panthera tigris tigris* 孟加拉虎). Historically, the Indochinese tiger is found in Southeast Asia countries of Myanmar, Thailand, Cambodia, Vietnam, Laos, and Southwest China. Today, most of the tigers that are smuggled into China (dead or alive) are believed to be Indochinese in origin. Their numbers in the wild are dwindling fast, however.

Although the Bengal tiger has been observed in the Tibetan mountain ranges – coming from India, this species is not considered to be a "native" species in China. Ironically today, it is the Bengal tiger that exists in the greatest numbers all thanks to the breeding programs in the tiger parks across China. This will be elaborated later in this chapter. The point here is that right up to the 1980s, there was very little state guidance or policy put in place to protect tigers in China. In fact, until recently, tigers were regarded as undesirable and a direct threat to human settlements in China, and their killing was hardly illegal.

The most important legislation protecting wildlife in China was enacted only in 1988 (中华人民共和国野生动物保护法) and subsequently revised in 2009, 2016, and 2018. The law, however, is not completely protective of wildlife, and has been criticized for "loopholes" that exist in its interpretation. For instance, it has been criticized for taking a very narrow definition of wildlife, giving protection only to species that are deemed "precious," endangered, or valuable. The corollary is that species which are not considered to be precious, valuable, or endangered are not accorded the same kind of protection. The law has two schedules (One and Two) which list the animals that are accorded protection, but these lists have been criticized as they are rarely updated (even the situation has changed in the aftermath of COVID). Under this law, wildlife can be bred if the technology is available, and bred wildlife can be sold and used under government licenses. It is under this provision by which many of the tiger parks are able to undertake "breeding" programs in their premises, and it is this provision that many conservationists are extremely unhappy about.

The Commodification of Tigers – The sad state of affairs of the wild tiger population in China is not only a direct result of poor animal protection, but also due to the fact that there is a demand for tigers – dead or alive. The consumption of tiger in China and wider Asia as a folk medicine/remedy is still a real phenomenon, and driven by mythical folklore engrained in Chinese and East Asian culture. Both Chinese court officials and British colonial officials (from Hong Kong) have been known to go to on tiger hunting expeditions in Southern China for egoistic reasons, while the Chinese and local Hong Kongers have a different set of beliefs for the consumption of the animal. According to folk beliefs, the consumption of tiger parts has important effects for the human body. Tiger can be consumed for its meat (exotic culinary), as processed medicinal products (preventive/restorative or enhancing medicine), as ornament (taxidermized exhibit or claws worn as lucky charms), as high-end decoration or as fashion accessories (tooth or claw as jewelry) or religious amulets. In particular, East Asian men (including Japanese and Korean men) have a penchant for the consumption of tiger parts, particularly the penis as a virility supplement prior to the arrival of Viagra. Tiger pelts were considered as luxurious household decorations since ancient times, often signifying wealth, power, or status. In Chinese folk medicine, almost every part of the tiger is believed to be beneficial for treating some ailment: fat (nausea, hemorrhoids, scalp diseases); flesh (malaria, spleen weakness); testicles (tuberculosis); tail (skin diseases); bile (convulsions in children); stomach (gastric troubles); gallstones (eye issues and hand abscesses); eyes (malaria, epilepsy, cataracts); nose (epilepsy; convulsions in children); teeth (rabies, asthma, penis sores); whiskers (toothaches); blood (increase strength), and brains (acne and overcoming laziness). This has resulted in an existing and significant black market today for tiger parts and it is met by the smuggling of poached animals, and by tigers bred in China and elsewhere.

The Illicit Political Economy for Tiger Parts

The consumption of tiger presents a real challenge for its conservation in China and Asia in three ways. First, the illicit political economy of tiger trade extends well beyond the borders of China. Protecting wild tigers in China does have some effect on this illicit trade on the supply side, but as long as the demand is there, tigers everywhere will always be at risk – whether in China or abroad. Second, legal and policy responses to protect tigers have so far only driven the trade further underground and increased the demand and consequently the prices of tiger-related commodities. This further incentivizes smugglers and illicit traders to secure their supplies from both domestic and foreign sources. This has also ironically created an industry where people sell "fake" tiger products (often alongside one or two genuine articles) made out of other animal parts (Figure 10.1). Third, outside pressure on China (or more accurately the Chinese government) might not necessarily be effective after a certain point – this demand cannot be curbed by policy and legal means alone. The Chinese government might not have the expertise, experience, funds, or capacity to protect tigers to the satisfaction of the NGOs and conservation establishments both inside and outside of China.

To have an idea of the illicit economy we are dealing with here, a brief survey will help. The black market for tiger parts is estimated to be USD 6 billion a year on international scale, with some research going as far to suggest that it is the third most profitable illicit industry after drugs and weapons trade (Frantz, n.d.). NGOs such as Traffic International have provided

Figure 10.1 Tibetan Vendors Selling Tiger products in Guangdong Province. Only a small percentage of items turn out to be genuine

(Photograph by the author).

estimates of how much each part of the Tiger is worth on their website. The most valuable part of the tiger is the pelt, and a good piece could fetch somewhere around USD 20,000, if marketed as high-end luxury décor. Tiger claws and tooth are used as lucky charms or religious amulets (Lam, 2018). Tiger bones go for RMB 14,000 a kilogram (AFP Beijing 29 March 2014). A cask of tiger bone wine fetches up to USD 30,000 (Frantz, 2019). As a National Geographic writer puts it, tiger is "walking gold" (Guynup, 2014). In 2007, a portion (plate) of braised tiger meat would cost RMB 588 (USD 77). On 30 December 2014, a wealthy businessman was arrested and jailed for 13 years for organizing the consumption of three tigers. He commissioned for three tigers to be caught and organized three tiger-eating parties for 15 people (SCMP, 30 Dec 2014). It is believed that these Tigers were smuggled into China. It was reported that he had a special hobby of "grilling tiger bones, boning tiger paws, storing tiger penis, eating tiger meat and drinking tiger-blood alcohol" (SCMP, 30 December 2014). Initial allegations were that this was a "visual feast" organized to entertain officials and rich business people, as these killings were organized as "eye-openers" to show off their social stature (AFP Beijing 27 March 2014). In 2015, three Chinese lawmakers from the eastern port of Qingdao were caught for keeping eight endangered Siberian tigers as pets. They were exposed after one of the tiger cubs jumped to its death from an 11-story building it escaped from during fireworks during the Lunar New Year holiday (SCMP 18 March 2015).

NGOs like the WWF have been actively lobbying the governments concerned to stem this illicit trade and help restore the wild tiger population in the relevant countries. To recap here, conservationists argue that the decimation of the tiger population results from the confluence of a few factors: (i) habitat loss; (ii) human–wildlife conflict; (iii) climate change, and (iv) tiger "farms" breeding captive tigers. The WWF estimates that tigers have lost 95% of their historical range, with their natural habitats destroyed through urbanization and other development activities. Human encroachment drove tigers to hunt domestic livestock and/or attack human settlements. Humans retaliated by hunting these tigers down for revenge or profit, often both. Tigers are also not immune to climate change, particularly those residing in areas (such as mangrove swarms) that see rising sea levels (WWF, n.d.).

The last item is of particular concern to this chapter. Tiger farms (or tiger parks), as well as captive breeding programs are being singled out as be a threat to tiger populations in the wild by the conservation camp. Out of the 7,000–8,000 tigers that are being held in 200 tiger centers across East and Southeast Asia, about three-quarters are located in China, that is, 5,250–6,000 tigers are estimated to be in China (WWF, n.d.). Tiger parks are considered a significant obstacle to the recovery and protection of wild tiger populations as they perpetuate the demand for tiger products, serve as a cover for illegal trade, and undermine enforcement efforts (WWF, n.d.). NGOs such as WWF advocate that national governments end breeding and phase out these farms. Understandably the NGOs who advocate these measures do so because they advocate for the animals. The "ideal" model, of course, is that the governments concerned should set aside large tracts of lands for tigers to be "rewild." The conservation camp has therefore labeled these tiger parks to be doing something against their stated mission, that is, to protect tigers. These places are therefore nothing but "fronts" that exploit tigers by allowing them to perform, be petted, be exhibited, and subsequently be quietly "farmed" when they die for parts to meet the illicit political economy.

Despite government regulations and conservation education, the "consumption" habits of the Chinese continue to fuel the demand for tiger-related products. This demand is rooted in superstition, mythology, and ill-informed beliefs "transmitted" across generations through folk culture and medicine. Ingesting animal parts to provide restorative and rehabilitative effects has almost no basis in science. The conservationists allege that there are only three possible sources that supply this illicit industry. The first source are the tigers found in the wild within China's borders. This is unlikely but not impossible. The second source pertains to tigers (or parts of them) that are smuggled from neighboring countries, most likely South and Southeast Asia. Most of the tiger pelts and bones used in Chinese medicinal wine tonic are smuggled from the Indian subcontinent or sourced from tiger farms based in Thailand, Vietnam, Myanmar, and Laos. Tiger parks and farms within China stand as a third possible source that continues to fuel this trade.

State versus Private Efforts in Tiger Conversation: Expectations, Realities, and Concerns

Judging from the conditions today, China probably has the least number of tigers in the wild compared to other Asian countries with native Tiger populations like India or Pakistan. Both Chinese and Western conservationists argue that only by enlarging the tiger's natural habitats, rewilding tigers, and reducing human–tiger interactions would the tiger population flourish. The Chinese government has largely gone all out to accommodate this view on tiger conservation. The Chinese government signed the CITES, agreeing to protect the animals listed in Appendix One and Two of the Treaty. In May 2013, Chinese President Xi Jinping put forward the idea of an ecological "redline" that the Chinese government should strictly observe. Ecological red lines, also known as "ecological conservation red lines," refer to the space boundaries and environmental management limits that must be strictly observed in ecological conservation services, environmental quality and safety controls, and natural resources utilization in order to ensure national and regional ecological safety, sustainable economic and social development, and people's health (China.org.cn, 30 May 2016). The government has set up approximately ten large national parks system and to have these reserves protected with the full force of law in order to preserve the animals in their natural ecosystem, including tigers. These nature reserves often cut across provincial lines and are being brought under the same management system and a harmonized legal infrastructure. For instance, one reserve is a huge panda reserve that spans Gansu, Shaanxi, and Sichuan, bringing pandas isolated on the mountains in these provinces together (Xinhua, 2017).

An important step forward in this initiative is the attention the central government has paid to the Siberian tiger. The Chinese government established the Northeast China Tiger natural reserve, a mega natural reserve that is actually a re-expansion of the Siberian (Amur) tiger's and leopard's natural habitat. The government has set aside a 5,790-square-mile expanse stretching across Jilin and Heilongjiang province (with Russia to the North and North Korea to the South) for the Siberian wild tiger population (Standert, 2018). This would allow the wild tiger population to enlarge their range and roam across national and provincial borders, thereby enlarging their natural habitat.[1]

The Chinese authorities are in the process of resettling 80,000 people out of the designated park area, introducing a logging ban, and repopulating the forests with the natural diets of tigers such as deers and wild boars into the park. Farmers in the area are asked to convert from cattle grazing to organic farming and tiger traps are removed (Standert, 2018). Authorities believe the higher the number of prey density, the better it is as it would mean the tiger needs only to move across a small range to have food. Today it is reported that there are about 40–50 tigers and 70 leopards in the area of the park. Chinese efforts are critical because out of the estimated 400–500 Siberian tigers living in the wild, most of them are on the Russian side of the border. If the Chinese park takes off, it would enlarge the tiger's natural habitat substantially, and convert 30,000 loggers to conservation jobs (Daley, 2017).

Yet, this project is feasible only if the geographic and economic conditions allow. While the projects look good on paper, the capital outlay of these projects is often huge and can be considered a substantial resource drain, especially for a developing country like China. The size of the parks makes implementation of the rules difficult especially when cross-provincial border issues arises. Additionally, there are no guarantees that poaching will be completely eradicated from these mega parks. Such an arrangement would likely to stir the resentment of locals who naturally feel that their rights might have been breached, particularly if their permission is not sought or if they are not compensated adequately. Persistent questions remain as to the extent the reserves can rejuvenate the wild tiger population, as does the persistence of the illicit economy involving tiger products. Practically speaking, there are limits to what the governments can achieve due to competing priorities of modernization and development. Therefore any plan to conserve tigers must realistically take into consideration of the interests of local residents and provincial authorities. The Siberian tiger population has got considerably lucky, but certainly not the South China tiger.

Private Tiger Parks and Farms: Detrimental to Conservation?

If the rejuvenation of tiger population is a metric for measuring the success of the state of conservation today, then the private sector in China clearly outperforms the state-related sector. Today, there are approximately an estimated 200 smaller tiger farms and zoological parks spread across China. The rise of tiger parks is an interesting, but controversial, experiment for the conservation of tigers in the Anthropocene epoch. Unlike the state-sponsored natural reserves, most of these tiger parks do not have the resources or clout to enforce a natural habitat spanning large tracts of territory like the central government can. Some are based in larger zoological premises in the provinces, while others look like farms with research conservation institutes, while others are located in a giant amusement and leisure entertainment complex. The phenomenon of the private zoos and tiger parks is an extremely controversial one, made all the more prominent by the recent Netflix series *Tiger King*, which features Joe Exotic, a tiger park keeper in Oklahoma who is accused of hiring a hit man to kill his competitor, Carol Baskin. The Netflix documentary portrays everything that is wrong with tiger parks: the owners were exploiting tigers as commercial assets: as performing animals; as road-show petting subjects, and as farm animals for hide/parts.

In China, these tiger parks are less controversial domestically even if they share some of the accusations hurled at these parks by environmentalists and conservationists in the United States. The more successful of these parks in China operate as private enterprises, often as zoological gardens, breeding and maintaining tigers in house. The largest of these venues can have staggering numbers of tigers bred at their zoo. In most of these venues, tigers are used as the main attraction to entice visitors (Figure 10.2). Ticket sales and other ancillary activities (e.g., theme parks, hotels, ticket and souvenir sales) would form the bulk of sales at these parks – and this money in turn sustains the park operations as well as the animals, with hefty profits going to the owners. Often they are well regarded by local governments for they provide a steady source of income stream via taxation, increase in tourist-related spending (e.g., hotels, transportation, goods, and services) and employment for the locals. Extremely popular with children and adults, these parks often provide an important social and therefore education node in promoting awareness about the importance of conservation in China.

Conservationists have always outraged at these Chinese tiger parks for a variety of reasons. In particular, they take issue with the zoological practices in these establishments. From the author's own experiences, common practices that irked conservationists abroad include: (1) having feeding stations where visitors purchase food for the animals, leading to situations where often the animals are fed nonstop on busy days or little to no food on days where there are no visitors; (2) petting stations for a fee; or (3) photo stations for a fee. Often, these stations are not just the odd monkey or parrot on show, but often involve baby tigers or leopards, often drugged to make the experience safe for patrons.

Figure 10.2 White Tigers Performing in Chimelong Park Daily for Visitors, Guangdong China

(Photograph by the author).

These irregular practices, however, are just the tip of the iceberg. The same concerns that have been leveled against caged farming for chickens are made against these private tiger parks/farms. Conservationists often express outrage that many of these so-called tiger parks are little more than just farms that exploit tigers, alive or dead. With inadequate conditions, often cramped and overcrowded, many of these smaller parks were often unable to provide adequate food and care for the tiger population in their charge, leading to starvation, stress, undue cruelty, and premature deaths. Due to the limitations of genetic pool within each tiger park, there have been many instances of handicapped or deformed tigers born due to inbreeding. Most of these parks have no intention to rewild the tigers; their inability to hunt in the wild often leads them back to create problems for human settlements. Given the numbers of tigers that are bred in these places, conservationists conclude that these outfits are nothing more than factories that meet the demands for the illicit demand for tiger products. Beyond this, some of the parks involved are accused of violating scientific ethics because of the way genetic modification and engineering is employed in the breeding process. The next section provides a snapshot of two important tiger parks in China that showcase some of the better tiger conservation management efforts in China. These two parks, however, have been criticized in their own ways, but nonetheless present two interesting case studies of how the private tiger parks can work with the central government to help conserve Tigers.

The Harbin Tiger Park (Heilongjiang Province) and Chimelong White Tiger Park (Guangdong Province)

The Northeast China (or Harbin) Tiger Park is one of the more successful enterprises in China. Located on the Sun Island (*Taiyang Dao*) in Harbin, the Northeast China (or Harbin) Tiger park occupies over 14,40,000 square meters and has pure Siberian tigers, alongside pumas, lions, and leopards, and receives about 400,000 visitors per year, including some of the most senior leaders in China such as President Xi Jinping (Northeast Tiger Park Website, n.d.). The park has its genesis in 1986 when the *Hengdao Hezi* animal reproduction center was born, making it the earliest tiger project in China. In 1996, the center was developed into a tiger park, making it a National 4A Tourism facility. The purpose of the park is to protect, research, breed, educate, and promote tourism. The Harbin Tiger Park is lauded by the people and provincial government of Heilongjiang as a top tourist destination and a major center for the conservation of Siberian (Amur) tigers. Via brochures, the park is being touted as one of North China's eminent zoological gardens, with a Siberian tiger population of 1,300, although the Heilongjiang tourism webpage suggests that the count is around 800, with 100 tigers in the exhibit areas. Regardless, the population of Siberian tiger here is staggering, considering that there are only about 450–500 wild Siberians left in the Russian Far East environs, with about 40-odd ones in the border area with China.

In the four different field visits to this park over the years, the operations and the physical arrangement of the park have remained relatively unchanged. The entire park is protected by an electric fence, and the park is segregated into two areas. The enclosed area with a segment of built-up areas of cement cells allows for visitors to walk around to observe tigers at close proximity. The facilities are spartan, and often the cats are locked in the bare cement cells. The number of tigers in this enclosed area is certainly way too many for them to live comfortably, and often leads to stress for these tigers. Visitors are allowed to buy meat and livestock to dangle them from rods to feed tigers, which is not a standard practice in zoos outside of China.

In the second segment of the park, visitors board a bus that is reinforced by cage metal bars over the window and drive into an open space where groups of tigers lounge about. Visitors who have paid for live feed at the entrance would be treated to the sight of another vehicle (SUV or dump-truck) driving along the visitor bus that would feed the tigers. An entire live cow, for instance, will cost RMB 2800. The feed vehicle will either throw (chicken or duck or small sheep) from an SUV or dump (in the case of the cow) from a dump truck, and visitors will be treated to the full gory of the tigers feeding on live animals. Again this has incurred the wrath and criticisms both within and outside of China. Videos of these controversial feeding sessions can be found on YouTube and have left some visitors outraged and shocked. Naturally such practices do not sit well with foreign audiences and conservation groups.

The park has been criticized in several respects. Over the course of the long Manchurian winter, as the visitor numbers dwindle, the park has come under pressure because of the high cost of upkeeping food for tigers. Without profits, it is hard to see how so many tigers can be kept and fed regularly. In order to substantiate the operating costs and the upkeep of tigers, the park needed visitors to subsidize the feed (through the purchases of live feed) as well as the sale of souvenirs and possibly tiger-related products. Moreover, despite the very successful breeding, it is almost certain that tigers lack the natural attributes of the wild tigers. To that extent, conservationists have criticized that such programs are unsustainable and to that effect does not contribute to the rewilding program that is necessary to increasing the "tiger" population in the wild. Today the park has tried to address its shortcomings by tying up her partnership with various scientific institutions and universities, and tried to start a "patron"-style adoption program where various tigers are being adopted by celebrities and corporations. Be that as it may, the park still faces substantial challenges as its tiger population still requires substantial resources to be committed in order to keep the park open. There are also criticisms that the park has been selling some sort of tonic (tiger bone wine in the gift shop) as a range of souvenirs popular with visitors.

The second park of interest here is the Chimelong Safari Park in Guangzhou. The Chimelong group runs the Chimelong Safari Park in Pany district, Guangzhou, and is one of the most successful tiger theme parks in Southern China. Founded in 1989, the owner started a Xiangjiang restaurant that grew into a hotel in 1994. Today, the group consists of a theme park which includes a safari park (biggest in Asia), a crocodile park, a bird park, an international circus, three hotels, and an amusement park. The group also owns the International Ocean Resort on Zhengqing Island off the Zhuhai–Macau border. The most important attraction at the Chimelong Safari Park and the White Hotel are the "anchor" exhibits: white tigers. Chimelong today has about 50% of the world's white tiger population. White Tigers are actually Bengal Tigers with white colour coats due to a rare recessive gene. Within the hotel lobby, guest can drink next to a huge enclosure in the lobby where white tigers lounge, as is the case in the white tiger restaurant where meals are served with tigers seated right next to diners separated by the glass. Again these practices are relatively unconventional zoological practices, and have been criticized for their effects on the well-being of the tigers. This park holds tremendous importance in the local economy of Pany district. Without exaggeration, the employment provided by theme park and the attendant restaurants and hotels provides significant economic vitality to the local district. It has emerged as a leading tiger breeding center in China, and is very much supported by the people and the government. Its success has had a very positive economic benefit for Panyu and the businesses around the area.

Like the Sun Island Tiger Park, the Chimelong White Tiger park has come under attack for their zoological methods and their treatment of tigers. The criticisms are somewhat similar in the respect that these parks are profiting from the human–animal interactions in the park. For instance, like the Sun Island Park, the Chimelong Park until relatively recently offered feeding stations at many animal exhibits. One can simply stroll up to respective exhibits and purchase food from the staff at hand. Fish for the pelicans, sweet potato for the monkeys, bananas for the elephants, branches of leaves for giraffes, and chucks of meats for tigers. Unlike the Sun Island Park, live animals are not used as feed, at least for the safari park. While this is a good model perhaps to subsidize the food costs of the animals, the regularity of feed and the volume fed to the animals vary tremendously. During public holidays such as Chinese New Year and weekends, as thousands of visitors throng the park, the animals might be overfed, while on days without visitors, it is not immediately clear if the animals are fed adequately (although the zoo keepers on duty assured the author that they are). There are also opportunities to pet and take photos with animals such as koalas, chameleon lizards, and naturally baby tigers. Again, during weekends and public holidays, the animals can be stressed as thousands of Chinese wait in line to pet and to have their photo taken and pet the animals for a fee (CNY 50–100 for a photo). No animals, particularly the baby tigers, would appreciate being stoked, pet or manhandled by thousands of enthusiastic Chinese adults and children for photos. To be fair, in the course of repeated visits over a decade, the author has seen the zoological practices in the Chimelong Park improve. The decision by Chimelong to scale down feeding and stop the photo-taking opportunities during the last fieldwork visit showed that the park management is slowly but surely responding to these criticisms.

Taken collectively, the above criticisms of the day-to-day operations pertaining to the issue of overcrowding and inadequate facilities, feeding practices, and making the animals perform shows for visitors pale are not just a critique of subpar zoological techniques in China. Conservationists argue strenuously that these parks are only interested in profits, not animals or conservation. Their concerns are, of course, not without merit. The evaluation of these criticisms must be taken seriously by the Chinese government and the people, in particular by those who support the existence of this park.

Pragmatism and Enhancing Domestic–Wild Balance

Like most things that work in the world, a healthy combination of idealism and pragmatism would work best for bringing back species on the brink of extinction. The dwindling number of (wild) tigers of China was a big concern. There is no question that these private tiger parks are doing extremely well from a numeric standpoint to boost the endangered population. There are many who argue that the animals in these parks are not "real" tigers, since many are born in captivity, and they cannot adapt to the wild. These same conservationists have called for the closure of these private zoological/tiger parks to help save the "real" tigers. While these advocations are not entirely without merit, it raises the question as to why some tigers are considered as "genuine" (the wild ones) while those born in captivity are considered "not so genuine" (and therefore less of a tiger than their wild counterparts). While this argument might make perfect sense, what if we were to substitute species with dogs? Would wild/stray dogs be more "natural" than the beloved dogs we keep at home? The answer is surely not. It therefore raises the problematic manner by which we categorize animals when it comes to conservation efforts.

A more nuanced and incremental approach might be more effective to enhance the good work that has been done so far with regard to the conservation and rejuvenation of the tiger population. One way is perhaps for us to recognize that these tiger parks and farms do not all operate on the same plane, intensity, methodology, and interests. It is also important to note that despite what conservationists think, there are actually staff members in the parks, as well as the management that do care about the animals deeply. Many actually do put the welfare of the animals well before profits. It is also important to recognize that many of these parks actually care about sustaining their businesses in the long run. Even though most of the business operation is widely different, if there is one central commonality between them, it is their inability to survive without profits or some sort of government funding.

The question before us is whether the closure of these tiger parks would necessarily improve the overall conservation of tigers in China. Even if the assumption that these tiger parks/farms exist essentially to provide the raw materials for the illicit tiger product industry holds true, would the eradication of these farms/parks necessarily reduce the demand for tiger-related products? This, of course, might be possible, but the reverse is true as well. It might instead drive the price of these products, further pushing up the demand (due to egoistic consumption) and thereby endangering the population in the wild, within and outside of China. The closure of all the private zoological gardens and parks with tigers will not be possible under any circumstances, not even in authoritarian China. Even if it were, where would all the animals currently in these parks go? Conservation alone cannot just be an ideology – and one cannot expect the interests of the conservation camp to triumph other competing priorities in the real world. As mentioned in Chapter one, today the world is largely facing the ghosts (or mistakes) of the past era. In terms of biodiversity and particularly in the case of tigers in China, things are not in a good shape at all. The way forward is perhaps to reconsider how the existing system can be tweaked to address the concerns of the conservation camp and at the same time advance the interests of other societal groups with a stake in this matter. For the tiger parks, being sensitive and responding to the criticisms and suggestions of the conservation movement are paramount. They ought to be seriously considered by both the public and private sector in their search for the best way forward for China in order to enhance the conservation work of tigers in China.

Greater Governmental Supervision and Proposed Policy Enhancements

The Chinese government has done well in its conservation work in some respects. The Communist state has no problems forcing thousands of people to move out of the confines of the natural reserves they want to establish (aforementioned in Northeast China). Such a move is not possible in many societies. To the delight of conservationists, very harsh punishment is meted out to poachers if they are caught. Notwithstanding this, problems persist – local-level corruption, weak political will in enforcement in some provinces; the persistence of cultural tendencies. There is, of course, much more to be done as there are various counter accusations that the government has not done enough, ranging from a lack of political will to inefficient or lackluster operationalization to outright collusion between government officials and poachers. Most important of all is the livelihood issues that involve the resilience and expansion of agricultural settlements that intrude into the tiger's living habitat remain. These are not easy issues to resolve and will not go away.

Without a doubt, it is the still the central government that has the ability to legislate and implement measures that will further strengthen conservation of Tigers in China. The central government's power to regulate and compel enforcement is the single most important instrument conservationists can have in advancing their cause. Given that the existing laws in China are extremely in favor of conservation and the protection of tigers, the conservation movement actually already has the support of the most important player in its corner. What it needs to do is to help facilitate and tweak the existing mechanisms to further its purposes. There are a number of important items of critical importance that need to be addressed.

a. **Number Reduction and Quality Enhancement of Tiger Parks** – The first proposal would be for the Chinese government to consolidate the number of facilities that are licensed to breed tigers. These facilities must have clear objectives, and a properly set up governance board overseeing its operations and finances before they can be licensed. Facilities which breed tiger for purely commercial profit should be closed down if they are unable to provide the appropriate and acceptable standards for keeping tigers in captivity. Captive tigers bred and grown in concrete enclosures have little in common with their wild counterparts. As China modernizes, the country continues to learn and implement "best practices" in many fields, zoology included. Therefore, the expectations and standards of what is "acceptable" for a zoo or animal park should correspondingly change. Those facilities that are unable to provide stress-free conditions for animals should be closed down. The facilities that demonstrate competence and interests in animal welfare, in this case in looking after tigers professionally, should be allowed to establish a unit to focus on breeding tigers. This unit that has clear guidance and good facilities to be able to help tigers reproduce in captivity (Figure 10.3). The same

Figure 10.3 "The Importance of Artificial Rearing" Exhibit in Chimelong Tiger Park, Guangdong Province China

(Photograph by the author).

considerations provided against horrific conditions in commercial chicken farms worldwide should be applied here too. Reducing the number of unfit and unsuitable parks would allow the central and provincial governments greater oversight to implement the regulations to protect and improve the welfare of the captive animals.

b.   **Adopting and Enforcing Scientific Research Ethics** – This has particularly to do with white tiger parks. Conservationists have argued that the white pigment in white tigers is in reality a birth abnormality that seldom occurs in nature. If any white tiger parks have been involved in intentionally manipulating genetics to breed white tigers, then this is ethically questionable. The researchers are intentionally distorting what Mother Nature has not intended. This allegation should be looked into more seriously by the Central government scientists, officials, and scholars. The government should investigate thoroughly if there should be an ethical guidance over what is permissible or not in the breeding of these animals. One aspect might be related to inbreeding. Due to the limited number of tigers at the beginning, some of these facilities might have undertaken to breed tigers from between members of the same family, leading to higher incidences of deformity among the tiger babies. The author has learnt from some of the park workers during the fieldwork that inbreeding is something they try to guard against, but the incidences of deformity and handicapped tigers are still seen. These tigers, however, often do not survive long.

This is one area that the state should pay closer attention to, and, if possible, have government experts join the work of the facility scientists to monitor what is being on the ground. The state should have a duty of care to ensure that these facilities do not overstep the bounds of scientific and ethics considerations. The state ought to step in and ensure the park staff understand that regulations exist for everyone's benefit and safety, and that adherence is in their best interest. At the same time, the government should also encourage the parks to come forward if they need help. Overregulation and excessive fines tend to inhibit park workers from coming forward if they need help as they are worried about punishment and recriminations that follow. It is this sort of fear that would further encourage unethical behavior in the scientific context (Figure 10.4).

c. **Focus on Sub-species Extinction Prevention** – An important unintended issue that is being raised by these Tiger parks and farms is the question of sub-species being bred. The parks have been mostly very successful in increasing the number of tigers in the country. Fundamentally, although the parks have reversed the possibility of extinction of tigers as a species, there is an imbalance between the variations of the sub-species. There are geographical constraints to be sure. The Sun Island predominantly breeds

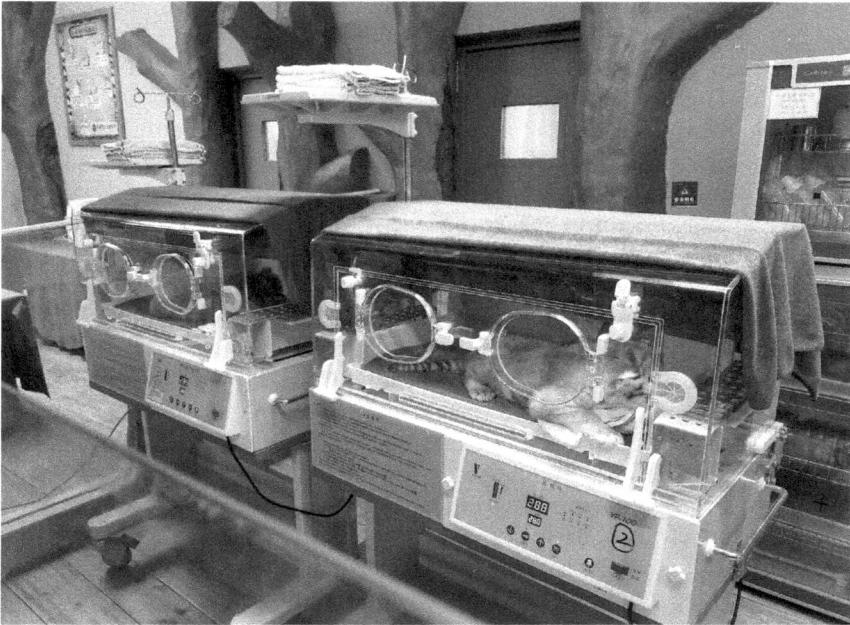

Figure 10.4 Baby Tigers in Incubators in Chimelong Tiger Park Guangdong Province China

(Photograph by the author).

Siberian tigers as it is located in Northeast China. The Chimelong Safari Park in the South, however, breeds white tigers. Conservationists argued that these white tigers actually belong to the Bengal tiger sub-species, and therefore the fixation on breeding the white tigers alone does not help preserve the tiger sub-species native to Southern China, which is the South China tiger. Reducing the number and improving the quality and the focusing on preventing and rejuvenation of endangered sub-species should be the focus of many of these parks. Government legislation and enhanced cooperation with state authorities would certainly enhance the conservation work. In short, these parks should be asked to refocus their efforts to breed more (and endangered) species that are native to China (South China tiger), and reduce their holdings on mutant Bengal tigers like the white tigers.

d. **Prioritizing Rewilding of Tigers** – There should be a central authority to ensure that breeding centers participate and contribute to the rewilding of baby tigers in their natural environment. This state authority would have the power to acquire land and set aside natural forests as natural preserves as in the case of Northeast China. They can also be empowered to mandate that the tiger parks set aside certain percentage of tigers they breed to be rewild in the natural environment, that is, Siberian tigers in the Northeast and South China tiger in the South. Captive tigers have little in common with their wild counterparts if they are not rewilded soon after birth. If the parks all have to partake and contribute to the rewilding work (such as meeting a state-mandated quota), the rejuvenation of wild tiger population in the nature reserves would be strengthened immeasurably. Rewilding a substantial portion of the best and most healthy tigers would provide added justification for other tigers (those weak, handicapped, or unsuitable to be rewild) to be kept in the zoological park for education and scientific purposes. The tiger parks could also be roped in to work with green groups, universities, and think tanks as well as state organs to ensure that such conservation work are conducted on a long-term basis as a private–public partnership.

e. **Conservation Outreach and Education** – Many of these tiger parks/zoological gardens have often received the enthusiastic support of local government and national conservation authorities. The better-run ones (such as the Sun Island Tiger Park in Harbin or the Chimelong Safari Park in Guangzhou) are important enterprises in the provincial tourism infrastructure, and play a critical role in the respective districts' employment and economy. While some aspects of their zoological practices might be different from international standards, these parks are nonetheless relatively immensely popular for tourists and locals alike. Their popularity enables them to play an extremely important role to promote and educate the public about animals and wildlife alike, particularly for future generations. Currently, from the number of zoos and parks the author has visited across China, the "educational" aspects of these parks are still lacking as tourism is still the main focus. This is a shame because better funded parks tend to have the resources to educate. More scientific explanations, less mythological narratives, and better regulated human–animal interactions and experiences will increase their appeal and expertise. Instead of closing down these tiger parks, converting them into education first centers helps the conservation cause. Like generational ban on smoking, educating children on the importance of wildlife will help defeat ignorance, superstition and poor taste. There might be a constant demand for the "exotic" such as using tiger skin for décor or those who live the lucrative profits in this illicit trade, but with proper education, this can be reduce over time with proper education.

f. **Control and Good Governance of the Tiger Trade** – The most pointed critique of the conservationist is that the missions of these tiger parks/zoos in China are tiger breeding farms that supply the illicit tiger trade in China under the guise of conservation. Reducing the number of and tightening regulations for parks that can undertake breeding is critical to fight this. The state can make it an onus for the parks to demonstrate that they provide a reasonable standard of care to the tigers, in terms of living space and food, and exercise the best standards in their breeding programs. Surprise or stealth inspections by cross-border teams (as they do for corruption investigations) could be employed in the supervision of the zoos. The key message to take away is this: it is not right for tigers to be bred like livestock, and this must be regulated.

In his interviews with the keepers at the fieldsites, the author learnt that each tiger is fed regularly with 10–15 kilograms of meat a day (costing USD 60–80 each day). The upkeep of tigers is therefore extremely high. To keep the parks running, funding is required. Conservation cannot just rely on charitable donations or state handout. The tiger parks, if managed well, will be a key way of sustaining public interest and broadcasting the message on conservation, as well as generating funding for further conservation and scientific programs.

Notwithstanding this, it is possible to consider letting these parks process bred tigers that die naturally within these parks to meet the demands of those who want to consume tiger-related products. Legalizing this trade will allow the authorities to control this instead of driving it underground. Capping the number of tigers that are harvested for profits per year will prevent the zoos from engaging in tiger farming as its main effort. Hardcore conservationists might balk at this suggestion, but perhaps there is a prima facie case with state backing and careful incremental collaboration, illicit trade can be further suppressed. This will help reduce poaching in neighboring countries to meet the demand.

g. **Stiffer Penalties for Importation, Smuggling, Sale, and Consumption of Wildlife**

In 2009, an Indochinese tiger in China was killed and eaten by a man, Kang Wannian, who was subsequently sentenced to a twelve-year prison term – ten years for killing a rare animal and two years for possessing an illegal firearm (Platt, 2009). The stiff penalties due to the animal protection law in China have deterred poaching of tigers in China, but this has stimulated an increase in the poaching cases across the bordering Indochinese countries of Vietnam, Thailand, Cambodia, and Laos. The key here is for authorities to legislate and increase the penalties for trafficking of the Indochinese (or Malayan) tigers coming from Southeast Asia as well as for the sale or consumption of these tigers for food. Further reinforcing the law would certainly help to reduce the demand destined for the Chinese market. Authorities should also consider levying heavy fines on zoological gardens that are found to be in breach of the principles and ethics with regard to breeding tigers for consumption and sale. If profit is the main motive behind such activities, then levying a financial disincentive would surely help in the deterrence of such activities.

Conclusion

If cultural inertia, uneducated beliefs, and egoistic consumption drive the illicit trade, suppression techniques alone will not eradicate this demand. Rethinking what the parks can do for conservation is quite important. Only by education and persuasion supported by the rigorous and intelligent application of the law would we be able to gradually reduce the demand across generations.

Tiger parks should be used as a primary tool to educate the next generation to overcome superstition, myths, unfounded beliefs as well as insulate against egoistic consumption. In Africa, many of the safari parks are too in reality large "zoological" parks, but with fenced up areas so large that it would take more than a couple of hours to drive from one end to the other. It is also common practice for these parks to give up their old and dying animals to be "sold" for hunting (even though it is not advertised) within the safari grounds as the revenues from the parks admissions and well as the sales help fund the rest of the animals in the park. In a personal conversation with some of the park rangers in South Africa, the author was told in uncertain terms that even though this practice of allowing some of their animals to be hunted was frowned upon by the conservation community, it was a practical one for the parks. The key issue here is how one old lion's death could be made worthwhile so that the well-being of all the other animals in the park can be improved. While this line of thinking might be controversial, it is certainly something that should be given a second thought with regard to the tiger parks discussed here.

In China, the numbers of tigers in the wild remain low, while the numbers in captivity are high. The next step for these tiger parks is to come under the regulation of a state body if this is not already the case. This body could actually compel all the existing big parks to partake in a national program to help collectively tackle the tiger conservation problem in China. A priority perhaps is these parks should be harnessed to help in two critical issues. The first is to increase the number of South China tigers, and, the second, to assist in the rewilding of the tiger population in say Northeast and Central/

Southwestern China. If human activities (as well as climate change) have been the primary contributing factors in the decimation of the wild tiger population, then surely human beings can undo this to a certain extent. With the legal infrastructure in place, the key now is to further regulate and incorporate the existing tiger parks and zoos as critical actors in conservation practice, rather than cast them as the opponent.

The key therefore is for the conservation camp in China to try and work with existing actors, including tiger parks. This will enable further conservation work to ensure that the next generation of young children understand and appreciate the fact that tigers (alongside other animals) are our cohabitants of the planet. The tiger parks and zoos are the most effective way to reach the next generation. If our next generation cannot understand and experience our fellow planetary cohabitants in the first place, how can expect them to protect them in the future? For that reason alone, conservationists in and out of China should try and recruit the government and the tiger parks on their side in their fight for tiger conservation. Only the state can compel consolidation of these parks, demand minimum standards, fight abuse, and extract certain portion of their revenue for public goods such as conservation.

Exotic goods (e.g., Tiger products) become very sought-after if the supply is diminished completely; that is, the more stringent the ban, the more exotic these products become, and the greater the profit and thus supply. If the decimation of the wild tiger population in mainland Southeast Asia is directly linked to the vibrancy of the illicit tiger trade in China, then the tiger parks can certainly help to reduce this demand if the deceased tigers are allowed to be harvested for this trade. Allowing the tiger parks to trade might also allow the state to increase its governance and control of this trade.

Misguided cultural beliefs and superstition lie at the root of the demand for tiger parts that drives the illegal trade. The state, the tiger parks, and the conservationists can work to reduce the "exotic" imagery of Tiger products, by educating consumers about the "correct" nature of these products. The educational efforts should be focused on young children as this is a generational effort. Tiger parks, if properly run, would be able to play a greater role in the conservation of the species in China.

Note

1 Modern technology has enabled the Chinese and Russian governments to keep track of wild tigers that have been tagged. Indeed, there was a human interest story that a tiger named *Kuzya*, released into the wild by President Vladimir Putin (nicknamed Putin's Tiger), had crossed into Northeast China in October 2014. Newspapers and social media were sounding alarm and making jokes that Putin's Tiger was at great peril. Chinese and Russian officials were in fact concerned that the death of the tiger on Chinese soil might cause a political and diplomatic incident and were in fact scrambling to find it. Jokes abound whether the tiger was trying to escape authoritarianism, or whether it was invading, spying, or a defecting, were trending on social media at that time.

Bibliography

AFP Beijing, "Tigers Slaughtered in Show of Social Stature for Guangdong Businessmen", 27 March 2014, https://www.theguardian.com/world/2014/mar/27/tigers-killed-entertainment-guangdong-businessmen

Alesha and Jarryd, "The Gruesome World of Harbin Tiger Park", Blog, 19 April 2018, https://www.nomadasaurus.com/harbin-siberian-tiger-park/

BBC Science, "White Tiger's Coat Down to One Change in a Gene", BBC News, 23 May 2013, https://www.bbc.com/news/science-environment-22638341

Cell Press, "White Tiger Mystery Solved: Coat Color Produced by Single Change in Pigment Gene", ScienceDaily, 23 May 2013, www.sciencedaily.com/releases/2013/05/130523143342.htm (accessed November 17, 2019).

Chen, Min, "The Problem with China's White Tiger Obsession", Six Tone, 17 January 2019, https://www.sixthtone.com/news/1003460/the-problem-with-chinas-white-tiger-obsession

Chris, Quan, "Siberian Tiger Park", ChinaHighlights, 26 January 2022, https://www.chinahighlights.com/harbin/attraction/siberian-tiger-park.htm

CITES (Convention on International Trade in Endangered Species of Wild Fauna and Flora), 70th Meeting of the Standing Committee, Sochi (Russian Federation), 1–5 October 2018, https://cites.org/sites/default/files/eng/com/sc/70/E-SC70-51.pdf

Fouts, Matt, "6 Truths about White Tiger" Conservation, 28 July 2018, https://www.twpark.com/blog/conservation/white-tiger-truths

Frantz, Lauren, n.d. "The Illegal Trade in Tiger Parts" Crownridge Tigers Feature, https://crownridgetigers.com/the-illegal-trade-in-tiger-parts

French, Paul, "Hunter Who Hastened Demise of the South China Tiger, and How Mao's Assault on Nature Finished It Off", South China Morning Post, 9 November 2019, https://www.scmp.com/magazines/post-magazine/long-reads/article/3036511/american-hunter-who-hastened-demise-south-china

Guynup, Sharon, "Illegal Tiger Trade: Why Tigers Are Walking Gold", National Geographic Blog, 12 February 2014, https://blog.nationalgeographic.org/2014/02/12/illegal-tiger-trade-why-tigers-are-walking-gold/

Jacobs, Andrew, "Putin's Tiger in a Territory Grab All His Own, Swims to China". The New York Times, 10 October 2014, https://www.nytimes.com/2014/10/11/world/asia/putins-tiger-crosses-into-china-prompting-a-diplomatic-rush.html

Jonathan, Watts, "US Diplomat posed as Korean Tourist in Undercover Visit to China Tiger Farm", The Guardian, 25 August 2011, https://www.theguardian.com/environment/2011/aug/25/us-diplomat-china-tiger-farm

Koehler, Gary M., 1991. *Survey of the remaining population of South China tigers.* Unpublished Report, WWF Project 4512/China, Washington, DC.

Launders, Jackson, "Why White Tigers Should Go Extinct", Slate, 13 December 2012, https://slate.com/technology/2012/12/white-tiger-controversy-zoos-shouldnt-raise-these-inbred-ecologically-irrelevant-animals.html

Lydia, Lam, "8 Cases of Illegal Tiger Parts Trade, Smuggling in Singapore since 2015: AVA on World Wildlife Day", The Straits Times, 3 March 2018, https://www.straitstimes.com/singapore/8-cases-of-illegal-tiger-parts-trade-smuggling-in-spore-since-2015-ava-on-world-wildlife

Ming, Zongdian, 2003. "The Tiger Trouble and Its Relevant Problems in Ming and Qing Dynasty". *Ancient and Modern Agriculture* (古今农业) 16(1): 17–23, https://scjg.cnki.net/kcms/detail/detail.aspx?DbCode=CFJD&dbname=CFJD9411&filename=GJNY200301002&uid=

Mo, YaBo, 2008. "百虎围村50年前的人虎大战" (Hundreds of Tiger Surrounds the Village - The Human-Tiger War Fifty Years Before) http://news.sina.com.cn/c/2008-03-28/115715246199.shtml

Nuwer, Richael, "The Key to Stopping Wildlife Trade: China", The New York Times, 19 November 2018, https://www.nytimes.com/2018/11/19/science/wildlife-trafficking-china.html

Platt, John, "Rare Siberian Tigers Face Potential Genetic Bottleneck", Scientific American, 8 July 2009a, https://blogs.scientificamerican.com/extinction-countdown/rare-siberian-tigers-face-potential-genetic-bottleneck/

Platt, John, "Man Convicted for Killing and Eating China's Last Indochinese Tiger", 25 December 2009b, https://blogs.scientificamerican.com/extinction-countdown/man-convicted-for-killing-and-eating-chinas-last-indochinese-tiger/

Richard, Michael Graham, "Fewer than 50 Wild Tigers Left in China", Says Wildlife Conservation Society, 11 October 2018, https://www.treehugger.com/natural-sciences/fewer-than-50-wild-tigers-left-in-china-says-wildlife-conservation-society.html

Scientific American, "Save the White Tigers", 16 October 2014, https://www.scientificamerican.com/article/save-the-white-tigers/

South China Morning Post, "From Poacher to Ranger: Saving China's Siberian Tigers", 6 October 2017, https://www.scmp.com/news/china/society/article/2114237/poacher-ranger-saving-chinas-siberian-tigers

Standert, Michael, "New Park: China Creates a Refuge for the Imperiled Siberian Tiger", Yale360 Online Edu, 1 February 2018, https://e360.yale.edu/features/china-carves-out-a-park-for-the-imperiled-siberian-tiger

Straits Times, "Chinese Businessman Jailed for 13 Years for Buying and Eating Three Tigers", 30 December 2014, https://www.straitstimes.com/asia/east-asia/china-businessman-jailed-for-13-years-for-buying-and-eating-three-tigers

The Straits Times, "China Tiger Farms Put Big Cats in the Jaws of Extinction", 12 February 2015, https://www.straitstimes.com/asia/east-asia/china-tiger-farms-put-big-cats-in-the-jaws-of-extinction

The Straits Times, "Profit-hungry Tiger Breeders behind Push to Lift China's Trading Ban, 8 January 2019, https://www.straitstimes.com/asia/east-asia/profit-hungry-tiger-breeders-behind-push-to-lift-chinas-trading-ban

The Sydney Morning Herald, "11 Starve to Death at Zone", 12 March 2010, https://www.smh.com.au/environment/conservation/tiger-tragedy-11-starve-to-death-at-zoo-20100312-q3rj.html

WWF, "The Truth About White Tigers", 13 July 2021, https://www.worldwildlife.org/stories/the-truth-about-white-tigers

WWF, "Tiger Species Page", n.d.-a, https://www.worldwildlife.org/species/tiger

WWF, "Restoring Asia's Roar: 12 Ways to Make Tigers Made a Comeback in 12 Years", n.d.-b, https://www.worldwildlife.org/stories/restoring-asia-s-roar-12-ways-tigers-made-a-comeback-in-12-years

Xinhua, "China to Set Up National Park System", 29 September 2017, http://www.chinadaily.com.cn/china/2017-09/27/content_32557996.htm

Xinhua, "China to Restore Siberian Tiger Population from Captive-bred Stock, 30 July 2018, http://www.china.org.cn/china/2018-07/30/content_57881997.htm

Xu, Xiao, Gui-Xin Dong, Xue-Song Hu, Can Xie, Ruiqiang Li and Shu-Jin Luo, 2013, "The Genetic Basis of White Tigers." *Current Biology* 23(11): 1031–1035, http://www.cell.com/current-biology/fulltext/S0960-9822(13)00495-8. 31 July 2017.

Zhao, Ying, "From 6 to 200: When the Tiger Came Down the Mountain" CGTN Feature, 7 January 2021, https://news.cgtn.com/news/2021-01-07/From-6-to-200-When-the-tiger-came-down-the-mountain-WIu9N3voKA/index.html

Index

Note: Page numbers followed by n indicate notes.

For Product Safety Concerns and Information please contact our EU
representative GPSR@taylorandfrancis.com
Taylor & Francis Verlag GmbH, Kaufingerstraße 24, 80331 München, Germany